Effective Communication for the Technical Professions

Jennifer MacLennan

University of Saskatchewan

Prentice Hall

Toronto

This book is dedicated to my benefactor,
Daryl K. Seaman.

National Library of Canada Cataloguing in Publication

MacLennan, Jennifer
 Effective communication for the
technical professions/ Jennifer
MacLennan. — 1st ed.

Includes index.
ISBN 0-13-003717-6

 1. Business communication.
I. Title.

T11.M32 2003   808'.06665
C2002-901705-X

Statistics Canada information is used
with permission of the Minister of
Industry, as Minister responsible for
Statistics Canada. Information on the
availability of the wide range of data
from Statistics Canada can be obtained
from Statistics Canada's Regional
Offices, its World Wide Web site at
www.statcan.ca, and its toll-free access
number 1-800-263-1136.

ISBN 0-13-003717-6

Vice-President, Editorial Director:
 Michael J. Young
Marketing Manager: Toivo Pajo
Developmental Editors: Lise Creurer,
 Susanne Marshall
Senior Production Editor: Joe Zingrone
Copy Editor: Susan James
Proofreader: Nancy Carroll
Senior Production Coordinator: Peggy
 Brown
Page Layout: Silver Birch Graphics
Art Director: Mary Opper
Cover/Interior Design: Silver Birch
 Graphics
Cover Image: Lenscape Incorporated

1 2 3 4 5   06 05 04 03

Printed and bound in Canada.

Contents

To the Instructor

As a professor of engineering for over 30 years, my "lifeline" has always been communication of one form or another. I have always have taken pride in my written publications, and considered myself a good lecturer. Until recently, however, I didn't realize how much more there was for me to learn. This last year, I undertook to teach our newly designed course in Oral and Written Communication, developed by Dr. Jennifer MacLennan. This course—like the book you are reading—is based on the premise that an understanding of basic rhetorical theory can enhance one's practical communication skills. In particular, two essays by Lloyd F. Bitzer and Wayne C. Booth (both of which appear as readings in this book) formed the theoretical framework of the new course.

As a result of participating in the instruction of this course, I made a series of discoveries that have profoundly changed my whole attitude toward communication. Like my students, I was given new tools for understanding what made my communication successful; I also discovered ways in which I could improve on my previous efforts. Teaching this course has influenced the way I approach both oral and written communication in my own life, even as I watched the skills and attitudes of my students increasing. The significant changes I have seen in them (and indeed, in myself) as they progress from the beginning to the end of the course are testimony to the power of the approach presented in this book. I have found the readings by Booth and Bitzer to be particularly influential in shaping the understanding of our students in the College of Engineering.

As you approach the teaching of your technical students, I encourage you to give more than lip service to these foundational articles; they can open a whole new world of understanding to you and your students, and provide a usable, practical framework for developing concrete skills in speaking and writing.

Dr. Richard Burton,
Professor of Mechanical Engineering
and Assistant Dean,
Undergraduate Programs
College of Engineering,
University of Saskatchewan

Preface

I am pleased to write this note to accompany the first edition of *Effective Communication for the Technical Professions*. The book has been a challenge to write, drawing as it does on several different disciplines, from engineering to rhetoric to business to communication theory. The task has been huge, but I have been grateful for the opportunity to bring these disparate fields together in what I hope is a comprehensive approach to the study of communication.

There are many fine books in technical communication already on the market; why, then, do we need another one? Good question: after six months of fourteen-hour days, I have often asked myself the same thing. One answer is surely that, while most technical communication texts offer only writing practice, this book incorporates the skills of critical reading and analysis that many employers note are missing in our graduates. The variety of writing samples offered here, some of which violate the principles of good writing that the students are learning, can be used to help develop skill in assessing the appropriateness and effectiveness of different communicative strategies. By comparing effective examples with ineffective or flawed examples, students will develop greater discernment and critical skill that, in turn, will assist them in editing their own work.

In my role as technical and professional communication expert in a College of Engineering, I work with engineering students and professionals every day. I see first-hand the need for effective communication skills, and am frequently told by practising engineers how important these skills are in the new graduates they hire. I have also seen a like appreciation among the students, who have been hungry for an approach that could help them build the kind of discretion and effective judgement they need to respond to any communication situation. My goal has been to make this book do that job. Like my previous texts dealing with strategies for effective communication, the approach in this text has been grounded in the ancient art of rhetoric—the systematic study of human communication—but more explicitly so. As a result, this book concentrates on more than format and content in technical writing: it approaches communication as a thoroughly human process, focussing not only on clarity of purpose, but also on developing a strong sense of audience, an engaging and competent projection of the writer or speaker, and an effective relationship between them.

In short, this book attempts to respond to the needs that have been articulated to me by engineering deans and faculty, by the employers who hire technical graduates, and by the students themselves. The specific features that will distinguish this book from its competition are as follows:

- it is supported by my experience as a communication specialist housed in an engineering college, using a proven and successful approach for teaching courses in the area

- it avoids cookie-cutter formats; instead, it is designed to inculcate in the students the judgement they need to respond appropriately to a variety of communicative demands in the workplace

- it offers assignments and situations for writing that, while addressed to a technical audience, are accessible and understandable to students who have not yet amassed years of professional experience

- it provides a solid and usable theoretical foundation that is accessible to non-specialists

- it features assignments designed to develop the skills of reading carefully and analytically as well as those of thinking and writing strategically and effectively

- it includes numerous samples of both effective and ineffective communication, and will help students develop the discrimination they need to edit their own work for clarity and effectiveness

- it offers a comprehensive treatment of the job application process, aimed squarely at its technical student audience and based on years of practical research

- it includes several Critical Readings, along with strategies for reading technical and other materials efficiently and critically

- its treatment of ethical communication is comprehensive and practical

- its contexts and examples are Canadian

My goal has been to produce a book that will serve both professors and students in communication classes; I have also tried to make it practical enough that its readers will continue to refer to it even after the course in which they first used it is over.

SUPPLEMENTS

For the Student

We are pleased to offer a *Text Resource Site* (**www.pearsoned.ca/text/maclennan**) for this edition of *Effective Communication for the Technical Professions*. The website provides "enhanced chapters," featuring online exercises and additional resources for students.

For the Instructor

We are pleased to offer an *Instructor's Manual* (ISBN 0-13-030718-1) to accompany this text. This supplement is also available for downloading from the Instructors' Resources section of the Text Resource Site.

Acknowledgements

The making of a book such as this one, as anyone knows who has attempted the task, is a huge undertaking, and it depends on the assistance and good will of many people. My initial thanks therefore go to those whose comments and questions initiated the project: the students and faculty in the College of Engineering at the University of Saskatchewan.

I would also like to thank the many folks at Pearson Education Canada who contributed to the book along the way; first, of course, is David Stover (former editor-in-chief), whose presence has been such a positive influence for me over the years. Thanks also to Sophia Fortier, senior acquisitions editor. I could not have survived the six months in manuscript development without the contributions of Lise Creurer, who has been in every way a delight to work with. I am also grateful to those whose hard work has contributed so indispensably to the finished book: Peggy Brown, Joe Zingrone, Susan James, Nancy Carroll, Imogen Brian, and Monica Kompter. Finally, I would like to thank Bonnie Benoit, Richard Burton, George A. Tripp, and Douglas Perry White, who took time out of their busy schedules to review the early manuscript.

I cannot begin to adequately thank those in the College of Engineering who have helped me out so much, both directly and indirectly: Ron Bolton, Professor of Electrical Engineering, who generously supplied diagrams for Chapter Six, and who gave me helpful feedback on the segment on log books; Richard Burton, Assistant Dean and Professor of Mechanical Engineering, whose advice on formal reports I found very helpful; Roy Billinton, acting dean and professor of electrical engineering, whose support has been unwavering; Gord Putz, associate professor of civil engineering, who shared with me several examples; these people have also supported me in innumerable ways as members of the D.K. Seaman Chair Advisory Committee. I am also indebted to those who have allowed me to include their work in the book: Julian Demkiw, Allan Dolovich, Curtis Olsen, and Burton Urquhart.

The staff of the Dean's Office in the College of Engineering also deserve a special mention: Brenda Bitner, who is so consistently helpful and positive, and who always knows the answer, no matter what I ask her — thank you so very much. I am also grateful for the same reasons to Brenda Rowe, Evelyn Laird, and Pernel Noble. As well, but for the computer wizardry of Ian MacPhedran, our systems genius, I might have given up entirely on the project when I accidentally overwrote a late draft of Chapter Eight. I have also been very thankful to Randy Hickson, Faculty and Staff Support, Computer Services, for his cheerful and always life-saving attentions to the techno-monster in my office. I would also like to thank my research assistant,

Laura Patterson, who has been an immeasurable help in this and other projects.

I would very much like to acknowledge some other marvellous folks at the University of Saskatchewan who have been very supportive of my efforts here: Vera Pezer, VP of student affairs and services; John Thompson, professor of sociology at St. Thomas More College; Jeremy Bailey, associate dean (academic), Western College of Veterinary Medicine; Lloyd Moker, coordinator of student services, Western College of Veterinary Medicine; Jana Danielson, director of retention, Student Affairs and Services; Roxanne Vandeven, coordinator of finance and development, College of Engineering; Colleen Teague, coordinator of personnel and facilities, College of Engineering; and Cathy McKenna, coordinator, academic programs and internships, College of Engineering; Eileen Herteiss, program director, Gwenna Moss, Teaching and Learning Centre; and Ron Marken, director.

Finally, of course, I cannot begin to express my debt to David Cowan, without whose support, understanding, and generosity I could not continue my work. Thank you, as always, from Meestie, and Feral, and me.

Jennifer M. MacLennan

CHAPTER 1

Understanding the Communicative Situation

LEARNING OBJECTIVES:

1. To understand the nature of communication.
2. To recognize the two principles upon which all effective professional writing is based.

WHY COMMUNICATION MATTERS

It may surprise you to learn that, of all the daily tasks performed by a technical professional, none is more important than effective communication. On the job, you may interact daily with dozens, or even hundreds, of people communicating by letter, memorandum, telephone, and electronic mail.

How important are communication skills? By some estimates, as much as seventy-five percent of the average person's day is spent communicating in some way. For example, while you are a student, you will spend sixty-nine percent of your communication time on speaking and listening, seventeen percent on reading, and fourteen percent on writing.[1] As one expert has put it, "we listen a book a day, we speak a book a week, read the equivalent of a book a month, and write the equivalent of a book a year."[2]

Your need to communicate clearly doesn't end when you graduate; indeed, it will likely escalate. In many technical jobs, for example, fifty percent or more of your work can involve communicating, persuading, and cooperating with others. For this reason, nearly all practitioners in the technical professions list communication among the essential skills for new graduates and experienced engineers alike.[3] For instance, a recent survey by the Association of Professional Engineers and Geoscientists of Saskatchewan (APEGS)[4] asked practising engineers to identify important areas for career development under the guidelines for "continuing competence" provided by the Canadian Council of Professional Engineers.[5] In response, nearly all the engineers surveyed identified communication skill development as a main priority.

The survey also asked respondents to list qualities they would use to judge each other's professional competence. As we will see, nearly all the categories identified by the engineers involved skill in communicating effectively. A separate study by the U.S. Department of Labor showed similar results, identifying sixteen qualities for high job performance in any field. Ten are communication skills: listening, speaking, creative thinking, decision making, problem solving, reasoning, self-esteem, sociability, self-management, and integrity/honesty.[6]

Research and anecdotal reports alike affirm that skills in writing, speaking, listening, and critical reading are valued by employers,[7] by the professions themselves,[8] and by society at large.[9] For example, consider some of the skills required for a recently advertised, entry-level position with a multinational firm:

> Ability to read, analyze, and interpret common scientific and technical journals, financial reports, and legal documents. Ability to respond to common inquiries

---

[1] L. Barker, R. Edwards, C. Gaines, K. Gladney, and F. Holley. "An Investigation of Proportional Time Spent in Various Communication Activities by College Students." *Journal of Applied Communication Research* 8 (1980) 101–109.

[2] Walter Loban, quoted by Roy Berko, Megan Brooks, and J. Christian Spielvogel, *Pathways to Careers in Communication*. Annandale, VA: Speech Communication Association, 1995: 1.

[3] Canadian Council of Professional Engineers, "Continuing Competence." On line. Available URL: http://www.ccpe.ca/ccpe.cfm?page=ContinuingCompetence

[4] Larry Hein, "Are We Competent?" *Professional Edge* 54 (February-March 1998) 8–9.

[5] Sir Graham Day, "The Engineering Profession: Some Thoughts on the Implications of Globalization and the Need for Continuing Education." *Proceedings of the Canadian Conference on Engineering Education* (Ottawa, ON: Canadian Council of Professional Engineers, 1998) 2–8.

[6] U.S. Department of Labor, *What Work Requires of Schools: A SCANS Report for America 2000.* Washington DC: Secretary's Commission on Achieving Necessary Skills, U.S. Department of Labor, June 1991: vii.

[7] Joseph Hoey, "Employer Satisfaction with Alumni Professional Preparation." *North Carolina State University, University Planning and Analysis* (1997). On line. Available URL: http://www2.acs.ncsu.edu/UPA/survey/reports/employer/employ.htm Accessed 15 September 1999; Isa Engleberg and Dianna Wynn, "DACUM: A National Database Justifying the Study of Speech Communication." *Journal of the Association for Communication Administration* 1 (1995) 28–38.

or complaints from customers, regulatory agencies, or members of the business community. Ability to write speeches and articles for publication that conform to prescribed style and format. Ability to effectively present information to top management, public groups, and/or boards of directors. Ability to define problems, collect data, establish facts, and draw valid conclusions. Ability to interpret an extensive variety of technical instructions in mathematical or diagram form and deal with several abstract and concrete variables.

The education specified for a job demanding such advanced and highly specialized communication and reasoning skills was not, as you might expect, a major in communication; instead, it was an undergraduate degree in engineering.[10] Unfortunately, many graduates of engineering programs are ill-prepared in skills like the ones listed, partly because technical students often don't realize the extent to which their career success will depend on their ability to communicate effectively. Even worse, some faculty, assuming that students will automatically pick up such skills along the way, fail to emphasize the importance of acquiring them.

However, engineering employers like the company that ran this ad recognize that, in our highly complex and demanding workplaces, skill in communication at all levels is essential; they know that clear and easy-to-understand messages will be dealt with efficiently and promptly, while those that are unclear, incomplete, or discourteous will be set aside for later—or possibly disregarded entirely. Similarly, employers know that people who cannot accurately and efficiently interpret the messages of others will be unable to fulfill the requirements of many engineering jobs. For this reason, significant effort has been spent in documenting the need for strong communication skills among engineering professionals,[11] and in discussing ways in which such skills might be implemented across the engineering curriculum.[12] Those who can bring such skills to the workplace have an edge in the job market.

[8] Burton Urquhart, *Communicative Demands of the Technical Workplace.* Thesis. University of Saskatchewan, 2002; Canadian Engineering Qualifications Board, "Admission to the Practice of Engineering in Canada." (Ottawa: Canadian Council of Professional Engineers, nd) 3. See also Larry Hein, "Are We Competent?" *The Professional Edge* 54 (February-March 1998); V. DiSalvo, D.C. Larsen, and W.J. Seiler, "Communication Skills Needed by Persons in Business Organizations." *The Speech Teacher* 25 (1976) 269–275; Elton L. Francis, "When the Engineer Speaks." *Today's Speech* 8 (1960) 6–9; Gary T. Hunt and Louis P. Cusella, "A Field Study of Listening Needs in Organizations." *Communication Education* 32 (1983) 393–401.

[9] Stephen Strauss, "Avoid the Technical Talk, Scientists Told." *The Globe and Mail* (27 August 1996) A1; The Royal Bank of Canada, "The Practical Writer." *The Royal Bank Letter* 62 (Jan/Feb 1981).

[10] Trane, Inc, "Sales Engineer." Position advertisement. On line. http://www.trane.com Accessed 15 May 2001.

[11] Marion G. Barchilon, "Writing for Engineering Fields." *The Practice of Technical and Scientific Communication* (Stamford, CN: Ablex Publishing, 1998) 37-48; R.A. Buonopane, "Engineering Education for the 21st Century." *Chemical Engineering Education* 31 (1997) 166-67; David Beer and David McMurrey, "Engineers and Writing." *A Guide to Writing as an Engineer* (New York: Wiley and Sons, 1997) 1-10.

[12] For example, D. MacIsaac and L. McLean, "Teaching Engineers How to Communicate Effectively." *Proceedings of the Canadian Conference on Engineering Education* (1998) 289–96; Katherine Staples and Cezar Ornatowski, eds., *Foundations for Teaching Technical Communication: Theory, Practice, and Program Design* (Stamford, CN: Ablex Publishing, 1997); R.J. Flemming and G. Wacker, "Communications Training in an Engineering College" *Proceedings of the Canadian Conference on Engineering Education* (1980) 122–133; Jennifer MacLennan, "Banishing Speak and Spell: A New Approach to Teaching Communication to Engineers." Keynote address. Victoria: Canadian Council on Engineering Education, August 2001.

Exactly how are graduates measuring up? A study conducted by the University of North Carolina asked the employers of recent university graduates to assess their new employees in both technical preparedness and communication skill, and then to rate the importance of both kinds of skills to the job being performed. The study showed that in specialized and technical skills, the new graduates consistently exceeded employer expectations and job requirements. However, their communication skills were not as well developed. In the four areas of writing, reading, listening, and public speaking, the employers consistently rated the actual ability of the students significantly lower than the level of skill required by their jobs.[13]

Engineers who hope some day to advance to management have added reason for attending carefully to communication skills. Another recent survey listed managerial communication skills as the single most important factor in creating an effective workplace environment. Managers who are good communicators aid employee retention by helping employees to understand how their work contributes to the organization's goals, thereby increasing their sense of value and commitment to the organization.[14] Since recruiting and training new employees is much more expensive than retaining those already on the job, hiring managers who communicate effectively is a sound investment.

Most people recognize that an engineer's ability to communicate highly technical subjects in a clear and understandable way is important, but they may not realize that good communication involves more than simple information transfer. It requires careful attention not only to the clarity of the message itself, but also to satisfying the needs of the audience and to establishing the speaker's credibility. This is a greater challenge than it may appear, since most of those with whom the engineer regularly communicates are non-specialists whose technical expertise varies widely—clients, support staff, managers, and administrators. While communicating with others who share the same level of technical training is usually quite straightforward, it is actually the exception for most professionals, who communicate more often with lay people than with others in their field of specialization.

The successful engineer must therefore be able to present specialized information in a manner that will enable non-specialist readers to make informed policy, procedural, and funding decisions.[15] In order to do this, an engineer's communication, like that of any other professional, must establish and maintain credibility and authority with those who may be unfamiliar

---

[13] Joseph Hoey, "Employer Satisfaction with Alumni Professional Preparation." North Carolina State University, University Planning and Analysis, 1997. Available URL: http://www2.acs.ncsu.edu/UPA/survey/reports/employer/employ.htm> Accessed 14 June 2001.

[14] "Three Out of Four Say Better Communication Equals Greater Employee Retention." KnowledgePoint, Press Release via *Business Wire* (8 December 2000).

[15] Hudspith, R.D. "Education for Social Responsibility: Two Approaches." *Proceedings of the Canadian Conference on Engineering Education* (Ottawa, ON: Canadian Council of Professional Engineers, 1998) 468–475.

with technical subjects. Because such lay readers cannot directly judge technical skill, they will instead rely on the clarity and confidence of a professional's communication as a basis for judging technical competence. Thus, skill in communicating specialized information often becomes the measure of an engineer's competence, irrespective of his or her actual technical expertise.

The course in which you are reading this book is intended to enhance your practical communication skills in four areas of communication: writing, speaking, critical reading, and listening. It will help to improve your ability to communicate by building a foundation for making effective judgements in the various kinds of communication situations you face on the job. For this reason, before considering the specific techniques of professional and technical communication, we will take time to explore some theoretical principles that form the foundation of communication effectiveness.

WHAT IS COMMUNICATION?

Although "communication" is frequently thought of as simply the "transmission of information by speaking, writing, or other means,"[16] it is actually a much more complex process than this simple characterization would lead us to believe. It is also a longstanding subject of systematic study; one of the oldest books in existence (dating to 2675 BCE) is a treatise on effective communication.[17]

One of the most comprehensive contemporary definitions characterizes communication as a process of "adjusting ideas to people, and people to ideas."[18] This definition recognizes not only that messages must be adapted to their intended audiences, but also that audiences in turn are affected and influenced by the messages that they hear. Subsequent messages must be further adapted to accommodate these changes in audience knowledge or awareness; these messages further affect the audience's understanding in ways that must be accommodated by future messages, and so on. For this reason, communication is a dynamic process of observation, judgement, and adaptation that cannot rely on cookie-cutter templates. It is an interaction, a process, rather than a simple transaction or exchange of static information.

Three Keys to Effective Communication

The three keys to effective communication that make possible the process of adjustment just described grow out of the three components of every

---

[16] Katherine Barber, ed., *The Canadian Oxford Dictionary* (Toronto: Oxford University Press, 1998) 287.

[17] Known as the *Precepts*, it was composed in Egypt by Ptah-Hotep. See James McCroskey, "A Rhetorical Tradition." *An Introduction to Rhetorical Communication* 6e (New Jersey: Prentice Hall, 1993) 2.

[18] Donald C. Bryant, *Rhetorical Dimensions in Criticism* (Baton Rouge, LA: Louisiana State University Press, 1973) 19.

Three Keys to Effective Communication:

- Sound and logical arguments
- Effective audience appeal
- Personal credibility

communication interaction: a speaker (or writer), an audience, and a message or meaning to be communicated. To communicate effectively, a speaker or writer must pay attention to all three: building sound and logical arguments, creating effective audience appeal, and establishing personal credibility.

The dynamic relationship among the three elements means that, though some typical situations tend to recur (job applications, incident reports, or performance evaluations, for example), no two communication interactions are precisely identical. We must be able to rely on trained judgement and sensitivity to nuance to help us communicate appropriately for the specific demands of each situation, finding a proper balance among audience, message, and speaker.

While at first glance we might be inclined to think that the most important element of any communication is the quality of the message itself, a communicator who wishes to be understood, respected, and heeded must pay careful attention to all three key components of effective communication. These principles have been recognized at least since Aristotle wrote his practical handbook on public communication, known as the *Rhetoric* (circa 330 BCE). Like many current communication textbooks—including this one—Aristotle's book combines theory and application, offering an analysis of the principles of communication along with practical advice on how to apply them. As different as today's study of communication is from the book Aristotle wrote, it still deals with the practical reasons why people communicate and the strategies they use to make their communication effective.

Aristotle understood, for example, that it is frequently the speaker's credibility that most determines the way a message is received. This feature of effective communication, which he named "ethos," refers to the way a speaker's or writer's character is revealed in the style and tone of the message. Effective ethos depends on communicating good will toward the audience, good judgement of the situation and issues, and good character–that is, integrity and credibility. As Aristotle pointed out, an audience will be unlikely to trust a speaker who does not demonstrate these qualities, and will therefore be inclined to doubt or even reject such a person's message, no matter how accurate it may otherwise be.

As an indicator of the importance of ethos in technical communication, consider once more the APEGS survey mentioned above. When asked to identify "categories used to describe competence," the professional engineers who took part in the survey identified seven qualities that reflect the same qualities of good judgement, good will, and good character that Aristotle noted as important to effective communication.[19] These seven measures are quickly recognizable as aspects of ethos. For example, the top

---

[19] An eighth category of "Miscellaneous" traits was also generated. Hein 8.

two categories, "experience" and "knowledge," assess the candidate's demonstration of good judgement, as does the category "publications, patents, awards." "Approach" and "attitude" are measures of good will, while "professional association," "references," and "clean record" offer evidence of good character.

In addition to ethos, or the speaker's character, Aristotle also recognized the importance of the other two components of communication. He emphasized the need to pay attention to the audience's needs, expectations, experience, and concerns, and the need to develop a coherent, well-argued message. A successful communicator, he said, must establish a bond with the audience. To do so, a writer or speaker must consider and respond to the things that the listeners value, need, hope for, fear, or care about, and must demonstrate respect and concern for the audience. Aristotle called this aspect of communication "pathos." Finally, an effective communicator must construct the message with care, being sure to provide reasonable and logically sound arguments. A speaker or writer must demonstrate a thorough knowledge of the issues being discussed, show a command of both style and subject matter, and avoid logical inconsistencies. This aspect of effective communication Aristotle called "logos."

In the remainder of this chapter, and indeed in the remainder of this book, we will explore the way in which the three forms of appeal identified by Aristotle—the good will, good judgement, and good character of the speaker (ethos), a clear and well-constructed message (logos), and a recognition of and accommodation to audience (pathos)—provide a foundation for building effective professional communication in any situation. Let's begin with nine axioms that can help us to understand exactly how effective communication functions.

NINE AXIOMS OF COMMUNICATION

These axioms about the practice of communication arise out of Aristotle's three principles of communication effectiveness. Some of these axioms may surprise those who think of communication as primarily information exchange, but they make sense once we recognize the relationship among the three components of the communication dynamic—audience, message, and speaker.

1. Communication is not simply an exchange of ideas or information, but an interaction between people.

The economic metaphor of communication as the "exchange" of a commodity of information is one we have come to take for granted, but because it tends to obscure the human element of the interaction, it is a flawed understanding of communication. You will quickly recognize how important this element of human contact is to communication if you've ever been frustrated in trying to communicate with the impersonal face of a bureaucracy where your concerns are ignored or unrecognized.

Communication has been described as a process of "building community by exchanging symbols."[20] This definition rightly emphasizes the relationships that are formed when we communicate, rather than the information exchanged. Our social success and our professional effectiveness alike often depend on such relationships, and every time we communicate, whether in face-to-face exchanges, by telephone, through letters or reports, or via e-mail, we should make human contact our first consideration. Thinking of communication primarily as interacting with other people will improve your communication skills immediately and markedly.

Emphasizing the human interaction at the heart of communication also leads directly to our second axiom.

2. All communication involves an element of relation as well as content.

When we understand communication as an interaction between people rather than simply an exchange of information, we realize that all communication—on the job and elsewhere—involves building or maintaining human relationships. This element of a message is called the *relation,* and it is frequently as important as—and sometimes more so than—the content or information that the message contains. In fact, ignoring this quality of relation is one reason that many workplace messages fail.

It is true that many professionals, and especially organizations, may attempt to depersonalize their communication and neglect its relational component. However, although this practice is all too common, it actually displays a failure to communicate, because it refuses to really engage the reader or hearer. Many writers and language specialists, among them the noted Canadian cultural critic Northrop Frye and the essayist and social commentator George Orwell, have condemned this bureaucratic use of language. Frye calls it "dead, senseless, sentenceless, written pseudo-prose that surrounds us like a boa constrictor, which is said to cover its victims with slime before strangling them."[21] For example, compare your typical response to a form letter with how it feels to receive a letter genuinely addressed to you by someone in an organization who has taken the time to deal directly with your request or situation.

---

[20] Roderick P. Hart, *Modern Rhetorical Criticism*, 2e. (Boston: Allyn & Bacon, 1997) 36.

Part of the function of every professional message, no matter how routine, is to create and maintain an effective working relationship between the writer and the reader. Thus, as you write, you need to think not only about the content of your message but about how you are building and shaping your professional relationships through your communication with your clients, colleagues, and managers. In the long run, neglect of this function will undermine the firm's relationships with its clients, and eventually will compromise its professional standing. E-mail, as we will later see, is especially prone to failures of relation-building because of the depersonalizing effect of the technology, and needs to be handled with particular care.

3. All communication takes place within a context of "persons, objects, events, and relations."[22]

This axiom reminds us that all our messages are products of their social, historical, professional, and personal context. If they are to be effective, we must consider not only the information requirements, but also the background, history, interpersonal qualities, professional constraints, and social consequences of the situation into which they are introduced. Communication is always situated; no message stands entirely on its own, and no communicator who hopes to be effective can afford to ignore the situation in which any message must function. As we create our messages, we must consider the needs of the audience and the impact of our own self-presentation, as well as the professional, social, political, historical, and interpersonal context that shapes both the message itself and the way in which that message is interpreted.

4. Communication is the main means through which we exert influence.

Influence—the ability to gain cooperation and compliance from others—is a fundamental requirement for our survival, comfort, and success. Few of us are skilled enough to provide all the necessities of life for ourselves. The food we eat, the homes we live in, the clothes we wear, the roads we travel, the vehicles we drive, and many other goods and services, large and small, are largely provided for us by others. To fulfill these needs, we must rely on our ability to influence others.

Our social structures depend on cooperation among individuals, and cooperation, in turn, depends on effective and ethical persuasion, an issue we will explore in greater detail throughout this book. With the exception of criminal behaviour and military combat, virtually all influence in the contemporary world is exercised through communication, whether it takes

---

[21] Northrop Frye, *The Well-Tempered Critic* (Bloomington: Indiana University Press, 1963) 37.

[22] Lloyd Bitzer, "The Rhetorical Situation." *Philosophy and Rhetoric* 1 (1968) 6. (Bitzer's essay can be found on page 74 of this book.)

place interpersonally, organizationally, politically, or through the mass media. We exert persuasive influence daily in many ways, large and small, and we do it through our interaction with others.

Consider a simple and immediate example. In order to acquire the qualifications needed for a professional position, you must convince a series of officials that you have mastered the fundamental skills you need. You do this by communicating your understanding in exams and assignments, thus persuading your examiners to grant you the degrees and credentials you require. After graduation, you will continue to exercise influence—an effective résumé, a well-prepared application form, and a successful interview will persuade an employer to hire you. Once on the job, you will exert influence through reports, e-mails, letters, phone calls, meetings, conversations, and presentations that must convincingly convey your professionalism and expertise to a series of audiences, causing them to respect and accept the advice, ideas, policies, or actions that you recommend.

The same is also true of all of our personal and social lives: satisfaction and success depend on our ability to enlist the cooperation of others. For this reason, our professional and personal success and fulfilment are directly dependent on the quality of our communication.

5. Communication is a principal way of establishing credibility.

Whether we like it or not, people do judge us by the way we communicate. Indeed, as we have already discovered, our ability to communicate is frequently the *only* basis on which others assess our professional competence and authority. Think for a moment about your own judgement of others' professional skill—your physician or dentist, your mechanic, your professors, your MP or MLA, your lawyer or clergyperson or veterinarian. Few of us are qualified to directly assess the skills of such specialists; instead, we form our impressions of their credibility and knowledge based on how they communicate, on the quality of their messages, and on the respect and good will they extend to us.

Even in cases where we feel we are sufficiently qualified to assess someone else's expertise, we rarely have the opportunity to do so directly. For example, your classmates have training and experience roughly equivalent to your own. You probably have an impression of how smart or how skilled they are—but ask yourself on what basis you have made these judgements. Since it is unlikely that you have actually studied their work, your assessments are more likely based on the way they have communicated to you in class, in labs, and in person. In exactly the same way, others judge our professional integrity and credibility based on how well we communicate our messages to them.

6. All communication involves an element of interpersonal risk.

This axiom recognizes the close relationship between credibility and communication. Because so much of our worthiness as professionals and as individuals is judged by how effectively we communicate, it is not surprising that every time we interact with others, our professional and personal reputations are at stake. In communicating with new audiences, we must demonstrate the good will, good judgement, and good character that will establish us as reliable, credible professionals. In interacting with those who already know us, we hope to reinforce those qualities that we have previously established. If we do not do so, we risk losing face, an experience that can shame and discredit us.[23]

For example, in a job interview where a potential employer is meeting you for the first time, you strive to create a positive impression in appearance, attitude, and knowledge, so that the employer will take you seriously as a potential employee. Once you have obtained the position, such assessments continue, both formally and informally, and you strive to maintain the good impression that you established in the initial contact. It is important to remember that all kinds of communication, not just interviews, involve similar risks of being judged and possibly rejected. Consider the "face risk" inherent in other frequent communication activities—seeking a favour from a friend, inviting an acquaintance on a date, requesting a raise, applying for a loan, phoning a stranger, selling an item door-to-door, canvassing for a charity, or giving a speech. Face risk is part of what makes such activities intimidating for many people. We wish to establish, and maintain, an air of competence and integrity; we wish to be accepted and respected. Each interaction presents a danger that this public image may be compromised, and thus each contact involves risk.

7. Communication is frequently ambiguous: what is unsaid can be as important as what is said.

Although our focus in this book is on verbal communication—written and spoken (oral) messages—it is important to remember that all communication also contains significant nonverbal elements. This is true especially, but not exclusively, in face-to-face interactions.

Nonverbal cues such as eye contact, facial expression, vocal tone, gestures, physical movement, stance and posture, manner of dress, personal hygiene, or overall attractiveness communicate personal traits. Your nonverbal expressions, commonly referred to as body language," can display

---

[23] For more on the importance of face in human interaction, see Erving Goffman, "On Face Work." *Interaction Ritual* (Garden City, NY: Anchor Books, 1967) 5–23.

assertiveness, confidence, self-possession, energy, enthusiasm, cooperativeness, reliability, and truthfulness—or their opposites. In written communication there is no opportunity for the reader to judge us on appearance or physical presentation, but such additional elements as diction, sentence structure, grammar, spelling, punctuation, handwriting, and document design combine with tone and style to provide plenty of nonverbal information.

Although the study of nonverbal communication can be quite complex, in general, you should remember that nonverbal cues are always present, and can either complement or contradict the verbal message. Nonverbal signals that do not seem to agree with verbal messages can discredit your communication. Although it is not possible to completely control your nonverbal messages, you can learn to pay closer attention to the way you colour your messages, both orally and in writing, through unconscious nonverbal cues.

8. Effective professional communication is audience-centred, not self-centred.

Another way of putting this axiom is that effective professional communication is rhetorical. This means both that it has a practical purpose and that it is addressed to an audience. Rhetoric is the word we use to describe communication intended to "produce action or change in the world"[24] by appealing to those who are in a position to make this change; it is the means by which we exercise influence in order to get things done. For this reason, rhetorical communication depends on engaging the audience, motivating their concern, and enabling them to make decisions or take an action that the speaker believes to be essential or desirable. To do so, it must be audience-focussed rather than speaker- or writer-focussed.

Unfortunately, however, many of us habitually focus on what we want to say rather than on what the audience needs to hear, concentrating on ourselves instead of our readers or listeners. In fact, many messages fail for exactly this reason, because they focus on the speaker's or writer's interests at the expense of the audience's needs and expectations. Serving our own needs and interests while ignoring those of the audience is acceptable in messages that are strictly self-expression. But because it is essentially self-centred, such expressive communication typically fails to inspire audience confidence or collaboration, and is useless in situations where we need their cooperation.

If we are interested in engaging others, in moving them to acceptance or action, we cannot ignore the needs, expectations, or perspective of our audience. We must train ourselves to take into account the audience's point of view as we design oral and written messages that require the cooperation of others.

[24] Bitzer 4.

9. Communication is pervasive: you cannot *not* communicate.

This final axiom recognizes the pervasiveness of communication and of the human need to interpret symbolic meanings. Interestingly, even non-messages are typically read as having communicative meaning, often because of the nonverbal cues discussed in Axiom #7.

For example, imagine you have interviewed for a job. If a length of time passes without your hearing from the employer, you will undoubtedly interpret this lack of communication as a negative message in itself. This simple example demonstrates that once a channel of exchange has been opened, anything that occurs—even if it is nothing at all—will be read as a message. As soon as we have entered into any form of interaction with others, we are engaged in a permanent relation in which non-communication becomes impossible.

Channels can be opened not only by choice, as in the interview scenario, but also by coincidence—as long as you live, work, attend school, socialize, or shop where there are other human beings, you are immersed in a social environment filled with messages. Participating in this pervasive environment is inevitable, because each action of yours will be read by others for its communicative meaning. Even a refusal to participate in communication is a form of communication that others will interpret, just as we read meaning in the non-messages of others.

For example, suppose that the apartment across the hall from you is occupied by a man with whom you have never spoken. You see each other every day as you come and go from the building, and yet he never greets or acknowledges you. Soon you will begin to infer meaning from his uncommunicative actions: you might consider him shy, unfriendly, a loner, or a snob. You might even begin to wonder whether he is antisocial, possibly even dangerous. Your inferences may be accurate or inaccurate, but they will be inescapable, because the man's resistance to normal social interaction itself constitutes a message, whether or not he intended it to do so.

Once a context for communication exists, there is no possibility of non-communication. Every action you perform will send messages that will be understood—or misunderstood—by others.

Using the Axioms to Improve Your Writing

The nine axioms we have just considered give us some insight into the way communication functions. However, as you will learn in the next chapter, they are also *useful* because they provide a foundation for devising effective practical strategies to ensure that our communication is effective. Whenever we are communicating with others on practical issues, we will enhance our skill if we remember to consider the expectations, needs, and interests of those we hope to engage, and if we can keep the purpose of our communication clearly in view. The rest of this book is devoted to helping you to do both of these things by developing some concrete methods for improving the quality of your professional messages.

SHARPENING YOUR SKILLS

1. Think about the last time you dealt with a representative of a bureaucracy: a clerk in a government office, an administrative functionary in the university or college where you go to school, a representative of an insurance company or a bank. Did you feel satisfied with the way your concerns were handled? If so, what elements of the communication made you feel this way? If not, why not? In what ways do the nine axioms of communication help to explain what took place?

2. Consider the following communication challenges: which involve the greater risks? Make a list of these experiences (and add others that occur to you), ranking them in order from the highest to the lowest level of risk.
   ___ returning an item to a store
   ___ attending a job interview with one interviewer
   ___ attending a job interview with a panel of interviewers
   ___ sending back a poorly cooked meal in a restaurant
   ___ asking a neighbour you do not know to turn down loud music
   ___ asking a neighbour you know to turn down loud music
   ___ telephoning a stranger
   ___ telemarketing
   ___ asking an acquaintance for a favour
   ___ asking a friend for a loan
   ___ asking a professor to reconsider a grade on your assignment
   ___ inviting an acquaintance on a first date
   ___ arranging a date with your regular boyfriend or girlfriend
   ___ making a television appearance

__ making a speech in front of family and friends
__ making a speech in front of strangers
__ going on a blind date
__ giving a presentation in class
__ persuading someone to take an action you have taken
__ persuading someone to take an action you have not taken
__ asking for a raise
__ canvassing door-to-door for charity
__ canvassing for charity by telephone
__ receiving an award in public

Compare your list with that of others in the class. How much consensus was there? What differences did you discover? To what extent does this exercise suggest that communication apprehension (the fear of face risk) is situational, cultural, or individual? Summarize the results of the class discussion and write a short memo report for your instructor. Submit the report via e-mail if your instructor directs you to do so.

CRITICAL READING

Below you will find an article on effective writing by the rhetorical communication expert Wayne Booth. "The Rhetorical Stance" discusses the relationship among speaker, audience, and message that creates the dynamic known as the "rhetorical triangle." Although the article deals with issues similar to those discussed in this chapter, and makes similar arguments, it differs significantly from the chapter in its treatment of the subject matter. As you read, consider why these differences might exist, and use the principles you have learned about communication to understand how the article and the chapter function differently. Questions about audience, purpose, and author credibility will help you understand both this reading in particular, and communication in general. Appendix A, "How Texts Communicate: Some Suggestions for Reading Critically," might be of some assistance as you approach this task.

After reading the article, respond as directed by your instructor to one or more of the following assignments.

1. Compare Booth's discussion to the treatment of similar material in this chapter. Write a memo to your instructor in which you describe the central principles emphasized by both the article and the chapter materials, and describe at least two major differences that you see in the treatment of their subject matter. Given what both the chapter and the article say about the nature of successful communication, why do you think the two are so different in approach?

2. After writing your memo, think about the following and be prepared to discuss it in class: which reading—Booth or the chapter—did you find more accessible? Why? For whom was each written? What is the purpose of each? Do these factors make a difference?

3. Answer the questions following the essay as directed by your instructor. Submit your responses via e-mail or in memo format.

THE RHETORICAL STANCE

Wayne C. Booth

Last fall I had an advanced graduate student, bright, energetic, well-informed, whose papers were almost unreadable. He managed to be pretentious, dull, and disorganized in his paper on *Emma*, and pretentious, dull, and disorganized on *Madame Bovary*. On *The Golden Bowl* he was all these and obscure as well. Then one day, toward the end of term, he cornered me after class and said, "You know, I think you were all wrong about Robbe-Grillet's *Jealousy* today."* We didn't have time to discuss it, so I suggested that he write me a note about it. Five hours later I found in my faculty box a four-page polemic, unpretentious, stimulating, organized, convincing. Here was a man who had taught freshman composition for several years and who was incapable of committing any of the more obvious errors that we think of as characteristic of bad writing. Yet he could not write a decent sentence, paragraph, or paper until his rhetorical problem was solved—until, that is, he had found a definition of his audience, his argument, and his own proper tone of voice.

The word "rhetoric" is one of those catch-all terms that can easily raise trouble when our backs are turned. As it regains a popularity that it once seemed permanently to have lost, its meanings seem to range all the way from something like "the whole art of writing on any subject," as in Kenneth Burke's *The Rhetoric of Religion*, through "the special arts of persuasion," on down to fairly narrow notions about rhetorical figures and devices. And of course we still have with us the meaning of "empty bombast," as in the phrase "merely rhetorical."

I suppose that the question of the role of rhetoric in the English course is meaningless if we think of rhetoric in either its broadest or its narrowest meanings. No English course could avoid dealing with rhetoric in Burke's sense, under whatever name, and on the other hand nobody would ever advocate anything so questionable as teaching "mere rhetoric." But if we settle on the following traditional definition, some real questions are raised: "Rhetoric is the art of finding

Wayne Booth is George M. Pullman Professor Emeritus of English at the University of Chicago. His best-known work is *The Rhetoric of Fiction* (1961). He is also the author of *A Rhetoric of Irony* (1974), *Modern Dogma and the Rhetoric of Assent* (1974), *Critical Understanding: The Powers and Limits of Pluralism* (1979), *The Company We Keep: An Ethics of Fiction* (1988), and *The Art of Deliberalizing: A Handbook for True Professionals* (1990). Our selection holds the distinction of being the most frequently reprinted article from *College Composition and Communication*, where it first appeared in 1964. In it, Booth probes the question of what it means to achieve an appropriate rhetorical balance in writing, a quality quite apart from correctness in grammar, sentence structure, and spelling. Booth is one of many contemporary theorists who have defined rhetoric quite broadly as encompassing the process of human communication.

*The words in italics are titles of novels.

and employing the most effective means of persuasion on any subject, considered independently of intellectual mastery of that subject." As the students say, "Prof. X knows his stuff but he doesn't know how to put it across." If rhetoric is thought of as the art of "putting it across," considered as quite distinct from mastering an "it" in the first place, we are immediately landed in a bramble bush of controversy. Is there such an art? If so, what does it consist of? Does it have a content of its own? Can it be taught? Should it be taught? If it should, how do we go about it, head on or obliquely?

Obviously it would be foolish to try to deal with many of these issues in twenty minutes. But I wish that there were more signs of our taking all of them seriously. I wish that along with our new passion for structural linguistics, for example, we could point to the development of a rhetorical theory that would show just how knowledge of structural linguistics can be useful to anyone interested in the art of persuasion. I wish there were more freshman texts that related every principle and every rule to functional principles of rhetoric, or, where this proves impossible, I wish one found more systematic discussion of why it is impossible. But for today, I must content myself with a brief look at the charge that there is nothing distinctive and teachable about the art of rhetoric.

The case against the isolability and teachability of rhetoric may look at first like a good one. Nobody writes rhetoric, just as nobody ever writes writing. What we write and speak is always *this* discussion of the decline of railroading and *that* discussion of Pope's couplets and the other argument for abolishing the poll-tax or for getting rhetoric back into English studies.

We can also admit that like all the arts, the art of rhetoric is at best very chancy, only partly amenable to systematic teaching; as we are all painfully aware when our 1:00 section goes miserably and our 2:00 section of the same course is a delight, our own rhetoric is not entirely under control. Successful rhetoricians are to some extent like poets, born, not made. They are also dependent on years of practice and experience. And we can finally admit that even the firmest of principles about writing cannot be taught in the same sense that elementary logic or arithmetic or French can be taught. In my first year of teaching, I had a student who started his first two essays with a swear word. When I suggested that perhaps the third paper ought to start with something else, he protested that his high school teacher had taught him always to catch the reader's attention. Now the teacher was right, but the application of even such a firm principle requires reserves of tact that were somewhat beyond my freshman.

But with all of the reservations made, surely the charge that the art of persuasion cannot in any sense be taught is baseless. I cannot think that anyone who has ever read Aristotle's *Rhetoric* or, say, Whateley's *Elements of Rhetoric* could seriously make the charge. There is more than enough in these and the other traditional rhetorics to provide structure and content for a year-long course. I believe that such a course, when planned and carried through with intelligence and flexibility, can be one of the most important of all educational experiences. But it seems obvious that the arts of persuasion cannot be learned in one year, that a good teacher will continue to teach them regardless of his subject matter, and that we as English teachers have a special responsibility at all levels to get certain basic rhetorical principles into all of our writing assignments. When I think back over the experiences which have had any actual effect on my writing, I find the great good fortune of a splendid freshman course, taught by a man who believed in what he was doing, but I also find a collection of other experiences quite unconnected with a specific writing course. I remember the instructor in psychology who penciled one word after a peculiarly pretentious paper of mine: *bull.* I remember the day when P. A. Christensen talked with me about my Chaucer paper, and made me understand that my failure to use effective transitions was not simply a technical fault but a fundamental block in my effort to get him to see my meaning. His off-the-cuff pronouncement that I should never let myself write a sentence that was not in some way explicitly attached to preceding and following sentences meant far more to me at that moment, when I had something I wanted to say, than it could have meant as part of a pattern of such rules offered in a writing course. Similarly, I can remember the devastating lessons about my bad writing that Ronald Crane could teach with a simple question mark on a graduate seminar paper, or a penciled "Evidence for this?" or "Why this section here?" or "Everybody says so. Is it true?"

Such experiences are not, I like to think, simply the result of my being a late bloomer. At least I find my colleagues saying such things as "I didn't learn to write until I became a newspaper reporter," or "The most important training in writing I had was doing a dissertation under old *Blank.*" Sometimes they go on to say that the freshman course was useless; sometimes they say that it was an indispensable preparation for the later experience. The diversity of such replies is so great as to suggest that before we try to reorganize the freshman course, with or without explicit confrontations with rhetorical categories, we ought to look for whatever there is in common among our experiences, both of good writing and of good writing instruction. Whatever we discover in such an enterprise ought to be useful to us at any level of our teaching. It will not, presumably, decide once and for all what should be the content of the freshman course, if there should be such a course. But it might serve as a guideline for the development of widely different programs in the widely differing institutional circumstances in which we must work.

The common ingredient that I find in all of the writing I admire—excluding for now novels, plays and poems—is something that I shall reluctantly call the rhetorical stance, a stance which depends on discovering and maintaining in any writing situation a proper balance among the three elements that are at work in any communicative effort: the available arguments about the subject itself, the interests and peculiarities of the audience, and the voice, the implied character, of the speaker. I should like to suggest that it is this balance, this rhetorical stance, difficult as it is to describe, that is our main goal as teachers of rhetoric. Our ideal graduate will strike this balance automatically in any writing that he considers finished. Though he may never come to the point of finding the balance easily, he will know that it is what makes the difference between effective communication and mere wasted effort.

What I mean by the true rhetorician's stance can perhaps best be seen by contrasting it with two or three corruptions, unbalanced stances often assumed by people who think they are practicing the arts of persuasion.

The first I'll call the pedant's stance: it consists of ignoring or underplaying the personal relationship of speaker and audience and depending entirely on statements about a subject—that is, the notion of a job to be done for a particular audience is left out. It is a virtue, of course, to respect the bare truth of one's subject, and there may even be some subjects which in their very nature define an audience and a rhetorical purpose so that adequacy to the subject can be the whole art of presentation. For example, an article on "The relation of the ontological and teleological proofs," in a recent *Journal of Religion*, requires a minimum of adaptation of argument to audience. But most subjects do not in themselves imply in any necessary way a purpose and an audience and hence a speaker's tone. The writer who assumes that it is enough merely to write an exposition of what he happens to know on the subject will produce the kind of essay that soils our scholarly journals, written not for readers but for bibliographies.

In my first year of teaching I taught a whole unit on "exposition" without ever suggesting, so far as I can remember, that the students ask themselves what their expositions were *for*. So they wrote expositions like this one—I've saved it, to teach me toleration of my colleagues: the title is "Family Relations in More's *Utopia*." "In this theme I would like to discuss some of the relationships with the family which Thomas More elaborates and sets forth in his book, *Utopia*. The first thing that I would like to discuss about family relations is that overpopulation, according to More, is a just cause of war." And so on. Can you hear that student sneering at me, in this opening? What he is saying is something like "you ask for a meaningless paper, I give you a meaningless paper." He knows that he has no audience except me. He knows that I don't want to read his summary of family relations in *Utopia*, and he knows that I know that he therefore has no rhetorical purpose. Because he has not been led to see a question which he considers worth answering, or an audience that could possibly care one way or the other, the paper is worse than no paper at all, even though it has no grammatical or spelling errors and is organized right down the line, one, two, three.

An extreme case, you may say. Most of us would never allow ourselves that kind of empty fencing? Perhaps. But if some carefree foundation is willing to finance a statistical study, I'm willing to wager a month's salary that we'd find at least half of the suggested topics in our freshman texts as pointless as mine was. And we'd find a good deal more than half of the discussions of grammar, punctuation, spelling, and style totally divorced from any notion that rhetorical purpose to some degree controls all such matters. We can offer objective descriptions of levels of usage from now until graduation, but unless the student discovers a desire to say something to somebody and learns to control his diction for a purpose, we've gained very little. I once gave an assignment asking students to describe the same classroom in three different statements, one for each level of usage. They were obedient, but the only ones who got anything from the assignment were those who intuitively imported the rhetorical instructions I had overlooked—such purposes as "Make fun of your scholarly surroundings by describing this classroom in extremely elevated style," or "Imagine a kid from the slums accidentally trapped in these surroundings and forced to write a description of this room." A little thought might have shown me how to give the whole assignment some human point, and therefore some educative value.

Just how confused we can allow ourselves to be about such matters is shown in a recent publication of the Educational Testing Service, called "Factors in Judgments of Writing Ability." In order to isolate those factors which affect differences in grading standards, ETS set six groups of readers—business men, writers and editors, lawyers, and teachers of English, social science and natural science—to reading the same batch of papers. Then ETS did a hundred-page "factor analysis" of the amount of agreement and disagreement, and of the elements which different kinds of graders emphasized. The authors of the report express a certain amount of shock at the discovery that the median correlation was only .31 and that 94% of the papers received either 7, 8, or 9 of the 9 possible grades.

But what *could* they have expected? In the first place, the students were given no purpose and no audience when the essays were assigned. And then all these editors and business men and academics were asked to judge the papers in a complete vacuum, using only whatever intuitive standards they cared to use. I'm surprised that there was any correlation at all. Lacking instructions, some of the students undoubtedly wrote polemical essays, suitable for the popular press; others no doubt imagined an audience, say, of *Reader's Digest* readers, and others wrote with the English teachers as implied audience; an occasional student with real philosophical bent would no doubt do a careful analysis of the pros and cons of the case. This would be graded low, of course, by the magazine editors, even though they would have graded it high if asked to judge it as a speculative contribution to the analysis of the problem. Similarly, a creative student who has been getting A's for his personal essays will write an amusing colorful piece, failed by all the social scientists present, though they would have graded it high if asked to judge it for what it was. I find it shocking that tens of thousands of dollars and endless hours should have been spent by students, graders, and professional testers analyzing essays and grading results totally abstracted from any notion of purposeful human communication. Did nobody protest? One might as well assemble a group of citizens to judge students' capacity to throw balls, say, without telling the students or the graders whether altitude, speed, accuracy or form was to be judged. The judges would be drawn from football coaches, hai-alai experts, lawyers, and English teachers, and asked to apply whatever standards they intuitively apply to ball throwing. Then we could express astonishment that the judgments did not correlate very well, and we could do a factor analysis to discover, lo and behold, that some readers concentrated on altitude, some on speed, some on accuracy, some on form—and the English teachers were simply confused.

One effective way to combat the pedantic stance is to arrange for weekly confrontations of groups of students over their own papers. We have done far too little experimenting with arrangements for providing a genuine audience in this way. Short of such developments, it remains true that a good teacher can convince his students that he is a true audience, if his comments on the papers show that some sort of dialogue is taking place. As Jacques Barzun says in *Teacher in America,* students should be made to feel that unless they have said something to someone, they have failed; to bore the teacher is a worse form of failure than to anger him. From this point of view we can see that the charts of grading symbols that mar even the best freshman texts are not the innocent time savers that we pretend. Plausible as it may seem to arrange for more corrections with less time, they inevitably reduce the student's sense of purpose in writing. When he sees innumerable W13's and P19's in the margin, he cannot possibly feel that the art of persuasion is as important to his instructor as when he reads personal comments, however few.

This first perversion, then, springs from ignoring the audience or over-reliance on the pure subject. The second, which might be called the advertiser's stance, comes from *under*valuing the subject and overvaluing pure effect: how to win friends and influence people.

Some of our best freshman texts—Sheridan Baker's *The Practical Stylist,* for example—allow themselves on occasion to suggest that to be controversial or argumentative, to stir up an audience, is an end in itself. Sharpen the controversial edge, one of them says, and the clear implication is that one should do so even if the truth of the subject is honed off in the process. This perversion is probably in the long run a more serious threat in our society than the danger of ignoring the audience. In the time of audience-reaction meters and pre-tested plays and novels, it is not easy to convince students of the old Platonic truth that good persuasion is honest persuasion, or even of the old Aristotelian truth that the good rhetorician must be master of his subject, no matter how dishonest he may decide ultimately to be. Having told them that good writers always to some degree accommodate their arguments to the audience, it is hard to explain the difference between justified accommodation—say changing *point one* to the final position—and the kind of accommodation that fills our popular magazines, in which the very substance of what is said is accommodated to some preconception of what will sell. "The publication of *Eros* [magazine] represents a major breakthrough in the battle for the liberation of the human spirit."

At a dinner about a month ago I sat between the wife of a famous civil rights lawyer and an advertising consultant. "I saw the article on your book yesterday in the Daily News," she said, "but I didn't even finish it. The title of your book scared me off. Why did you ever choose such a terrible title? Nobody would buy a book with a title like that." The man on my right, whom I'll call Mr. Kinches, overhearing my feeble reply, plunged into a conversation with her, over my torn and bleeding corpse. "Now with my *last* book," he said, "I listed 20 possible titles and then tested them out on 400 business men. The one I chose was voted for by 90 percent of the businessmen." "That's what I was just saying to Mr. Booth," she said. "A book title ought to grab you, and *rhetoric* is not going to grab anybody." "Right," he said. "My *last* book sold 50,000 copies already; I don't know how this one will do, but I polled 200 businessmen on the table of contents, and . . . ."

At one point I did manage to ask him whether the title he chose really fit the book. "Not quite as well as one or two of the others," he admitted, "but that doesn't matter, you know. If the book is designed right, so that the first chapter pulls them in, and you *keep* 'em in, who's going to gripe about a little inaccuracy in the title?"

Well, rhetoric is the art of persuading, not the art seeming to persuade by giving everything away at the start. It presupposes that one has a purpose concerning a subject which itself cannot be fundamentally modified by the desire to persuade. If Edmund Burke had decided that he could win more votes in Parliament by choosing the other side—as he most certainly could have done—we would hardly hail this party-switch as a master stroke of rhetoric. If Churchill had offered the British "peace in our time," with some laughs thrown in, because opinion polls had shown that more Britons were "grabbed" by these than by blood, sweat, and tears, we could hardly call his decision a sign of rhetorical skill.

One could easily discover other perversions of the rhetorician's balance—most obviously what might be called the entertainer's stance—the willingness to sacrifice substance to personality and charm. I admire Walker Gibson's efforts to startle us out of dry pedantry, but I know from experience that his exhortations to find and develop the speaker's voice can lead to empty colourfulness. A student once said to me, complaining about a colleague, "I soon learned that all I had to do to get an A was imitate Thurber."

But perhaps this is more than enough about the perversions of the rhetorical stance. Balance itself is always harder to describe than the clumsy poses that result when it is destroyed. But we all experience the balance whenever we find an author who succeeds in changing our minds. He can do so only if he knows more about the subject than we do, and if he then engages us in the process of the thinking—and feeling—it through. What makes the rhetoric of Milton and Burke and Churchill great is that each presents us with the spectacle of a man passionately involved in thinking an important question through, in the company of an audience. Though each of them did everything in his power to make his point persuasive, including a pervasive use of the many emotional appeals that have been falsely scorned by many a freshman composition text, none would have allowed himself the advertiser's stance; none would have polled the audience in advance to discover which position would get the votes. Nor is the highly individual personality that springs out at us from their speeches and essays present for the sake of selling itself. The rhetorical balance among speakers, audience, and argument is with all three men habitual, as we see if we look at their non-political writings. Burke's work on the *Sublime and Beautiful* is a relatively unimpassioned philosophical treatise, but one finds there again a delicate balance: though the implied author of this work is a far different person, far less obtrusive, far more objective, than the man who later cried *sursum corda* to the British Parliament, he permeates with his philosophical personality his philosophical work. And though the signs of his awareness of his audience are far more subdued, they are still here: every effort is made to involve the *proper* audience, the audience of philosophical minds, in a fundamentally interesting inquiry, and to lead them through to the end. In short, because he was a man engaged with men in the effort to solve a human problem, one could never call what he wrote dull, however difficult or abstruse.

Now obviously the habit of seeking this balance is not the only thing we have to teach under the heading of rhetoric. But I think that everything worth teaching under that heading finds its justification finally in that balance. Much of what is now considered irrelevant or dull can, in fact, be brought to life when teachers and students know what they are seeking. Churchill reports that the most valuable training he ever received in rhetoric was in the diagramming of sentences. Think of it! Yet the diagramming of a sentence, regardless of the grammatical system, can be a live subject as soon as one asks not simply "How is this sentence put together," but rather "Why is it put together in this way?" or "Could the rhetorical balance and hence the desired persuasion be better achieved by writing it differently?"

As a nation we are reputed to write very badly. As a nation, I would say, we are more inclined to the perversions of rhetoric than to the rhetorical balance. Regardless of what we do about this or that course in the curriculum, our mandate would seem to be, then, to lead more of our students than we now do to care about and practice the true arts of persuasion.

Things to Consider

1. What is Booth's essay about? What is its purpose?

2. This essay started out as a conference talk. To what extent does it retain elements of its oral nature? What does this do to the style of the passage?

3. Who are the intended readers of this paper? How can you tell?

4. What devices does Booth use to capture and retain the audience's interest?

5. What are three components of any communication situation, according to Booth? How may these be perverted?

6. What is "rhetorical stance" as Booth describes it? Define it in your own words.

7. Based on this passage, what kind of person is Wayne Booth? How does he gain authority or credibility?

8. What values or beliefs of the audience does Booth address? What assumptions does he make about the reader's needs, expectations, prior experience, or concerns? Given his intended audience, are these assumptions reasonable?

9. How difficult was it to read this passage? Compare it with the information presented in the chapter; which is more difficult? Can you suggest why that might be?

10. Is this passage formal, informal, or casual writing? How can you tell? How suitable is its chosen style—the language, the structure, the tone—to the purpose and audience for which it was written?

CHAPTER 2:

Style in Professional Writing: An Overview

1. To recognize the two principles upon which all effective professional writing is based.
2. To learn the Seven Cs of professional writing and incorporate them into all your writing.
3. To examine several common causes of unclear professional writing and learn how to edit them out of your own writing.

The nine axioms of communication introduced in Chapter One are interesting ways to reflect on human behaviour, and they give us some insight into the way communication functions. They also provide a foundation for practice, since they give us important cues as to how we should proceed when we are communicating in any situation. Since our interest in this book is communication on the job, most of our discussion and examples will therefore involve professional communication, but the principles and practices we will discuss can easily extend to other communication situations.

As the axioms make clear, there are good reasons for paying attention to the quality of our communication, both on the job and elsewhere. For example, professionals who have mastered both style and form in their letters, memos, and reports and who are clear and coherent in their electronic messages are more effective and successful. Their messages get dealt with more efficiently and they seem to get more done. They know the secret of effective writing: they think of their written works not as products but as tools, not as works of art but as communication between people. In other words, good writers know that all effective professional writing follows two basic principles: focus on your purpose, and remember your audience.

Remembering these two principles will help you to create clear messages and to acknowledge your audience's interests and concerns. Doing so will also enhance your credibility by establishing you as a person of good will and good judgement. For this reason, these two principles are understood to underlie all good communication, in the professions and elsewhere, because they are anchored in the nine axioms we discussed earlier, and because they make effective use of the three keys to communication identified by Aristotle: personal credibility, a coherent and sound message, and effective audience appeal.

Since so much of our professional contact with others takes place through the things we write, whether they are distributed in "hard copy" or by electronic mail, we will focus for most of this book on writing clearly, engagingly, and effectively. However, you will also find some attention to oral communication in the segments on public speaking and employment interviewing. As you read, you should remember that all your communication—not just the communication you do on the job—could be improved by understanding the axioms of communication and applying them through the principles of application we will learn in this and the subsequent chapters.

COMMUNICATING IN WRITING: TWO PRINCIPLES

If we are to be effective professional or technical writers, we must write clearly and courteously *for at least three reasons:* so that our messages are dealt with efficiently, so that we establish positive working relationships, and so that we maintain our credibility and professionalism. It is important to state our meaning coherently and edit our messages carefully, so that we can prevent misunderstandings or delays that can cost time and money, or cause embarrassment to the profession and ourselves. Achieving clarity and effectiveness in writing is not easy, but the good news is that with a bit of effort almost anyone can learn to write effective professional messages. As you will discover throughout this book, the formats and structures of professional writing are actually designed to help you do so.

Two Principles of Written Communication

1. Focus on your purpose and be able to state it clearly:
   - put the main point first
   - be specific
   - simplify your message
2. Remember your reader's
   - needs
   - expectations
   - concerns
   - interests
   - background knowledge
   - professional relationship to you

The bulk of this book will consider such standardized forms, but before we begin dealing with these, let's look more closely at the two important principles of communication that will help you to forge effective professional relationships:

1. Focus on your purpose and be able to state it clearly; and

2. Remember your reader's needs, interests, expectations, background knowledge, concerns, and professional relationship to you.

Focus Clearly on Your Purpose

Before you begin writing any e-mail, letter, memo, or report, ask yourself why you are writing. You can't make any message clear to a reader unless you know, before you begin to write, exactly what your purpose or objective is. What do you want your message to accomplish? What do you want the reader to do, or to know? What image of yourself, your company, and your profession do you hope to convey or maintain? Is your primary aim to pass on information or to persuade your reader to act or believe something? If you are primarily interested in informing your readers, you can state your facts clearly and simply. If, however, you want to move your readers to action, convince them of a point of view, or encourage them to accept a change in plans, you will need to use more persuasive techniques in your writing. These are covered in greater detail in Chapter Four.

A clearly defined purpose is equally important no matter what you are writing, from an ordinary letter of request to an application for a job, from e-mail to letters and reports. There are several specific steps you can take to ensure that your purpose will be clear to the reader.

Put Your Main Point First

Writing on the job differs from the other kinds of writing that you may be used to reading or that you have learned to produce yourself. In your work-related writing, you should begin with the main point of your message. Putting your main point first will seem awkward at the beginning, because it will seem a little like giving away the punchline before you've told the joke. We have all been conditioned to write in a more or less chronological order, which usually results in the main point coming last. This arrangement is ideal for a movie script or a novel, but inappropriate for a report. Think about your reader's situation: he or she may have many time demands, and may receive several dozen or more e-mail messages, memos, and letters daily, in addition to numerous reports. All of these messages require attention and action. You can't afford to waste the reader's time with a message that doesn't get to the point. Put the main message at the beginning of your communication, preferably in a subject (or "re") line. You can teach yourself to put your main idea first by beginning your rough drafts with, "The main thing I want to tell you is that...." Doing so will help you to focus on your main message. Be sure to delete this clause in your final draft.

Be Specific

Make sure you can identify exactly what you wish your reader to know and what response you expect. Be as concrete and specific as possible. For instance, if you wish to request a file or document, book a room, submit a tender, or propose a policy change, state this purpose clearly, identifying the document required or the policy you will discuss. Don't waste your reader's time; get to the point quickly, and provide as much concrete detail as is necessary to get the job done.

Simplify Your Message

Try to keep your letters, memos, and e-mail messages as simple as possible: don't clutter them with irrelevant information. If you can, stick to one main topic and include only what is necessary to the reader's understanding of your message. If you must deal with several topics in one letter or memo, be sure that each is dealt with fully before moving to the next issue. Cluster related information and get to the point as quickly as possible. Make sure the subject line of memos and e-mail messages clearly indicates what the message is about so that the reader can easily file and locate it.

Remember Your Reader

Whether it is primarily informative or persuasive, professional writing, like all effective communication, must catch and maintain a reader's attention and interest. You cannot do this if you don't understand the expectations and concerns of the person for whom you are writing. Why does this reader care about your message? What is the reader's interest in the information? What background does this person have? What will the reader need to know to make a decision? Since any professional communication is really an attempt to convince your intended reader that your position is valid and your recommendations necessary, you must present the information in the manner most likely to convince that specific reader. Good communication puts the reader's needs first; it focuses not so much on what the *writer* feels like saying, but on the things the reader needs to hear in order to make an informed decision. In order to do this, you need to analyze your reader as to needs, level of knowledge, expectations, and concerns. You must also consider the relationship your organization or company has had with this reader. Without this information you cannot hope to create a message that will effectively motivate and enable your reader to respond to your request, proposal, or notice.

Needs

Consider first the information that your reader needs in order to make a decision. What is the reader's interest in this subject? What will the reader be doing with the information? How much detail is needed? Leave out any information that is not immediately relevant to the reader, no matter how interesting it may seem from your point of view. If you are to communicate your point successfully, you must address the reader's need for the information you are providing.

Background Knowledge

Keep in mind your reader's level of expertise. If you are writing to someone who has no prior knowledge of your project or field of specialization, you will need to explain substantially more than you will if you are writing to someone who is well acquainted with them. On the other hand, it is just as inconsiderate to provide unnecessary detail to someone who knows a great deal about your work as it is to provide inadequate detail to someone who does not know very much about it. Also, the kind of prior knowledge your audience brings can influence the way you choose to present your information as well as the amount or kind of information to present.

Remember also that, even if your reader is generally familiar with the contents and context of your report, the specific details will not be as familiar to the reader as they are to you. It is courteous to help the reader along by contextualizing and clarifying as necessary, reminding the reader of pertinent details that may have been forgotten.

Expectations

What you say, and how you say it, will be very much affected by what the reader is expecting from your work. A reader who has been expecting a negative response will be relieved and delighted by good news. However, a reader who has been anticipating positive results is likely to be disappointed, frustrated, or angry if those expectations are not met. Ignoring or overlooking such a reader's frustrated expectations when you write will only aggravate the situation and may damage both your credibility and your relationship with your correspondent.

On the other hand, appropriately acknowledging a reader's legitimate disappointment can help to cushion the impact of bad news and emphasize your professional concern and interest in the reader's viewpoint. Such positive reinforcement can go a long way in cementing effective professional relationships.

Concerns

Hand in hand with reader expectations are the things that the reader is worried or concerned about. Perhaps the reader fears costs escalating out of control, or problems meeting deadlines; perhaps the company's reputation is at stake. Try to be attuned to the kinds of concerns that your intended reader—whether a client or a supervisor—might be expected to have. Although it may not be appropriate to address such fears directly, they should at least be indirectly acknowledged by providing appropriate information and assurances that will help the reader see that you are sensitive to the context and the situation into which the report or message will be delivered.

Relationship

A large part of the context in which the message will be read is the existing relationship with the audience. Is the reader of your message a long-time client who knows your firm and its reputation, or someone new, for whom your credibility is unknown? Have previous interactions been cooperative and pleasant, or contentious? Is this reader likely to be compliant or challenging? Is the message going to your own client with whom you have a personal relationship, or to a client of the firm with whom you have had no previous personal contact? Is the reader a colleague, a subordinate, or a superior? While you should be courteous and polite to all of these potential readers, you may require an additional element of deference when communicating with those higher in the organization.

In order to make the communication a success, it is the writer's responsibility to pay attention to the relational level of the message as well as to its content. To create and maintain effective relationships within your professional circle, you will need to think carefully about the needs and expectations of the reader who will receive your communication. If you do this each time you create an e-mail, a memo, a letter, or a report, you will not only communicate your content more effectively, but you will also maintain more successful interactions with clients and colleagues.

MASTERING THE SEVEN CS OF PROFESSIONAL WRITING

Good professional writing—like any other effective writing—exhibits a number of identifiable characteristics, which I have named the "Seven Cs." These principles show how the axioms of effective communication are actually displayed in the written work of an effective communicator.

1. Completeness

The first thing you must do in preparing a memo, letter, e-mail message, or report is to make sure that no important details have been overlooked. To ensure that you have included all of the information your reader will need in order to understand and act on your

The Seven Cs of Professional Writing

- Completeness
- Conciseness
- Clarity
- Coherence
- Correctness
- Courtesy
- Credibility

message, always ask yourself after you've written the first draft, "Have I said everything I needed to say?" Use the following questions as a guideline to ensure that you have included all the information your reader needs. Be sure you have answered any that are relevant:

- Who?
- What?
- How many?
- When?
- Where?
- How?
- Why?
- How much?

2. Conciseness

This means saying as much as you need to say in as little space as possible without being curt. Achieving conciseness without sacrificing completeness is more difficult than it sounds, because it involves more than simple brevity. Although you want to eliminate unnecessary information, you must at the same time preserve the details necessary to full understanding. Once you have completed your first draft, and have made certain that all essential information is included, you need to be sure to take out any *irrelevant* information that has crept into your message, and remove any unnecessary wordiness.

Ask yourself whether the reader really needs to know a fact you have included. If the answer is "no," then cut it. Similarly, avoid repeating yourself unnecessarily, and try to avoid wordy expressions and clichés, such as:

- at this point in time
- if this proves to be the case
- it is probable that
- it has come to my attention that
- until such time as
- please do not hesitate to

There are many more clichés common to professional writing; many of them can be found in the discussion of how to improve your writing style that appears later in this chapter. In it, you will learn to recognize these deadeners of style and clarity, and eliminate them from your messages.

3. Clarity

In addition to making sure you've included all the important information and eliminated details that are irrelevant to your reader's needs, you need to pay close attention to detail and organization. Be as concrete and specific as you can, identifying exactly what the issue or problem is and what you would like done. Try to avoid ambiguous phrasing: the message should be clear on the first brief reading. Your reader should never have to puzzle out your meaning and should have no unanswered questions after reading your correspondence.

You should also help to clarify your meaning for your reader by organizing your message in a logical way, for example, moving from problem to solution, from request to thanks, or from general to specific. You will find more detailed information on organizational patterns in Chapter Three.

4. Coherence

Any professional correspondence should "hold together"; the parts should logically follow one from the other so that the pattern of argument is easy to follow and clear to the reader. Coherence is partly a function of linking strategies, or devices of cohesiveness, which bind each sentence to the subsequent sentence, each paragraph to the subsequent paragraph, and each part of the correspondence to the overall purpose or focus of the message. Three elements provide this impression of coherence; they are:

- adherence to the conventional structure and format of a professional document;
- use of a sensible and appropriate organizational pattern in the discussion segment of the message; and
- effective use of connective, or linking, devices in each sentence.

The devices of cohesiveness that create coherence in a text are discussed at length in the next chapter.

5. Correctness

Check the accuracy of all information—names, dates, places, receipt numbers, measurements, lab results, prices—that you include in any correspondence. Correctness, of course, also includes correct spelling, grammar, and sentence structure, as well as attractive and appropriate formatting. Try to make it a habit never to send a written message of any kind—including e-mail—without proofreading it first and checking its tone. A work that contains several errors in spelling, grammar, or factual material suggests to the reader that the writer is unprofessional, unreliable, and sloppy. In turn, a poorly written letter makes inappropriate demands on the reader's patience and understanding, and is therefore also discourteous. For this reason, a failure to proofread for correctness and accuracy could be costly.

6. Courtesy

In professional life, as in all human interactions, things usually go more smoothly if people are pleasant and courteous to one another. Your relationships with clients and colleagues are very important and, as we have already discussed, every communication you send can enhance or undermine those relationships. Since your attitude is displayed clearly in your writing and will affect the relational level of your message, it's important to pay close attention to tone in what you write.

Tone refers to the attitude of the writer toward the audience and also toward the topic of the message. In general, you should try to make it a habit to be pleasant and positive in your professional communication. Be sure to thank your correspondents for any services or favours requested or received. In particular, avoid sarcasm, which—though it can satisfy our immediate expressive needs—is always read by the target as hurtful and offensive. It therefore has no place in professional correspondence.

Even if you are writing to someone you believe has done you wrong, give this person the benefit of the doubt, at least initially: allow the recipient to save face by taking the attitude that the error was the unintentional result of a misunderstanding. This approach will be much more effective in resolving difficulties than will a confrontational, accusatory, or sarcastic tone. To check this important aspect of your own professional communication, always read over your work carefully, putting yourself in the place of the reader. If you are uncertain whether you've achieved an appropriate tone, ask someone you trust to read the message for you.

7. Credibility

As we have already established, one of the most important features of effective communication is this projection of credibility—the quality that Aristotle called "ethos." As you know by now, the way others judge you as a professional depends on how you communicate with them, and your correspondents will judge your professionalism, expertise, and competence by the clarity and attention to detail revealed in your writing.

Credibility involves demonstrating to your reader your good judgement, good will, and good character, just as Aristotle outlined. You do this in part through the appropriateness of your style, diction, tone, emphasis, and sentence structure. You also demonstrate your judgement by the details you choose to include, your understanding of the issues, the reasonableness of your solutions, and your command of the facts, as well as through correct spelling, grammar, and punctuation. You demonstrate good will through the way you address your readers and attend to their needs. Attention to the first six Cs on the list will actually help to establish the seventh by contributing to the reader's perception of your competence, judgement, and good will.

Credibility also involves the presentation of the message in appropriate format, layout, and structure. Letters, memos, and reports are fairly standardized in their appearance, and a reader will expect yours to conform to these accepted standards. Letters, memos, or reports that do not observe the conventions of professional formatting may suggest a sloppy or unprofessional writer.

E-mail messages, of course, are far less consistent in their format: some writers, perceiving e-mail to be a casual medium of communication, habitually ignore all conventions of punctuation, spelling, and capitalization. However, you should know that not all readers appreciate *receiving* such unconventional messages, since they can hinder understanding by being much more difficult and time-consuming to decipher. It is much better, from the point of view of reader consideration, to observe the conventions of spelling, punctuation, and grammar than to take the chance of annoying or alienating a correspondent.

Professional documents, of course, are always word-processed (or typed), with a clear, dark printer cartridge on clean, good quality paper. They should also use an effective layout, one that makes use of generous margins, never looks too crowded or too widely spaced, and appears balanced on the page. Your message may also make use of other visual techniques to make it both more attractive and easier to read. For example, if it includes several important facts, lists steps in a process, or identifies items that the reader must include in a response, these may be indented in list form to set them apart from the rest of the information.

THE STRUCTURE OF AN EFFECTIVE PROFESSIONAL MESSAGE

Six Standard Parts of a Professional Message

- Summary
- Introduction
- Discussion
- Conclusion
- Recommendations (optional)
- Appendices (optional)

Once you have recognized that your job as a writer includes building effective professional relationships by considering the purpose and the audience for your message, and by paying careful attention to the Seven Cs of effective communication, you can begin to think about how professional communication is typically organized. Most professional writing can be divided into several parts, which are differentiated according to their function. These parts are presented in the same order in every message, forming a conventional structure that your readers will both expect and welcome.

The six parts of all professional communication may be indicated by section headings in lengthy or formal documents, or they may simply be indicated by paragraph breaks in shorter, informal correspondence. There are four essential parts and two optional parts; the more formal and lengthy the communication, the more likely it is to contain all six. The six standard parts of all professional communication are:

1. The summary, or main message statement ("the main thing I want to tell you is that ..."). In a report, this section will likely be named "summary" or "abstract." In an electronic mail message, a written memorandum, or even a letter, this information usually appears in the subject or "re" line.

2. The introduction contains a statement of the issue or problem, and provides any background information the reader may require. This information is placed in your first section, paragraph, or sentence, depending on the length of your correspondence.

3. The discussion, or body, of the work contains the full development of the report, letter, memo, or e-mail. This part provides any necessary details or the facts of the issue, usually organized in chronological, spatial, cause-to-effect, or other appropriate arrangement. In a report, particularly if it is lengthy, the discussion is sub-divided into sections with appropriate headings. Each section in turn deals with one aspect of the problem or issue being discussed.

4. The conclusion is the closing sentence or paragraph. It should remind your reader of your main point and should indicate any results you expect or intend. In a letter or memo where formal recommendations are not included, the conclusion may also suggest to the reader any appropriate action he or she should take in response to your communication.

5. The recommendations segment is optional and is usually reserved for more formal documents such as reports. Its role is to clearly spell out the action desired as a result of the information presented. In a letter, memo, or e-mail message, the recommendation for action may be added to the concluding paragraph rather than being presented in a separate section.

6. The appendices are additional documents that provide background information germane to the report (in a letter, "attachments" sometimes serve the same purpose). Generally, they contain facts, details, calculations, or research findings that were referred to in the body of the report but are not central to the reader's understanding. Appendices provide additional information that the reader may wish to consult, but they are not considered part of the body of the report itself. For example, if I am requesting a letter of recommendation from a former professor, I may as a courtesy provide that person with a copy of my grade transcripts or my résumé as an appendix to my request, to assist her in writing the letter.

Remember these parts by the acronym "SIDCRA." You can use this acronym when you are writing any professional correspondence, as a quick check to see that you have structured your message adequately. After all, you want your message to be communicated, and as the writer, it is your responsibility to help your reader understand that message. Following this structural outline and jotting down your main ideas first can make your written communication more effective and the writing process easier, because you will have organized your ideas about the message before actually composing a draft.

Layout as Self-Presentation

In addition to its conventional structure and adherence to the Seven Cs of professional communication, a professional message of any kind should also be arranged as attractively as possible for both clarity and readability. In the case of e-mail, effective presentation has become easier as advances in mail programs have made formatting and spell-checking more convenient. While systems and settings still vary, and it is difficult to control exactly how an e-mail message will appear to every recipient, there is a need to pay attention to format, as there is in all your professional communication. The casual air of e-mail is no excuse for sending a sloppy, disorganized message; though we think of the medium as ephemeral, in some ways it's as permanent as paper, and easier to circulate endlessly. We will explore e-mail in greater detail in Chapter Four.

Print messages, in particular, can and should be laid out carefully on the page. This means the writer should create an effective balance between the textual material (known as print) and the blank areas of the page where no print appears (known as white space). A page that is crowded from one edge to the other without visual breaks may intimidate the reader; on the other hand, generous margins, paragraphing, and standardized formats can help a reader understand your message more easily and quickly, and will also make a positive impression. Standardized formats and layouts are discussed in each of the subsequent chapters.

SHARPENING YOUR WRITING STYLE

It is one thing to know, in theory, that you need to put the main message first, accommodate the needs and expectations of your reader, and attend to your credibility. It's another thing entirely to accomplish all these things in your writing. Most writers experience difficulty in achieving the sharpness of style that distinguishes good professional writing and establishes credibility. Bad writing is surprisingly easy to produce, while good, clear, effective writing—which often looks simpler on the printed page—is actually much more difficult to achieve.

Writing concisely is a challenge, and almost every writer produces first drafts that need correction. Although you should give some thought to planning and organization when you write the first draft, don't imagine that it will be perfect. Expect—and be willing—to revise your work to make it more effective. One way to begin improving your writing is to become familiar with the most frequent violations of the Seven Cs and learn how to edit these common faults out of your own professional messages. With practice, this process will become quicker and easier.

COMMON FAULTS OF PROFESSIONAL WRITING

Common Faults of Professional Writing

- Did not identify the central issue
- Overused phrases and clauses
- Overused passive voice
- Used unnecessary repetition
- Did not cluster related points
- Did not identify desired action
- Provided incomplete information
- Writer referred to as "myself" or "this writer"

Once you have a full draft of your message, you must look at it critically to see whether it is as effective as it could be. Will it engage your reader's attention? Does it recognize and respond to the reader's needs and expectations? Will it contribute to an effective professional relationship? Does it portray you as a competent professional? Will it achieve its purpose? If the message fails to achieve its purpose, it will need to be improved so that it will more effectively engage its readers, motivate them to accept its arguments, and enable them to take the necessary action. You can begin the process of polishing your messages by avoiding common violations of the Seven Cs such as these troublemakers.

1. Failure to Identify the Central Issue

This is the most common flaw in professional writing. To avoid it, be sure that you can identify your purpose and that you understand, before you begin to write, exactly what you want your reader to know or to do. This information must come early, and it should be clearly expressed. Before you work on the body of the message, write down your main idea. You may even, as we discussed, begin your rough draft with the words, "The main thing I want to tell you is that...." (Remember to cross out this clause for your final draft.) In fact, you might even wish to put the point into a "re" line; doing so will not only force you to put it first, but will also help your reader to grasp your message more quickly.

2. Use of Clichés

Clichés make any writing uninteresting and deaden its human connection. They should not be confused with idiomatic phrases, which are set expressions that function as a single unit and cannot easily be expressed in any other way. Idiomatic expressions such as "My favourite jeans *wore out*," "The robber *held up* the bank," "John turned in for the night" or "He *pulled through* after the operation" have a different meaning from the combined meaning of the words themselves. "I pulled the cord through the keyhole" or "I wore my jeans out to the park," though they may appear similar, are not the same kind of phrases. The words that make up a true idiom cannot be split apart within the sentence. For instance, it would not make sense to say "John turned for the night in" or "Out my favourite jeans wore."

While idiomatic expressions are an indispensable part of the way we speak, clichés such as the ones shown in the checklist are a form of verbal padding, and frequently they are a signal that the writer is not thinking carefully about the message or about the reader who will receive it; instead she is reaching for ready-made phrases regardless of whether they capture her meaning precisely. Thus, clichés can usually be replaced by clearer, more direct phrases.

Despite their distancing effect, overuse of ready-made phrases is one of the most common faults in job-related writing. Inexperienced writers often imagine that professional correspondence is supposed to sound hackneyed, because so much of it does, and many writers, not knowing what else to do, fall back on the clichés they have seen in the writing of others. Because such clichés add words without adding meaning or clarity, because they say in several words what could more clearly be said in one, and because they frequently obscure rather than clarify your meaning, they are bad writing. You will do yourself, and your readers, a favour if you avoid this trap.

A complete list of all the clichés of professional writing would take up the rest of this chapter, but the examples in the list give you a taste of what to look for and avoid. As a general rule, any phrase that sounds as though it "ought" to be in a piece of correspondence should be avoided.

Most of these cumbersome and meaningless phrases can be replaced by much simpler language, often a single word that communicates much more forcefully and directly. For example, "at this point in time" could be replaced by "now"; "if this proves to be the case" could be written simply as "if"; "postpone until later" should simply be "postpone." Some of them, such as "it has come to my attention that," can be eliminated completely without any loss of meaning. See if you can translate these awkward and stuffy phrases and clauses into plain language. Always check your own writing for such phrases and replace them with more direct language.

3. Use of the Passive Voice

The passive voice expresses not action done *by* the subject, but action done *to* the subject, a crucial difference. For example, "Assistance would be greatly appreciated" is almost always better written, "I would appreciate your assistance."

There are some exceptions. In certain types of very formal writing, and in some (not all) scientific disciplines, the use of passive voice is considered appropriate, particularly in cases where an appearance of objectivity is the goal. For example, a chemist reporting laboratory results will likely write "The

Some Clichés of Professional Writing

- at this point in time
- if this proves to be the case
- in the amount of
- it is probable that
- postpone until later
- under separate cover
- it has come to my attention that
- reach a decision
- until such time as
- please do not hesitate to
- on or before
- whether or not
- send you herewith
- enclosed herewith find
- with reference to
- give consideration to
- at the present time
- due to the foregoing consideration
- in view of the foregoing
- in the near future
- due to the fact that
- in accordance with your request
- in the event that
- it will be our earnest endeavour
- in accordance with your request
- at your earliest convenience
- make a decision about
- in today's society

experiment was conducted under controlled conditions," instead of "My associates and I conducted the experiment under controlled conditions."

The conventions of scientific writing that call for the use of passive voice are usually not an indication of nefarious purposes. However, as you can imagine, passive voice is sometimes used (along with vague and clichéd wording) as a deliberate strategy by writers who want to confuse or obfuscate, or who wish to avoid acknowledging responsibility for an action: bureaucratic documents and administrative memos are often written in such opaque language. Consider, for example, the sentence, "The patient was given the wrong medication." There is no indication of who administered the medication, and therefore no indication of where responsibility for this mistake rests. In fact, because "the patient" is made the subject of the sentence, it may even seem that he or she bears responsibility for the error.

At times this sort of rhetorical sleight-of-hand is legitimate and desirable. For example, even though use of the passive voice should generally be avoided, at times it may be legitimately used to avoid assigning blame to someone. Keeping in mind the relational quality of communication sometimes means that we must take care to avoid causing someone else to lose face. For example,

George carelessly misplaced the Ferguson project file.

is more likely to cause loss of face to George than is

The Ferguson project file has been misplaced.

John's tardiness has caused a delay to the design phase of the project.

may unfairly blame John, while

There has been a delay in the design phase.

will avoid unnecessarily apportioning blame to anyone.

Remember that, while some uses of the passive voice are legitimate, in most cases it is unnecessary, and you can make your own writing more powerful and concise by eliminating it. In particular, it becomes a problem when writers render the majority of their sentences in the passive voice in situations that are not sensitive or contentious. Such deliberate obscurity or unintentional muddiness is not a desirable quality in writing designed to communicate. You want your writing to stand out for its clarity, directness, and human connection, and overuse of the passive voice in ordinary professional reports and correspondence distances writer from reader, deadens style, and often causes a loss of clarity. Consider how much more vivid and powerful, not to mention how much shorter, the following sentences are when they are written in the active voice.

Larry completed our report.

is more concise than

The report that we were working on has been completed by Larry.

Shirley conducted the required tests.

is less wordy than

The tests that were required have been conducted by Shirley.

Doug hired John Smith to complete the project tender assessment.

is more direct than

John Smith was hired by Doug for the completion of the assessment for the project tender process.

We have evaluated your application.

is more personal than

Your application has been evaluated by us, or your application has been evaluated.

Unless there is a legitimate, ethical reason for refocussing the reader's attention away from the agent in a sentence, you should generally prefer active to passive constructions in the things you write, in order to develop a more energetic, engaging style.

4. Overuse of Phrases and Dependent Clauses as Modifiers

Very often we find ourselves using several words where we could use just one. Believe it or not, long windy sentences are actually easier to create than clear, concise ones are, because they don't demand as much care or attention to meaning, and they require no concern for the audience's needs. But they also make dull, boring, tedious messages that remain unread or unheeded.

Excessive use of phrasal and clausal modifiers is one of the most common causes of this kind of wordiness, and when it is combined with the passive voice (as in the examples above) the result can be confusing to read. As a rule, you should not use several words when a few, or even one, will do the job. See how much more concise the following examples can be:

- the project I am working on (my project)
- the equipment that our department recently purchased (our recently-purchased equipment)
- the store on the corner (the corner store)
- the tender submission belonging to this consultant (this consultant's tender submission)
- the reason for which I am writing (my reason for writing)
- the project files that I recently acquired (my recently acquired project files)
- property that belongs to the government (government property)
- a friend whom I have known for a long period of time (a long-time friend)

Watch, in particular, for overuse of the words "of" (or other prepositions), "which," and "that." Rewrite such phrases or clauses into one or two words whenever you can do so without a loss of precision of thought.

5. Unnecessary Repetition of Ideas

Repetition can be a powerful tool for persuasion when it is used effectively and deliberately. It can engage and motivate an audience to action. However, two things must be said about repetition as a strategy: first, effective repetition is primarily an oral device and rarely achieves the same effect in writing. In fact, much repetition in writing is cumbersome and unworkable; it may even seem unnatural and overdone. In professional writing, where the goal is to communicate efficiently and clearly, unnecessary repetition can actually get in the way of your message.

In a memo, letter, or e-mail, you need to say what you mean in as little space as possible. To avoid unnecessary repetition, cluster related information and make each point only once. If you say it clearly the first time, you can eliminate the useless and often confusing repetition that weakens your writing and obscures your message. Use repetition only where necessary for clarity and cohesiveness; in general, you should avoid using it for emphasis.

6. Failure to Cluster Related Points

In organizing your message, you must be sure to place related points together: jumping back and forth is confusing to the reader and is one of the things that leads to the unnecessary repetition we have just discussed. If you find yourself writing the phrase "as I said above... " you probably need to do more to cluster related information.

If, for example, you are writing a letter or an e-mail to obtain registration information for a conference, and to inquire about submitting a paper to the convention program, you should cluster all information pertaining to the registration into one paragraph and your query about the paper submission into another. For added clarity and visual appeal, you can place the main details for each topic in indented lists.

7. Failure to Identify the Desired Action

A frequent function of persuasive professional correspondence is to prompt a specific kind of action from the reader. You are writing because you want results. Although as a writer you may feel that the appropriate course of action is obvious, what you want done may not be quite so clear to the reader, whose idea of a suitable solution may differ from yours. Don't expect your reader to come automatically to the same conclusion you have reached about what must be done. State directly, in clear and specific language, what you expect the reader to do. If there are several steps to be taken, list and enumerate them for the reader's convenience.

8. Incomplete Information

Before writing your final draft, check once more to be sure you have included all relevant information. To be sure that you have supplied all the necessary details, ask yourself, *Who? What? How many? Where? When? Why? and How?* Always double check that you have included any attachments (electronic or hard copy) that you intend to send with your message.

9. Using "Myself" or "This Writer" to Refer to Yourself

These are slightly different problems, but each one is equally ugly in its own way. The reflexive pronoun—the "-self" form—should never be used to replace the nominative (subject) pronouns I, you, he, she, they, or we, or the accusative (object) pronouns me, you, him, her, them, or us.

> I starved myself; I was punishing myself for my bad habits; You said so yourself; I didn't believe it until I saw it for myself.

are all correct, whereas

> He gave the file to myself; Bruce went on the outing with myself and the girls; You will receive a letter from myself in the next mail.

On a similar note, "on behalf of myself" is redundant, since of course you are speaking for yourself; you can only speak "on behalf of" others.

The construction "this writer" came about as a result of proscriptions against the use of the personal pronouns "I" or "we" in formal writing. Unfortunately, this and similar phrases sound pretentious in current usage, and impose an unnecessary distance between writer and reader. Contemporary usage sanctions the personal pronoun even in formal and some scientific writing, though some fields (such as mathematics) prefer "we" to "I" and some disciplines prefer only sparing use of the personal pronoun. If you are writing in a context in which the pronoun "I" is not typically used, arrange your sentences so as to avoid resorting to "this writer," "the author," or "this researcher." Figure 2.1 shows the first draft of a letter that violates the principles of communication we have been discussing. Read it through to see if you can think of ways to improve the clarity and professionalism of its message. An analysis of the errors in this first draft can be found later in the chapter, along with a corrected version of the letter.

Trish Trcka
PO Box 123
Drayton Valley, AB T5Y 7H8

9/28/04

Ministry of Economic Development
Province of Alberta
Parliament Buildings
Edmonton, Alberta

Dear Sir:

It has come to my attention that your ministry can offer some valueable information to potential entrepreneurs who are concidering opening a new fledgling business in the best province in the country.

I am writting you this letter because I am interested in starting my own business with a design that I developed myself. It would be greatly appreciated if you could provide me with some documents outlining your services and benefits, also general information for student entrepreneurs.

Thank you in advance for your assistance in this matter. Your immediate response will be appreciated.

Sincerly yours,

Trish Trcka

Trish Trcka

1. All effective professional communication is based on two primary principles:
   - Know your purpose and be able to state it clearly.
   - Understand your reader's needs, expectations, knowledge, and concerns.

2. Good professional style observes the Seven Cs:
   - Completeness
   - Conciseness
   - Correctness
   - Clarity
   - Coherence
   - Courtesy
   - Credibility

3. Good professional communication follows conventional formats and displays effective balance between print and white space.

SHARPENING YOUR SKILLS

Section A

Revise the following sentences, all taken from actual memos and letters, to make them clearer and more concise.

1. Please be advised that interested parties who wish to apply must consider that documents for application should meet the deadline of submission which is August 31.

2. I have a colleague of mine, Brian Quigley, who passed on to myself the facts and details contained in the information you provided.

3. In the event that any employee should be the final individual to exit the premises of this firm on the eve of any given working period, it would be greatly appreciated by management as a gesture of fiscal responsibility if such individuals should leave the offices in a state of darkness.

4. The full and complete application materials she submitted on behalf of herself have received a nod of acceptance.

5. Please be advised that I would like to express my interest in consideration to being interviewed for this very attractive opening that is currently available for hire in your firm.

6. Please accept this letter as formal notice of the fact that the scholarship award of which you had expectation of confirmation has received a negative decision in the form of a rescindment.

7. Due to the difficulties involved with the aforementioned request, the writer would like to take this opportunity to thank you in advance for your assistance in this difficult matter.

8. A cheque for the amount specified to cover the loss experienced due to the above incident of April 30 has been prepared by this office. The appearance of yourself is requested at your earliest convenience to complete the necessary paperwork and receive such payment. We trust this is in order.

9. With reference to your communication of the above-referenced date, enclosed herewith find the documents which you requested at that point in time.

10. In the event of circumstances beyond our control which affect delivery of this service, some alteration to the planned schedule may be required.

Section B

1. In his famous essay "Politics and the English Language," the writer George Orwell rewrites a familiar passage from *Ecclesiastes* into what he calls "modern" language.[1] Here is the original, followed by Orwell's rewritten version:

> *I returned and saw under the sun, that the race is not to the swift, nor the battle to the strong, neither yet bread to the wise, nor yet riches to men of understanding, nor yet favour to men of skill; but time and chance happeneth to them all.*

Compare Orwell's version, rewritten in modern English:

> *Objective consideration of contemporary phenomena compels the conclusion that success or failure in competitive activities exhibits no tendency to be commensurate with innate capacity, but that a considerable element of the unpredictable must invariably be taken into account.*

Of course, Orwell's point is to demonstrate how much less clear the modern language is than the original. As a class, compare Orwell's passage with the passage from *Ecclesiastes* and explain why the original is more effective.

2. Just for fun, try your hand at this reverse process by rewriting some clear passages into "modern" language, using the clichés and abstract phrases we have been discussing. Try rewriting some of the examples of good writing from this book in such a manner, or select a clear passage from one of your other textbooks. You can even apply the process to a familiar text—perhaps a fairy tale such as the "The Three Little Pigs" or

---

[1]George Orwell, "Politics and the English Language." *The Orwell Reader*, ed. Richard H. Rovere (New York: Harcourt, Brace, and World, 1956) 360.

Molson's famous "I Am Canadian" rant, a copy of which can be found on line at http://www.adcritic.com/content/molson-canadian-i-am.html.

Compare and discuss your choices with those of the other members of the class.

3. Read the following professional messages (Figures 2.2 and 2.3). Suggest ways they might be improved, keeping in mind the axioms of communication and the Seven Cs of style in professional writing.

FIGURE 2.2   Why might this e-mail message fail to resolve the problems the class has with Dr. Wolf Child?

To: Dr. Wolf Child

Department of Mechanical Technology

From: energy Systems lab

Date: september 30, 2003

Re: problems with last lab assignment

we would like to get together with you to discuss the marks we received on the last lab—everybody in the class is upset with their grades and we don't think you explained clearly enough what you wanted anyway we want to meet with you thurs. on your lunch hour to settle this problem. please give us an answer in tommorrow's class.

Jim Shenassa, esq.
class rep

FIGURE 2.3 In how many ways does Raied Shawaga's letter violate the Seven Cs?

Raied Shawaga
100 Membertou Road
Saint John, New Brunswick
E3H 4R7

March 3, 2004

Sheena Truman, Customer Service
Parts Department
Testek Supplies and Equipment Ltd.
64 Weniam Street
Vancouver, British Street
V1R 9V1

Dear Sir:

I am writting you a letter to ask you about the parts I ordered from you about six weeks ago. I still haven't received it even though you have already cashed my cheque.

If you don't send them to me right away I'll have to report you to the department of consumer affairs.

Yours truely,

Raied Shawaga

FIGURE 2.4 Here is the first draft of Trish Trcka's letter, along with an analysis and a corrected version. Errors in this version of Trish's first draft have been numbered for easy reference in the analysis. Compare it carefully with the corrected version (Figure 2.5), noting where the improvements have been made and why.

Trish Trcka **[1]**
PO Box 123
Drayton Valley, AB T5Y 7H8

9/28/04 **[2]**

Ministry of Economic Development
Province of Alberta
Parliament Buildings
Edmonton, Alberta **[3]**

Dear Sir: **[4]**

It has come to my attention that **[5]** your ministry can offer some valueable **[6]** information to potential entrepreneurs who are concidering **[7]** opening a new fledgling **[8]** business in the best province in the country. **[9]**

I am writting **[10]** you this letter because **[11]** I am interested in starting my own business with a design that I developed myself. It would be greatly appreciated **[12]** if you could provide me with some documents outlining your services and benefits, also general information for student entrepreneurs. **[13]**

Thank you in advance for your assistance in this matter. **[14]** Your immediate response will be appreciated. **[15]**

Sincerly **[16]** yours,

Trish Trcka

Trish Trcka

Analysis

[1] Since Trish's name appears under her signature, she need not include it here. Although we have not yet discussed letter format, this is such a common formatting error that it's worth noting here.

[2] As a courtesy to her reader, Trish should write the date out in full for ease of understanding.

[3] To complete the address, Trish should include the postal code, which is easily located in any telephone book in the province.

[4] Since Trish can't be sure that her correspondent is male, she should avoid the salutation "Dear Sir." In fact, she could leave out the salutation completely in this type of letter. As well, subject or "re" line would help to clarify what Trish seeks and could replace the entire first paragraph.

[5] "It has come to my attention that... " is a cliché that should be avoided. It is also unnecessarily wordy and pompous.

[6] "Valuable" is misspelled. This and other misspellings create a negative impression and make Trish appear incompetent.

[7] "Considering" is misspelled.

[8] Both "new" and "fledgling" mean the same thing in this case. Trish should eliminate at least one of them. "New business" is sufficient.

[9] This kind of overstatement is out of place in formal correspondence, and actually makes Trish sound insincere or manipulative. Since it serves no useful purpose and could actually compromise her ethos, Trish should eliminate the phrase completely.

[10] "Writing" is misspelled.

[11] "I am writing you this letter... " is clearly unnecessary: the reader, who has Trish's signed letter in hand, does not need to be told again that she has written it.

[12] "It would be appreciated... " is a cliché, and so probably does not sound sincere. Since Trish is making a personal request, she should also increase the sense of human contact in her letter by avoiding the passive construction.

[13] The phrases "some documents" and "general information" are vague; Trish has not identified clearly what she wants. She should be specific about her needs.

[14] "Thank you in advance for your attention to this matter" is another cliché; as such, it is wordy and sounds insincere and stuffy. A simple thank you would be better.

[15] "Your immediate response will be appreciated" is not only another unnecessarily wordy and tired phrase, but it borders on rudeness by implying that the people in the Ministry of Economic Development won't respond quickly enough unless she orders them to do so.

[16] "Sincerely" is misspelled.

In addition to the above mistakes, Trish has made another major error in this letter: she has not provided some of the specifics that would enable the people in the Economic Development office to help her most effectively. If she can, she should indicate the nature of her business, where in the province she plans to locate it, and exactly what information she needs from the government—is she interested in advice on drawing up a business plan, on securing financing, on government programmes and incentives, or on the kinds of businesses students have successfully run? Finally, Trish should take care to centre her letter vertically on the page. This kind of detail makes a letter more attractive and even more pleasant to read. Take a look at Trish's improved letter (Figure 2.5).

Analysis

1. Since Trish does not know the identity of her reader, nor does she know whether the reader is male or female, she has simply deleted the salutation, a practice that is increasingly common. A more old-fashioned approach, but one which is still courteous, is to use "Dear Sir or Madam" in place of "Dear Sir."

2. The "re" line specifies more closely the kind of information she is requesting.

3. The opening paragraph avoids clichés and gets right to the point: she describes the business she hopes to start so that her reader will best be able to assist her in choosing from the publications that the Ministry offers.

4. In paragraph two, Trish lists some of the specific documents that she would like to receive. Note that she includes the location of her proposed business, because it may affect what the Ministry people will send her.

5. A statement of thanks closes the letter effectively.

FIGURE 2.5 Compare the first draft of Trish's letter with the improved version. In what ways has she observed the requirements of effective professional writing?

PO Box 123
Drayton Valley, Alberta
T5Y 7H8

September 28, 2004

Ministry of Economic Development
Province of Alberta
Parliament Buildings
Edmonton, Alberta T7R 0B8

Re: Information on Entrepreneurship Opportunities

As a university student in mechanical engineering, I am planning to operate my own business next summer. I would appreciate receiving information on how to start and successfully run my own small manufacturing and retailing firm to market a series of topological puzzles that I have designed. I expect to be operating this business in Calgary.

Specifically, I would appreciate copies of the booklets "Can You Make Money with Your Idea or Invention?," "Checklist for Going Into Business," and "Learning about Your Market," but I would also be grateful for any additional information packets that you think would be helpful and appropriate.

I hope to get started with my planning well before summer, and would appreciate receiving the information as soon as it's available. Thank you very much for your help.

Sincerely,

Trish Trcka

Trish Trcka

CRITICAL READING

Did you know that effective communication is just as important in mathematical expressions as in other written work? Below is a set of guidelines for the submission of mathematical assignments in an engineering course, written by Allan Dolovich, a professor of mechanical engineering at the University of Saskatchewan, and Lisa Coley, his graduate teaching assistant. As you read, consider the ways in which Dolovich's advice corresponds to the guidelines given in this chapter.

After reading the article, respond as directed by your instructor to one or more of the following assignments.

1. To what extent do Dolovich and Coley's standards correspond to the Seven Cs? Which of the seven does he emphasize? Why? Write a memo to your instructor in which you show how Dolovich and Coley's guidelines offer some general advice on communicating effectively.

2. After writing your memo, think about the following and be prepared to discuss it in class: which reading—Dolovich–Coley or the chapter—did you find more accessible? Why? For whom was each written? What is the purpose of each? Do these factors make a difference?

3. Answer the questions following the essay as directed by your instructor. Submit your responses via e-mail or in memo format.

ASSIGNMENT SUBMISSION REQUIREMENTS FOR ME 316

Allan T. Dolovich and Lisa Coley

Industrial practice and the profession of engineering demand that all written work be of the highest quality, both in technical content and clarity of communication. Each submission, whether a report or other form of technical analysis, should be self-contained and effective in presenting the problem addressed and the method of solution. Consult the solution sets in the library as examples of application of the principles outlined in this document.

In keeping with this standard, marking of assignments will include evaluation of adherence to the following requirements:

- Start each question with a title, giving the problem number at the top of the page.
- Use only one side of each page.
- Solutions must be stapled together in the order in which they were assigned.

Allan Dolovich is on the faculty of the Department of Mechanical Engineering at the University of Saskatchewan. An outstanding teacher, he is known for his engaging classroom style and genuine commitment to undergraduate education. His research has been in the areas of numerical techniques in stress analysis and optical techniques for non-destructive testing. Lisa Coley, who was teaching assistant for the course ME 316 when this document was written, is currently completing a PhD in Mechanical Engineering in the same department.

- For each problem within an assignment, divide your presentation into three sections, each with its own clearly marked heading:
    Given
    Required (or Find)
    Solution
- Place significant space between sections.
- Under the heading "Given," a complete statement of the problem must be provided. This does not necessarily mean copying verbatim the problem statement in the textbook, but all the pertinent information and data must be listed, such as given angular velocities, angular accelerations, dimensions, angles, and so forth.
- The given information must include a diagram depicting the problem (except for the odd question for which no diagram is appropriate). This diagram must be
    large (use one-quarter to one-half page as a guideline)
    neat
    clearly labelled
    with clearly labelled coordinate axes.
- Quantities (such as angular velocities, angular accelerations, angles, etc.) should be labelled in your work as specified in the textbook description of the problem OR should be clearly defined otherwise.
- In your analysis, write all subscripts CLEARLY. Make sure they look like subscripts and not like part of the main variable or letter.
- Arrows must be placed over vector quantities or variables, except for unit vectors in which case the circumflex (^) should be placed over top.
- Specify the units of all angles throughout your analysis (that is, degrees or rads).
- If you start on a certain path (that is, velocity analysis or acceleration analysis), continue to completion or document any digressions.
- Present your analysis in a logical order which the reader can easily follow.
- Space out your work. Leave sufficient space between steps or equations to indicate a visual break. (The solution sets in the library provide a good example of appropriate spacing.)
- Final answers must logically follow from your calculations, and not be changed to match the answer at the back of the book. For example, 26.004 does not round off to 26.1.
- Final solutions must be expressed in the correct units.
- Place a neat box around your final answers.
- If you use a pencil and make a mistake, erase the mistake. Don't write over top of the mistake. If you use ink, cross out mistakes neatly. Don't just cross out one or two things in an equation; instead, put a line through the entire equation, leave a space, and write it again.
- YOUR HANDWRITING MUST BE LEGIBLE THROUGHOUT.
- NEATNESS, LOGICAL ORGANIZATION, AND CLARITY OF PRESENTATION COUNT.

Clarity in the communication of your results is as important as their correctness. For this reason, marks will be deducted for non-compliance with the above rules in the same way that marks would be deducted for incorrect calculations, incorrect equations, or other incorrect analysis.

Things to Consider

1. What is the purpose of Dolovich and Coley's document? Who are its intended readers?

2. What strategies do Dolovich and Coley use to capture and retain the audience's interest?

3. Dolovich and Coley recommend adherence to at least two of the Seven Cs of professional communication. How many do they observe in their own document?

4. How would you describe the writing style in this document? Is it formal or informal, personal or impersonal, direct or indirect, superior or equal? How suitable is its chosen style—the language, the structure, the tone— to the purpose and audience for which it was written?

5. Based on this passage, what kind of teachers are Allan Dolovich and Lisa Coley? How does he gain authority or credibility?

6. How well do Dolovich and Coley understand their intended audience? What assumptions do they make about the reader's needs, expectations, prior experience, or concerns? Given their intended audience, are these assumptions reasonable?

7. How difficult was it to read this passage? Compare it with the information presented in the chapter; which is more difficult? Can you suggest why that might be?

CHAPTER 3

Style and Strategies in Technical Communication

LEARNING OBJECTIVES:

1. To analyse and put into use an effective method for planning your professional correspondence.
2. To learn how to use a problem-solving cycle approach to message design.
3. To master the Three Ps of the writing process: Plan, Prepare, and Polish.
4. To become familiar with common organizational patterns and devices of cohesiveness.

PROOFREADING AND EDITING YOUR WRITING

Applying the Seven Cs (as discussed in Chapter Two) to achieve a clear, engaging writing style involves not only care in writing the first draft, but also careful editing after the first draft is completed. Very few writers, including professional writers, achieve a perfect message in one or even two drafts.

Instead, they understand the need for careful reflection and re-vision. They expect, and are willing, to take time to edit what they have written to improve its clarity, coherence, and tone. As I write this sentence, the chapter you are currently reading has been through more than a dozen complete re-writes; by the time you see it in its final version it will have gone through several more.

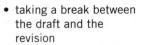

Tip: To help with the editing process, try

- taking a break between the draft and the revision
- reading your writing aloud to expose lapses in clarity

Most people proofread in some form or another. But good writers know that real proofreading—editing—is not simply rereading what you have written, or simply running the spell-check program on your word processor. Instead, editing means reading subsequent drafts with the clear intention of improving your written message. It means looking not only for errors in spelling and grammar, but also for any places where your message is unclear, unfriendly, or misleading.

Although you will often be short of time, you should try to take a break—even a short one—between writing the draft and proofreading it for the Seven Cs; try also reading your work out loud, listening for lapses in clarity or sense. Both of these strategies will help you to uncover mistakes you might otherwise miss. They give you a chance to show yourself at your best, and to communicate with your reader as you would like to be perceived.

Because the Seven Cs of professional writing are closely linked, improving one will often result in an improvement to the others. To achieve clarity and conciseness, for example, you must increase organization and reduce wordiness; at the same time that you attend to these qualities, you will also increase the coherence of your writing, and thus, because your work is easier to read and understand, the courtesy that it shows its intended reader is also increased. This process of improvement requires several drafts and careful attention to each of the Cs. The process can be mastered by anyone, and with practice it will become more automatic and less time-consuming. The remainder of this chapter introduces such a method for writing and editing your work.

PROFESSIONAL WRITING AS PROBLEM SOLVING

Writing well is hard work. As a form of systematic problem solving, it follows thoughtful steps of invention and assessment. Like other forms of practical trouble-shooting, designing an effective message cannot be done in one pass, except perhaps for the most routine communicative tasks. The more sensitive the situation or complex the task, the more demanding the writing process, as might be the case, for example, with a formal report to head office or an assessment of unsatisfactory work by a colleague. In such cases,

Communication Problem Solving

- State the problem and determine the purpose of the message.
- Consider all possible ways of understanding the situation.
- Identify constraints.
- Brainstorm ideas.
- Analyse the alternatives and eliminate unlikely or unnecessary details.
- Select the best approach and write the draft.
- Reconsider your message and edit until it is as effective as possible.

the writing task is usually more manageable and ultimately results in a more effective message if it is carried out in stages.

The stages of writing well are very like those you may have encountered in other forms of engineering problem solving. A framework known as the problem-solving cycle enables a practitioner in fields such as design, accounting, mediation, engineering, media production, or medicine to approach problems in a structured, repetitive pattern. You have likely learned a similar process in your technical courses. When all the steps in the problem-solving process have been completed and a preliminary solution developed, these practitioners do not consider their work finished. Instead, they typically start again at the beginning of the cycle, reconsidering the original problem in order to decide whether the draft solution is appropriate and complete. If it is not, they repeat the steps in the problem-solving process as many times as necessary to refine their solution until it is comprehensive and satisfactory. Writing, particularly in longer formats like reports, involves a similar process of "re-seeing"—or revising—until the proposed solution is as effective as it can be.

The steps in the process of problem solving can easily be adapted to the requirements of written communication. They can be expressed as follows:

- State the problem and determine the purpose of the message, making sure you understand exactly what is needed.
- Consider all possible ways of understanding the situation, the audience, the purpose of the message, and the relational context, and restate the purpose to include these.
- Identify any limitations or constraints facing you as you design your message; consider the context and history of the situation, previous attempts at solutions, limitations of time, funding, or personnel. Refine and identify specifications that will limit the kind of solutions that can be proposed.
- Brainstorm ideas to identify as many possible interpretations as you can, no matter how unlikely they may initially appear.
- Analyse the suggestions and ideas; separate relevant details from irrelevant information, and categorize relevant information into essential and peripheral details.
- Select the best approach from those you have identified and write the first draft of your message.
- Reconsider your message and edit until it is as effective as possible using the Three Ps of professional writing (discussed below).

As this list of steps suggests, the problem-solving approach is an appropriate way to understand and describe the process of writing well. Just as an engineer, a physician, or a profiler studies a problem and then draws upon theory and experience to design an appropriate intervention, a good writer considers the demands of the communication problem—its intended audience, the purpose of the message, and the best way to establish credibility—and then draws upon the principles of effective communication in order to create an appropriate message.

One of the greatest challenges for any communicator is finding the appropriate accommodation of audience needs, message development, and speaker credibility. Finding this appropriate balance means learning to understand the particular challenges of an individual communication scenario, and using your knowledge of effective communication principles to prepare a solution that is effectively adapted to the situation. A misplaced emphasis can render the communication ineffective by placing the focus or emphasis on the wrong element of the interaction. The problem-solving cycle can help to identify such imbalances so they can be corrected before the message goes to its intended readers.

There are three principal influences on what a writer or speaker can choose to say in any given situation. These are: the nature of the problem or issue that the message addresses; the particular characteristics of the audience for whom the message is to be prepared; and any additional limitations imposed by the physical setting in which the communication occurs, the historical, social, professional, or personal context, and the manner in which the message must be communicated. In addition, influences are always exerted by the audience, the form and structure of the message itself, and the speaker's character and experience. As in all human experience, some constraints also arise from moral, legal, and ethical considerations, as well as from technical or communicative limitations.

The problem or obstacle that prompts the communication, and the context or situation in which the communication occurs, together help to determine the kind of message that will be considered appropriate to the situation, just as the nature of any problem and the situation in which the solution must function will shape the design process. A problem presents itself, and the reflective practitioner draws upon experience and good judgement to create a solution that fits the constraints of the particular situation.

As in other kinds of problem solving, a technical writer or speaker does not stop at the first solution that occurs; instead, an effective writer reviews the problem and the written response several more times to be sure that the problem has been adequately understood and effectively addressed. In solving a design problem, the engineer will redefine the original problem so as to ensure that all constraints have been identified and considered; in creating an effective message, the writer edits and revises the document in light not only of the original request or situation but also with an eye to crafting a message that observes the Seven Cs of professional writing and appropriately addresses the intended audience. Seeing writing as a form of problem solving will help you to keep in mind all the elements of effective communication.

STEPS IN MESSAGE DESIGN: THE THREE PS

The writing process, like any design process, can be divided into three main components. These are a planning stage, a preparation or draft stage, and a polishing or editing stage.

During the planning stage, you define your task by analysing your purpose, your audience, and your own credibility requirements. At this stage also, you should sketch a brief outline of your main points.

A preparation or draft stage follows, during which you produce two or more rough drafts, reviewing each to ensure that you have addressed the issue adequately, that you have considered your audience's interests and needs, and that you have projected effective ethos.

The polishing, or editing, stage involves re-reading and revising the message for clarity, completeness, conciseness, coherence, correctness, courtesy, and credibility. This stage follows the completion of a more or less "finished" draft, and the whole process may be repeated several times until a satisfactory message has been produced.

Although you will rarely have unlimited time to devote to each writing task, you should make it a rule to allow yourself at least some time for proof-reading and editing. This can be managed more easily if you break larger writing tasks into stages and spread these over a longer period, rather than leaving the report until near the deadline and trying to finish the job at one sitting.

These three general stages can be broken down further into manageable steps. Though following these steps may initially seem more time-consuming

---

The Three Ps of the Writing Process

Plan	*Prepare*	*Polish*
1. Make a writing plan.	1. Start with the main message.	1. Let it sit for a while.
2. Generate ideas.	2. Keep the audience in mind.	2. Read it out loud.
3. Outline and organize.	3. Focus on your purpose.	3. Run speller and grammar checker.
	4. Choose an organizational pattern.	4. Seek help from others.
	5. Build cohesiveness.	5. Look for common flaws.
	6. Break up the task.	6. Ask style questions.

than what you currently do, this process will actually save you time by producing more efficient, competent messages. Careful planning, preparation, and polishing can help to ensure that your communication conforms to the Seven Cs. The next time you have to write a lengthy or complex message, try the following steps:

Stage I: Plan Your Message

1. Make a writing plan.

Before you begin to write, plan your communication carefully. Use a device such as the Professional Writing Planner (discussed below; see Figure 3.1) as a guide to sketch a preliminary overview of your task. Identify the important elements of your message and the probable needs and expectations of your reader.

2. Generate ideas.

If the writing task is short—a memo or letter—you may be able to hold in your head all the points that you hope to cover. However, most people find it useful to jot down their main points on scrap paper or on the computer screen before they begin writing. At this point, you should jot down everything that seems relevant; you can always eliminate later any information that comes to seem unnecessary.

Write down the major points you wish to cover and juggle them around to achieve the most logical order. Consider eliminating those that don't seem relevant; if you're unsure about any of them, retain them into the next writing stage. The jotted ideas need not be arranged into a formal outline; they are simply intended to help you remember your points and structure your message in an understandable way. This kind of "scratch outline" is particularly useful for planning longer messages. It can also help to give you an overview of the message as a whole, clarifying its purpose and its subject matter.

3. Outline and organize.

Once you've jotted down all the ideas you need to communicate, organize them into a logical progression of main message, background, full details, conclusions, and any recommendations or requests for action. Select a sequence that will make clear to the reader the details and intent of the message; distinguish between main ideas and supporting details. Some writers find that numbering the points in the scratch outline into the order in which they will be presented is enough at this stage, while others like to rewrite the list in the new sequence right away. Doing so gives a visual sense of the whole message, and will help you to identify ideas that are misplaced or superfluous. As you begin to order the points, you should also consider whether any elements can be eliminated and which need to be further supported.

As you sequence your thoughts, it may help you to think of the body of your report or letter as a kind of story which is meant to engage and motivate your reader. The action you require should make sense to the reader by the time he or she reaches the end of the message.

Stage II: Prepare a Draft

1. Start with the main message.

After developing your scratch outline, write your first draft beginning with "The main thing I want you to know is that . . . ." Beginning your rough draft in this way will help remind you to put the most important piece of information—your purpose—at the beginning, and will help you to keep it in view as you develop the draft. Be sure to cross out this clause in later drafts, especially the final draft.

2. Keep your audience in mind as you write.

As you already know, audience is one of the most important aspects of any communication. Although most people understand the principle of audience appeal, surprisingly few actually keep the audience in mind when they start the task of writing. Instead, they become preoccupied with the message itself or with their own writing challenges. Not just at the very beginning of the process, but throughout, each time you take a break, each time you read through your draft, think carefully about the person or persons who will be reading your message. Try to imagine their reactions to what you are telling them and to the way you are presenting the information. Reconsider all the important elements of your audience: What do these people need to know? What is their interest in the information you are providing them? What are they likely to be worried about, excited over, moved by? Do they have any concerns that need to be addressed? How are they likely to react to the message as you have written it? Is there anything that the reader might find confusing, offensive, or inappropriate? Is there any way to make your message clearer, more acceptable, less confusing?

3. Keep the focus on your purpose.

You started your draft with the main idea you wanted to communicate; this is a good beginning. But it's important to continue thinking consciously about what you want the message to accomplish. Do this not only at the beginning, but throughout the writing and editing process. As you re-read your drafts, keep clearly in mind your reason for writing in the first place. What do you want your message to accomplish? Has it achieved that purpose? Is there some way to make it better? What do you hope your audience will do with the information? Have you made it possible for them to make the decision or take the action that you want made or done? Does every statement in your draft contribute to fulfilling your intended purpose? As you read, look for ways to make your message clearer and more convincing, and check that you have achieved the purpose for which you are writing.

4. *Choose an organizational pattern.*

Although the general shape of your message as a whole should conform to the SIDCRA structure described in Chapter One, the body or discussion will need its own appropriate internal structure. Choose an organizational pattern that will assist you in presenting your specific information and that will help your reader to follow your argument. Some common patterns of organization include: chronological or first-to-last; least-to-most important or most-to-least important; spatial (top to bottom, bottom to top, left to right, inside to outside); cause to effect or problem to solution; general to specific or specific to general; comparison or contrast; size, function, or levels of complexity.

Your choice of organizational pattern will depend on the subject matter and purpose of your message. For example, a technical description may be best written using a top-to-bottom or left-to-right order; an incident report may employ a chronological or cause-to-effect arrangement; a process analysis is usually presented chronologically, with greater or lesser detail depending on whether the reader is expected to understand the process or to reproduce it; a site analysis may move from general to specific or from perimeter to centre.

In some cases, two or even more modes of organization may be combined. A technical description of a complex piece of equipment may be organized from general to specific, and also by function. In such a case, the device is described initially according to its general or overall operation; then each of the parts in turn is treated as to its specific function within the whole. For example, a guitar may be described as to type (acoustic or electric; bass, rhythm, or steel), overall appearance (size, shape, material, colour), and general performance or range of sound. Then its sound box, finger board, or string configuration can be described, along with the contribution that each makes to the overall function and sound of the instrument. Similarly, a site analysis may compare or contrast two potential building sites, using both spatial and general-to-specific patterns. Select the pattern, or combination of patterns, that best accommodates your purpose and the needs or interests of your audience.

5. *Build in cohesiveness.*

Although a sensible and appropriate organizational pattern will go a long way toward developing an understandable structure for your message, you also need to attend to its cohesiveness—the way the individual parts of the message are connected, so that the whole makes sense to your reader. Cohesive texts are more coherent and easier to understand than fragmented texts, and readers generally experience them as better written (and their authors as more competent). Devices of cohesiveness help to ensure that your message will be regarded as credible. There are two primary principles of cohesiveness, and several specific linking devices you can use to build this quality into your messages.

- Place familiar information before new information in sentences, paragraphs, and documents.
- Connect every sentence in some explicit way to the sentence that went before, and connect every paragraph in some explicit way to the paragraph that went before.

The first principle of cohesiveness is to observe reader expectation by adhering to something called the "known-new contract." In any argument, information that is known to the reader is placed both logically and chronologically before that which is new. Most messages involve a combination of the known and the new, and if they are to be understood by their readers, they are typically arranged so that familiar information comes first. This ordering can be seen at the sentence level, so that the known information is presented in the subject of the sentence and the new information in the predicate; it can also be seen at the paragraph level, so that the familiar information is located at the beginning of a paragraph, followed by the new data. Finally, individual segments of the document (summary, introduction, discussion, and so on), as well as the document as a whole, also follow this general ordering.

Here's how the known-new contract operates in a document. Suppose you are introducing your reader to a new concept or idea. The first time you use the term you will need to define it and provide synonyms or examples. Subsequent uses, especially if they are accompanied by brief reminders or quick summative statements, reinforce the explanation. Finally, the new term itself becomes familiar enough to be considered known information, and can in turn be used to introduce other new information.

For example, recall the introduction of the term "ethos" in Chapter One. You will remember that it was defined as the speaker's character revealed in the style and tone of the message. Synonyms such as "credibility" further illustrated the meaning of the concept. Examples drawn from common experience—the way we assess the professionalism and expertise of our colleagues, for instance—helped to reinforce this newfound knowledge, as did continued repetition of the idea through the chapter. Finally, your familiarity with the concept of "ethos" was used to introduce two other forms of appeal—"logos," the quality of the message, and "pathos," adapting the argument to the needs and expectations of the audience. If you look at any of your textbooks—including those from mathematics or engineering courses—you will find a similar pattern of development. Mathematical arguments, for example, begin with what is given (the known information) and proceed to the proof of a theorem (the new information). Writers in all fields who ignore the known-new contract typically confuse and alienate their readers.

Once you have established a general pattern of organization and ensured that you have abided by the known-new contract, you can further enhance the cohesiveness of your message by linking every sentence in some explicit way to the sentence that went before it, and connecting every paragraph in some explicit way to the paragraph that went before. There are several specific means you can use to link clauses, sentences, and paragraphs, including key word repetition, explicit linking words, and structural devices. All of these devices are discussed fully later in the chapter.

6. Break up writing sessions into smaller periods, and divide a large task into smaller ones.

Do not try to write the entire report at one sitting. For example, if your deadline is two weeks away, devote an hour or more per day to the writing, rather than leaving everything to the day before the report is due. If your deadline is only two days away, try writing for shorter periods of half an hour or forty-five minutes; then take a short break of five or ten minutes so that you can return to the task refreshed. Dividing up the time spent is a useful strategy for more effective writing; similarly, dividing a large task into smaller tasks will make your job easier. For example, write the introduction and the conclusion as separate tasks; write the discussion in several sections; write the recommendations as the last stage before you write the summary.

If you can get started early enough, intersperse the writing with other regular tasks. That way, each time you return to the job of writing you will bring a fresh perspective that will help you to understand the problem anew. Spreading out the writing task also means that you will come to a fuller understanding of how to achieve your purpose. Some nuances of a problem won't be clear to you until you have thought about the challenges through several drafts.

Step III: Polish Your Written Message

1. Let it sit for a while.

If you can, take a longer break from the task once you have finished each full draft, or, if the document is complex, between drafts of major segments. Set your work aside and turn your attention to something else. When you return, you will be able to see the document as the reader will see it, and you may be able to see areas of weakness in what you have written so that they can be corrected before you send the message out. The length of this break will depend on the time available and the complexity of the message, but if you can, try to let your document sit at least overnight so you can approach it with a fresh perspective.

2. Read it out loud.

Sometimes the process of speaking the message, or hearing it out loud, will reveal flaws in reasoning or wording that you are less likely to catch when you read it silently. Your eye tends to see what you wanted to say rather than what you did say, and reading aloud will help you to locate weaknesses before the message goes to your audience.

3. Run the spell checker and grammar checker.

Use both, but recognize their limitations. Although helpful, these programs do not catch mistakes in diction, and can even introduce new errors into your writing. The spell checker will change specialized words it does not recognize into similar words that are in its dictionary, turning sense into nonsense; the grammar checker, by its limited list of common grammatical rules, will frequently recommend changes that are simply wrong. Discerning good advice from bad requires informed judgement on the part of the writer, and if you are unfamiliar with grammar principles, you may end up following recommendations that actually compromise the clarity or correctness of your writing. On the positive side, when they are used judiciously, these programs can help you to catch typos and other small errors that you might otherwise miss, but be sure to proofread on your own as well. Don't automatically assume that any changes these programs suggest are for the best; use your own judgement as well. If you're uncertain about a spelling or grammar point, you can always check in your dictionary and your handbook.

4. Seek help from others.

If you can, have someone else read your work for you. Be sure to choose someone who will appraise it honestly rather than someone who will simply flatter your ego by telling you that what you've written is fine as it is. The more experience you gain in having someone else review your work, the more distance you will develop and the better you will become at editing your own work and seeing it as a reader might see it.

5. Read for common flaws of professional writing.

Some of the more common flaws of on-the-job writing can be removed before you get to the final draft if you know what to look for. Several of these were identified later in the previous chapter, and you should take a look at them to familiarize yourself with the kinds of awkward phrases and wordy constructions that tend to creep into professional writing. Remove these wherever you see them in your work, and you will improve its style significantly.

6. At last, when you think you have reached the final draft of your work, read it again as you ask yourself the following questions:

a) Will my opening catch the attention of the person with whom I want to communicate?

b) Is my argument clear and direct?

c) Is the material presented in a logical manner?

d) Does my writing have a clear beginning, middle, and end?

e) Is it easy for a reader to follow along?

f) Have I put myself in my reader's place?

g) Is the report interesting enough to make people want to read all the way through?

h) Is my main point or request obvious?

i) Does my conclusion follow from what I've presented, or does it seem to come out of left field?

j) Would people who have read this report likely want to take the action I've recommended, and be willing to encourage others to do so?

Use the Professional Writing Planner (Figure 3.1) below as a first stage in developing a draft of your written messages. It will assist you in understanding the context in which your messages will be received.

FIGURE 3.1   A Writing Planner can help you in all your written communication. Your instructor may be able to give you a copy of this form on a handout. A similar planner for writing reports can be found in Chapter Four.

PROFESSIONAL WRITING PLANNER

Before beginning your letter or memo, consider these points carefully.

1. What is the topic of this letter or memo?

2. What is my focus or purpose? Am I providing information? Promoting an idea? Proposing an action?

3. What is my main point? (This will appear in a subject or "re" line.) "The main thing I want to say is that... "

4. Who is my reader? What is his or her interest in this subject? What does my reader already know and what further information will be needed or wanted? What is the reader's likely attitude to this information?

5. What background information does my reader need as preparation for what I am going to say?

6. What are my primary supporting points? Which details are important? Have I answered any Who, What, When, Where, Why, and How questions the reader might have?

7. What, if any, action do I wish my reader to take after reading my letter or memo? Have I made it possible for her or him to do so?

8. How do I wish to be perceived by the reader of this message?

DEVICES OF COHESIVENESS

Devices that create cohesiveness

- Key words
- Pronouns
- Synonyms
- Antonyms
- Commonly paired words
- Connecting words
- Parallelism

One of the most important principles of good writing that you will ever learn is this one: always explicitly link every sentence to the sentence that preceded it, and explicitly link every paragraph you write to the paragraph that preceded it by placing a linking device close to the beginning of the sentence or paragraph. The linking device helps the audience to pick up the thread of your argument by helping to fulfill the known-new contract. All good writing follows this principle, a fact that you can easily check by turning to any piece of prose that you consider well written. You will find at least one, and frequently more than one, of the following devices in every sentence in the text. Interestingly, though these devices are easy to find once you know what to look for, and may strike you as cumbersome when you are learning to use them, they pass unnoticed by the average reader, who simply experiences the text as particularly clear. You can make your own writing similarly lucid by mastering these strategies.

The examples below show linking strategies from sentences in this book; you can find others if you look for them. Some of the examples show connections between clauses rather than separate sentences, but the principles of linkage are the same in both cases.

1. *Repeat a key word from the previous clause or sentence in the new clause or sentence.* Do not overuse the device of repetition, but employ enough "planned redundancy" to make your text readable.

   > *Nonverbal* cues—such as eye contact, facial expression, vocal tone, gestures, physical movement, stance and posture, manner of dress, personal hygiene, or overall attractiveness—communicate personal traits. Your *nonverbal* expressions can display assertiveness, confidence, self-possession, energy, enthusiasm, cooperativeness, reliability, and truthfulness—or their opposites.

2. *Use a pronoun in the new clause or sentence to refer to a specific noun or noun phrase in the previous clause or sentence.* Avoid using pronouns such as "it" or "this" unless they clearly point to a preceding noun; pronouns with no identifiable antecedent can actually undercut the cohesiveness of your text.

   > Your *need to communicate* clearly doesn't end when you graduate; indeed, *it* will likely escalate.

3. *In the new clause or sentence, use a synonym for a key word or phrase in the previous clause or sentence.* Make sure that the synonym fits the context, and that your reader will connect the two; remember to place the synonym near the beginning of the second clause or sentence.

Unfortunately, many *graduates of engineering programs* are ill-prepared in skills like the ones listed, partly because *technical students* often don't realize the extent to which their career success will depend on their ability to communicate effectively.

4. *In the new clause or sentence, use an antonym for a key word or phrase in the previous clause or sentence.* The principle of contrast helps your audience to link two thoughts and to follow the thread of your argument. Contrast may be combined with parallelism (device #7) for greater effect.

> Although "communication" is frequently thought of as *simply* the "transmission of information by speaking, writing, or other means,"[1] it is actually a much more *complex* process than this characterization would lead us to believe.

5. *Use a word in the new clause or sentence that is closely associated with a word in the previous clause or sentence.* The link appears similar to the synonym, but differs in that the paired items are words that have become yoked by habit or convention, rather than synonyms in meaning: cats and dogs; salt and pepper; trains, planes, and automobiles; lions and tigers and bears.

> *Ethos* helps to build the relationship with the reader. *Pathos* performs the same function.

> *Readers* will expect a text to be clear. *Writers* need to respond to this expectation.

6. *Use a specific connecting word or phrase.* If you have organized your ideas well, cohesiveness (and hence coherence) can be enhanced by adding explicit connectives. Some that you may wish to use include:

since	as well as	in addition to
therefore	however	on the other hand
naturally	of course	as a matter of fact
also	nevertheless	for instance
for example	once again	furthermore
moreover	thus	first this; second, this
as a result	then	for this reason
because	after	following

There are many more, of course—a complete list would be much too long to include here. Any word that explicitly signals a movement in your thought from cause to effect, problem to solution, first to last, front to back, and so on, is considered a connective. Here are some examples of their use from the previous chapter:

---

[1] Katherine Barber, ed. *The Canadian Oxford Dictionary* (Toronto: Oxford University Press, 1998) 287.

In many technical jobs, for example, fifty percent or more of your work can involve communicating, persuading, and cooperating with others. *For this reason*, nearly all practitioners in the technical professions include communication in the list of essential skills for new graduates and experienced engineers alike.

Because lay readers cannot directly judge technical skill, they will instead rely on the clarity and confidence of a professional's communication as a basis for judging technical competence. *Thus*, skill in communicating specialized information often becomes the measure of an engineer's competence, irrespective of his or her actual technical expertise.

7. *Place ideas in a parallel structure*, which means that phrases, clauses, sentences, paragraphs, or sections of the document should be arranged in the same structural or organizational pattern. (For more on parallel structure within sentences, see Appendix B.)

In many technical jobs, for example, fifty percent or more of your work can involve *communicating*, *persuading*, and *cooperating* with others.

This example demonstrates parallel structure in a list, but it can also be used to unify sentences, paragraphs, or even sections of a document. When you are writing your résumé, for example, you create sections covering education, volunteer activities, and work experience. If you want to help your reader to a clear picture of your background, you will do well to organize the information within each section in the same pattern, perhaps by date, institution or firm, job title or course of study, or duties. Parallel structure assists the reader in making sense of unfamiliar information.

Use one or more of these devices for every sentence you ever write in your professional correspondence. Using explicit writing strategies to create cohesiveness will help to keep you from drifting or jumping around, and will ensure that your readers perceive you as a competent and considerate writer.

Learning strategies for planning, developing, and effectively editing your own work will improve your writing and make your messages easier to understand and act upon. Approaching your message design in a systematic way, using the problem-solving cycle, may seem cumbersome when you first try it, but if you stay with it, the method will become second nature to you, and will ultimately help to ensure that you have fully understood and addressed the issues involved. As well, planning your message using an aid such as the Writing Planner will assist you in focussing on both purpose and audience, thereby enhancing your credibility as a professional. Taking the time to master these strategies now is a small investment that will pay large dividends when you're faced with writing tasks on the job.

1. Break up the writing into smaller tasks, and take breaks.

2. Always be prepared to write more than one draft.

3. Always begin with your main idea; put it into a subject (or "re") line where possible.

4. Cluster related points together.

5. Check wordiness by eliminating unnecessary repetition, excessive use of clauses and phrases, clichés, and the passive voice.

6. Include all necessary information and eliminate irrelevant details.

7. Identify desired action.

8. Be courteous—always consider your reader's feelings and watch your tone.

9. Re-read and edit for clarity, conciseness, coherence, and correctness.

10. Repeat the process until the message is as effective as you can make it.

SHARPENING YOUR SKILLS

1. Following your instructor's directions, select a page or two of text from this or another of your textbooks, or from any source that you consider to be well written. Following the list on page 66, identify all the linking strategies that are used in the text. To what extent do these strategies affect your understanding? How easy would the text be to read if these strategies were removed?

2. Turn to the "Sharpening Your Skills" segment of Chapter Four. Read assignment #8 in Section A, pages 116–117. Using the Writing Planner on page 65, strategize the way in which you would respond to the sensitive communication situation described there. Be sure to consider all aspects of the problem, and in particular, to evaluate the needs and expectations of your audience. If your instructor directs you to do so, write a memo justifying the approach you have taken to the problem. Be prepared to discuss your responses with the class.

3. The following real examples contain weaknesses similar to those we've discussed above. As you read, look for ways in which each could have been improved. Ask yourself whether the writer in each case has attended to both content and relation, established credibility and professionalism, or developed an understandable, clear message. Identify any ways in which the writers violate the Seven Cs. What is your impression of each? Be prepared to discuss the flaws in each example and to offer suggestions as to how to correct them. If your instructor asks you to do so, edit the samples to improve their ethos, clarity of content, and audience relation.

FIGURE 3.2 If you were Mr. Loblaw, what would you make of this letter? Has this writer established credibility? Is he respectful of his audience? Is his message clear? If not, why not?

PARK ESTATES DEVELOPMENT INC
Your Home is Our Business

25 April 2003

Mr. Bob Loblaw, MPP
2345 Eglinton Avenue, East
Toronto, Ontario
M3R 5T6

Dear Bob,

Hi, You don't know me but Thanks for an insightful and interesting speech this afternoon at the Bakery Road—Service Club whose guest I was of John Cookman of the American BAnk of Canada at Pope and Lottery Avenue.

As I listened to some of the questions especially Development that were directed at you after your speech, the thought occured to me to ask about this issue which is herewith included and alarmed me when I first read about it recently. An excerpt from Harold Blossomb's April Fool Edition of the Saturday World. I could not remember all the details at the time during the luncheon speech you were making, but I do remember tearing this section of the leftover World while I patiently awaited a haircut the other day at the Pickborough Mall, and the thought still remained to whom could I take this isue if in fact there was any truth to it*

Although I work in a field directly related to Devlopment, Construction, the imediate thoughts that go through ones mind when reading something like this. First of all every once and awhile I personally enjoy using our national parks and especially in the summer they make good picnic grounds for large clubs and family outings and company gatherings and personally in my opinion I would not like to see even one of these wonderful natural resources tampered with in any negative way!

The second thing is I very much respect your "Lowly Politicians Job" as you so aptly put it but hope you will continue to carry on the superb responsibilities you have so admirably taken on that I am wondering why you would not be seriously interested in being willing to take over where Mr Harrass is leaving off? I hope you would be willing to answer these questions either personally or or a province wide level-entirely your choice!

Thanks Again.
Yours Professionally,

FIGURE 3.3 As government funding diminishes, many colleges and universities increasingly rely on fundraising appeals such as this recently-circulated letter. What would your reaction be to this message? Would you donate to the communication fund? If not, why not? The letter was very likely written for the dean by a member of his administrative staff. How responsible is the dean for the quality of this letter?

Waskasoo College
Office of the Dean of Student Services,
Division of Science and Technology
PO Box 1977, Forestville ON P6R 2T7
www.waskasooc.on.ca/sci-tech/htm

The "Communication for the Future" Challenge
April 2003

Dear Waskasoo Alumni:

Exciting, challenging, and innovative are the words to best describe the "Communication for the Future Challenge." At the close of this first campaign for the CFFC fund we have reached the $45,000 mark of donations matched by contribution from an alumni who wishes to remain anonymous. I would like to take this opportunity to thank you for your gift which enabled us to get this fund off the ground.

As we pass the close of an old century and the beginning of a new millennium, we are taking positive steps necessary with the Communication for the Future Challenge Fund to ensure that each and every one of our students receives the appropriate training in communication that they so richly deserve and which are our continuing priority. This year, we would like to build the "Communication for the Future Challenge Fund" into a stable source of funding. The accumulated interest income o this endowment fund, which will not only allow the Division to have some flexibility, but ensures delivery of the best possible educational experience to our students.

The Curriculum Committee of the Division of Science and Technology has been working labouriously to implement a new communication program revisions into the curriculm at all levels. The need for program changes is necessary as academic and continuing competence demands escalate. The size of student enrollment has increased at the same time as allocated resources decreasing. We all face the challenge of doing more with less, putting our shoulders to the wheel, and striving toward the future. Whjle we acknowledge with undying gratitude, support from our alumni, industry and friends combined with the commitment to dedication, hard work and loyalty demonstrated by our instructors and staff, the circumstances we confront every year signal us to continue the vigil of trying to keep up with ever-changing societal demands.

The Division of Science and Technology continues to maintain its commitment to provide each and every student the highest possible quality education and the best possible working conditions and learning environment for students, faculty and staff. Some of the support we require comes from those who accredit their success as a technical professional to the exceptional level of education they received here; thereby giving back to their college a small measure of support to enhance the future for those science and technology students of today and tomorrow.

It is people like you who have played a major role in the succes of this initiative to date. Every single donation counts* Please consider your choices and support the education of future technologists. I encourage you to fill out the enclosed Donor card. Please keep in mind that receipts will be issued for donations of $10 or more. My heartfelt gratitude and thanks to you for your undying support.

Very sincerely yours,

Mikhail Bruin, Dean of Student Services.

Many people believe that only fussy writers notice others' mistakes, and that their own readers will forgive any weaknesses in style. The following reply to the dean's letter shows that even those who do not write well themselves can notice and criticize the mistakes of other writers. Consider your own reaction to the letter, then compare it with the response that follows.

FIGURE 3.4 How effective is this letter from U.V. Ray? Is he a better writer than the person who wrote the letter for the dean? How might his letter be made more effective? Do you think the dean can repair the ethos he has lost with Ray and with other alumni who did not directly express their concerns? Brainstorm with your class a possible response to this letter.

513 – 2 Avro Place,
Erin Mills ON M4D 5Y7
May 12, 2003

Dr. Mikhail Bruin, Dean of Student Services
Division of Science and Technology
Waskasoo College
PO Box 1977
Forestville ON P6R 2T7

Dear Dean Bruin:

In today's world of business and industry, technical people have always been accused of poor English and not being able to write good letters and reports. An unfair criticism when considered generically but true when some individuals are assessed. I'm sorry, Mr. Dean, but you do fall in the latter category vis a vis your letter regarding the Communication for the future Fund (which by the way I supported last year as one of the contributors). My marked up copy of your letter is enclosed to illustrate my position of bad writing. If you are concerned about standards for technical grads please give a little more consideration to english and reports/letters in their curriculum.

In the education system these days more stress seems to be given to the facts you are conveying in your writing and less on communicating clearly and correctly. Perhaps in writing to friends on the Internet this approach might be "acceptable." However I think more is expected of instructors, staff *and Deans* at centres of higher learning. Maybe you should spend some of the fund money on someone who can thoroughly check the letters coming out of your office to insure that they are an *effective* communication. (Only "Kidding!" but serious too.)

Please do not consider the previous comments except in the sense of constructive criticism both to yourself and to the staff member who wrote this letter for you. They are written a bit "tongue-in-cheek" and facetiously but the concern is still there.

Sincerely,

U.V. Ray
Waskasoo Alumnus '89

CRITICAL READING

Below you will find Lloyd Bitzer's "The Rhetorical Situation," which discusses the ways a message is shaped by the context in which it is spoken or written, by the participants in the interaction, and by limitations encountered through the interaction of physical, political, social, personal, or professional considerations. Bitzer's article has been very influential in the study of communication and provides a rich source of understanding for novices as well as experts.

Bitzer's discussion, like that of Wayne Booth in Chapter One, covers some of the same ground that we have covered in our discussion. However, you will find significant differences between what Bitzer says and what the chapter says. Just as you did for Booth's article, consider as you read why these differences might exist, based on the principles of communication you have learned. If you have not already read it, Appendix A, "How Texts Communicate: Some Suggestions for Reading Critically," might be of some help in identifying elements to look for.

After reading the article, respond as directed to one or more of the following assignments. Your instructor may ask you to submit your work in memo or e-mail form, or to bring your responses for a class discussion.

1. Compare Bitzer's discussion to the treatment of similar material in this chapter. What are the central principles emphasized by both the article and the chapter materials? Describe at least two major differences that you see in the treatment of their subject matter. Given what both the chapter and the article say about the nature of successful communication, why do you think the two are so different in approach?

2. After writing your memo, think about the following and be prepared to discuss it in class: which reading—Booth, Bitzer, or the chapters of this book—did you find more accessible? Why? For whom was each written? What is the purpose of each? Do these factors make a difference?

3. Compare the two articles by Booth and Bitzer. Which did you find more readable? What significant similarities do you see in subject matter, purpose, tone or style, and structure? What is the purpose of each article? Who are their intended audiences? In what respects do these factors influence the shape of the message?

4. Answer the questions following the essay as directed by your instructor. Submit your responses via e-mail or in memo format.

THE RHETORICAL SITUATION

Lloyd F. Bitzer

If someone says, That is a dangerous situation, his words suggest the presence of events, persons, or objects which threaten him, someone else, or something of value. If someone remarks, I find myself in an embarrassing situation, again the statement implies certain situational characteristics. If someone remarks that he found himself in an ethical situation, we understand that he probably either contemplated or made some choice of action from a sense of duty or obligation or with a view to the Good. In other words, there are circumstances of this or that kind of structure which are recognized as ethical, dangerous, or embarrassing. What characteristics, then, are implied when one refers to "the rhetorical situation"—the context in which speakers or writers create rhetorical discourse? Perhaps this question is puzzling because "situation" is not a standard term in the vocabulary of rhetorical theory. "Audience" is standard; so also are "speaker," "subject," "occasion," and "speech." If I were to ask, "What is a rhetorical audience?" or "What is a rhetorical subject?"—the reader would catch the meaning of my question.

When I ask, What is a rhetorical situation?, I want to know the nature of those contexts in which speakers or writers create rhetorical discourse: How should they be described? What are their characteristics? Why and how do they result in the creation of rhetoric? By analogy, a theorist of science might well ask, What are the characteristics of situations which inspire scientific thought? A philosopher might ask, What is the nature of the situation in which a philosopher "does philosophy"? And a theorist of poetry might ask, How shall we describe the context in which poetry comes into existence?

The presence of rhetorical discourse obviously indicates the presence of a rhetorical situation. The Declaration of Independence, Lincoln's Gettysburg Address, Churchill's Address on Dunkirk, John F. Kennedy's Inaugural Address—each is a clear instance of rhetoric and each indicates the presence of a situation. While the existence of a rhetorical address is a reliable sign of the existence of situation, it does not follow that a situation exists only when the discourse exists. Each reader probably can recall a specific time and place when there was opportunity to speak on some urgent matter, and after the opportunity was gone he created in private thought the speech he should have uttered earlier in the situation. It is clear that situations are not always accompanied by discourse. Nor should we assume that a rhetorical address gives existence to the situation; on the contrary, it is the situation which calls the discourse into existence. Clement Attlee once said that Winston Churchill went around looking for "finest hours." The point to observe is that Churchill found them—the crisis situations—and spoke in response to them.

Until his retirement in the early 1990s, Lloyd Bitzer taught rhetoric and communication at the University of Wisconsin, where he is now Professor Emeritus. Our selection, a portion of Bitzer's famous and influential essay, appeared in the inaugural issue of *Philosophy and Rhetoric* in 1968. Here, Bitzer explores the role of context in the creation and understanding of pragmatic human communication. He proposes that such pragmatic uses of language—rhetorical uses—must be understood as the product of a particular context or situation, and sets out the constituents that form the background to all purposive communication.

I hope that enough has been said to show that the question—What is a rhetorical situation?—is not an idle one. I propose in what follows to set forth part of a theory of situation. This essay, therefore, should be understood as an attempt to revive the notion of rhetorical situation, to provide at least the outline of an adequate conception of it, and to establish it as a controlling and fundamental concern of rhetorical theory.

I

It seems clear that rhetoric is situational. In saying this, I do not mean merely that understanding a speech hinges upon understanding the context of meaning in which the speech is located. Virtually no utterance is fully intelligible unless meaning-context and utterance are understood; this is true of rhetorical and non-rhetorical discourse. Meaning-context is a general condition of human communication and is not synonymous with rhetorical situation. Nor do I mean merely that rhetoric occurs in a setting which involves interaction of speaker, audience, subject, and communicative purpose. This is too general, since many types of utterances—philosophical, scientific, poetic, and rhetorical—occur in such settings. Nor would I equate rhetorical situation with persuasive situation, which exists whenever an audience can be changed in belief or action by means of speech. Every audience at any moment is capable of being changed in some way by speech; persuasive situation is altogether general.

Finally, I do not mean that a rhetorical discourse must be embedded in historic context in the sense that a living tree must be rooted in soil. A tree does not obtain its character-as-tree from the soil, but rhetorical discourse, I shall argue, does obtain its character-as-rhetorical from the situation which generates it. Rhetorical works belong to the class of things which obtain their character from the circumstances of the historic context in which they occur. A rhetorical work is analogous to a moral action rather than to a tree. An act is moral because it is an act performed in a situation of a certain kind; similarly, a work is rhetorical because it is a response to a situation of a certain kind.

In order to clarify rhetoric-as-essentially-related-to-situation, we should acknowledge a viewpoint that is commonplace but fundamental: a work of rhetoric is pragmatic; it comes into existence for the sake of something beyond itself; it functions ultimately to produce action or change in the world; it performs some task. In short, rhetoric is a mode of altering reality, not by the direct application of energy to objects, but by the creation of discourse which changes reality through the mediation of thought and action. The rhetor alters reality by bringing into existence a discourse of such a character that the audience, in thought and action, is so engaged that it becomes mediator of change. In this sense rhetoric is always persuasive.

To say that rhetorical discourse comes into being in order to effect change is altogether general. We need to understand that a particular discourse comes into existence because of some specific condition or situation which invites utterance.

Hence, to say that rhetoric is situational means: (1) rhetorical discourse comes into existence as a response to situation, in the same sense that an answer comes into existence in response to a question, or a solution in response to a problem; (2) a speech is given *rhetorical* significance by the situation, just as a unit of discourse is given significance *as* answer or *as* solution by the question or problem; (3) a rhetorical situation must exist as a necessary condition of rhetorical discourse, just as a question must exist as a necessary condition of an answer; (4) many questions go unanswered and many problems remain unsolved; similarly, many rhetorical situations mature and decay without giving birth to rhetorical utterance; (5) a situation is rhetorical insofar as it needs and invites discourse capable of participating with situation and thereby altering its reality; (6) discourse is rhetorical insofar as it functions (or seeks to function) as a fitting response to a situation which needs and invites it. (7) Finally, the situation controls the rhetorical response in the same sense that the question controls the answer and the problem controls the solution. Not the rhetor and not persuasive intent, but the situation is the source and ground of rhetorical activity—and, I should add, of rhetorical criticism.

II

Let us now amplify the nature of situation by providing a formal definition and examining constituents. Rhetorical situation may be defined as a complex of persons, events, objects, and relations presenting an actual or potential exigence which can be completely or partially removed if discourse, introduced into the situation, can so constrain human decision or action as to bring about the significant modification of the exigence. Prior to the creation and presentation of discourse, there are three constituents of any rhetorical situation: the first is the *exigence;* the second and third are elements of the complex, namely the *audience* to be constrained in decision and action, and the *constraints* which influence the rhetor and can be brought to bear upon the audience.

Any *exigence* is an imperfection marked by urgency; it is an obstacle, something waiting to be done, a thing which is other than it should be. In almost any sort of context, there will be numerous exigences, but not all are elements of a rhetorical situation—not all are rhetorical exigences. An exigence which cannot be modified is not rhetorical; thus, whatever comes about of necessity and cannot be changed— death, winter, and some natural disasters, for instance—are exigences to be sure, but they are not rhetorical. Further, an exigence which can be modified only by means other than discourse is not rhetorical; thus, an exigence is not rhetorical when its modification requires merely one's own action or the application of a tool, but neither requires nor invites the assistance of discourse. An exigence is rhetorical when it is capable of positive modification and when positive modification requires discourse or can be assisted by discourse. For example, suppose that a man's acts are injurious to others and that the quality of his acts can be changed only if discourse is addressed to him; the exigence—his injurious acts—is then unmistakably rhetorical. The pollution of the air is also a rhetorical exigence because its positive modification—reduction of pollution—strongly invites the assistance of discourse producing public awareness, indignation, and action of the right kind. Frequently rhetors encounter exigences which defy easy classification because of the absence of information enabling precise analysis and certain judgment — they may or may not be rhetorical. An attorney whose client has been convicted may strongly believe that a higher court would reject his appeal to have the verdict overturned, but because the matter is uncertain—because the exigence *might* be rhetorical—he elects to appeal. In this and similar instances of indeterminate exigences the rhetor's decision to speak is based mainly upon the urgency of the exigence and the probability that the exigence is rhetorical.

In any rhetorical situation there will be at least one controlling exigence which functions as the organizing principle: it specifies the audience to be addressed and the change to be effected. The exigence may or may not be perceived clearly by the rhetor or other persons in the situation; it may be strong or weak depending upon the clarity of their perception and the degree of their interest in it; it may be real or unreal depending on the facts of the case; it may be important or trivial; it may be such that discourse can completely remove it, or it may persist in spite of repeated modifications; it may be completely familiar—one of a type of exigences occurring frequently in our experience—or it may be totally new, unique. When it is perceived and when it is strong and important, then it constrains the thought and action of the perceiver who may respond rhetorically if he is in a position to do so.

The second constituent is the *audience*. Since rhetorical discourse produces change by influencing the decision and action of persons who function as mediators of change, it follows that rhetoric always requires an audience—even in those cases when a person engages himself or ideal mind as audience. It is clear also that a rhetorical audience must be distinguished from a body of mere hearers or readers: properly speaking, a rhetorical audience consists only of those persons who are capable of being influenced by discourse and of being mediators of change.

Neither scientific nor poetic discourse requires an audience in the same sense. Indeed, neither requires an audience in order to produce its end; the scientist can produce a discourse expressive or generative of knowledge without engaging another mind, and the poet's creative purpose is accomplished when the work is composed. It is true, of course, that scientists and poets present their works to audiences, but their audiences are not necessarily rhetorical. The scientific audience consists of persons capable of receiving knowledge, and the poetic audience, of persons capable of participating in aesthetic experiences induced by the poetry. But the rhetorical audience must be capable of serving as mediator of the change which the discourse functions to produce.

Besides exigence and audience, every rhetorical situation contains a set of *constraints* made up of persons, events, objects, and relations which are parts of the situation because they have the power to constrain decision and action needed to modify the exigence. Standard sources of constraint include beliefs, attitudes, documents, facts, traditions, images, interests, motives and the like; and when the orator enters the situation, this discourse not only harnesses constraints given by situation but provides additional important constraints—for example his personal character, his logical proofs, and his style. There are two main classes of constraints: (1) those originated or managed by the rhetor and his method (Aristotle called these "artistic proofs"), and (2) those other constraints, in the situation, which may be operative (Aristotle's "inartistic proofs"). Both classes must be divided so as to separate those constraints that are proper from those that are improper.

These three constituents—exigence, audience, constraints — comprise everything relevant in a rhetorical situation. When the orator, invited by situation, enters it and creates and presents discourse, then both he and his speech are additional constituents.

III

In the best of all possible worlds, there would be communication perhaps, but no rhetoric—since exigences would not arise. In our real world, however, rhetorical exigences abound; the world really invites change—change conceived and effected by human agents who quite properly address a mediating audience. The practical justification of rhetoric is analogous to that of scientific inquiry: the world presents objects to be known, puzzles to be resolved, complexities to be understood—hence the practical need for scientific inquiry and discourse; similarly, the world presents imperfections to be modified by means of discourse—hence the practical need for rhetorical investigation and discourse. As a discipline, scientific method is justified philosophically insofar as it provides principles, concepts, and procedures by which we come to know reality; similarly, rhetoric as a discipline is justified philosophically insofar as it provides principles, concepts, and procedures by which we effect valuable changes in reality. Thus rhetoric is distinguished from the mere craft of persuasion which, although it is a legitimate object of scientific investigation, lacks philosophical warrant as a practical discipline.

Things to Consider

1. According to Bitzer, what is rhetoric? What does he mean by his term "rhetorical situation?"

2. What are the three constituents of the rhetorical situation?

3. Who are the intended readers of this article? How can you tell?

4. What is the purpose of this article? What does Bitzer hope to accomplish?

5. If all communication is a product of its situation, as Bitzer contends, is his own essay also a product of its particular context? Can you point out any evidence that this is so?

6. What assumptions does Bitzer make about the reader's needs, expectations, prior experience, or concerns? Given his intended audience, are these assumptions reasonable?

7. How difficult was it to read this passage? Compare it with the information presented in the chapter; which is more difficult? Can you suggest why that might be?

8. Is this passage formal, informal, or casual writing? How can you tell? How suitable is its chosen style—the language, the structure, the tone— to the purpose and audience for which it was written?

CHAPTER 4

Letters, Memos, and E-mail Messages

LEARNING OBJECTIVES

1. To learn the common types of professional correspondence and when to use them.
2. To learn the standard parts and formats for letters and memos.
3. To learn judicious care in dealing with electronic mail.
4. To learn to select writing formats appropriate to audience and context.

Whether they are sent by surface mail, by facsimile machine, or through electronic mail, the business letter and the memo remain the most common forms of business communication. Of course, information can also be exchanged by telephone, especially since the cost of long-distance calls has decreased significantly in recent years, but for many practical reasons, written communication may be preferred. For example, it allows a degree of precision not possible in oral exchanges and creates a permanent record for your own files.

Although many messages are now communicated by electronic mail (e-mail), the letter has traditionally provided the appropriate vehicle for exchanging written information with someone in another organization. Within the company, written communication has commonly been conducted via the memorandum. Nowadays, traditional post and interoffice mail systems continue to be used for official and formal messages; as well, facsimile transmissions have made the movement of information in hard copy much more convenient.

E-mail provides a quick and inexpensive means of communicating with people within an organization or with other people and organizations around the world; generally, communication by e-mail is less formal than by letter or memorandum. Although its use is widespread for routine exchanges both within and outside an organization, e-mail is still considered largely a casual medium, not suitable for conveying sensitive, significant, or official information. As a result, it has not yet replaced the business letter or the memo as the dominant form of correspondence, particularly for the formal and legal demands of professional communication.

Despite this fact, the implications of its increased use are significant for professional communicators. First, as many of you know, e-mail is not a secure forum for discussing personal or sensitive issues, as it may be monitored periodically by your employer or accessed by someone else on the system. As well, although it retains the casual air of conversation, e-mail is far from being a transient or fleeting form. In fact, it is in many ways as permanent as hard copy and is infinitely easier to distribute more widely and indiscriminately. Personal messages, once sent, can easily be passed on by the recipient to readers you never imagined would see your words. For this reason, you should always exercise caution in your electronic comments. Ironic, humorous, or off-hand comments may easily be misread, particularly if they are seen by someone other than the intended reader.

Finally, e-mail remains unsuitable for conveying official or important information simply because of sheer volume. If you have an e-mail account through your college or university, you already know that you can easily receive thirty or more messages per day, many of them from unwanted sources —advertising clutter; stale jokes that some acquaintances habitually circulate to larger and larger lists; routine mailings to everyone in the organization that may have no direct relevance to you. Many professionals automatically delete e-mails of this kind and may inadvertently delete important messages as they routinely clear their queues.

Part of communicating your message effectively is choosing an appropriate method or medium of communication, and the principles that follow are applicable to all your written communication, including e-mail. Here are some general guidelines for choosing between oral and written communication, or between a letter or memo and e-mail.

Generally, you should use a letter or memo when:

- an official copy of the information you are sending must be retained on file;
- the information concerns organization policy;
- the information is important; and
- copies of the document will be formally distributed to someone other than the individual to whom it is addressed.

You should use a letter or a memo rather than e-mail or a telephone call for any information that might be considered "official" or formal, and for any information which is to be retained on file. A call or even a personal note will suffice if the information is for one person only, if no copies will be distributed elsewhere, and if it is not for the record.

E-mail, on the other hand, may be used for a whole variety of communication situations, is handy if you have to send the same message to a number of people, and can be quite casual in its set-up. However, because it cannot as yet carry your signature, it is not usually considered suitable for formal needs, and busy people who are inundated with dozens of e-mail messages a day may not give an important piece of communication the attention it deserves.

You should not use e-mail communication for confidential messages unless you can ensure that they will be securely encrypted and decoded only by the intended recipient. Even then, you should remember that anything you send via e-mail may be retained indefinitely by the recipient or forwarded without your knowledge or permission to someone else virtually anywhere in the world.

Because letters and memos are used so frequently in professional situations, consistent standards of content, format, and style have evolved to help make the task of writing—and reading—easier. Observing conventional standards in your writing will not only help to make your message more easily understood, but will also create a positive impression and help to establish your credibility as a professional. Eventually, as electronic mail replaces more and more functions of "snail mail" for professional communication, style requirements will evolve for its use as well. All your written messages—memos, letters, and e-mails—should observe the same rule: get to the point and don't waste the reader's time. To be effective, they must be accurate, clear, and written in an acceptable format. In the remainder of this chapter we will consider general guidelines for the content of your professional correspondence, and then we will learn the standard formats used in writing them.

Types of Professional Letters, Memos, and E-mail Messages

Letters, memos, and e-mail messages can carry both positive and negative messages, and may serve a multitude of purposes, from providing information to soliciting financial support. Whether they carry good news or bad news, these messages can usually be divided into two main categories: request and response. Whichever form your message takes—letter, memo, or e-mail—it will follow the same general guidelines.

Making a Request

A *request* may be written for any number of purposes: to order merchandise; to request information or documentation (printed materials, specifications, user's manuals) or to book an appointment; to reserve a conference or hotel room; to apply for a job; to ask for favours (a reference, for instance); to submit a tender or funding application; or to invite donations. These days, professional contacts may be initiated via e-mail as well as by letter, and the following guidelines also apply to these exchanges. Basically, a request is any message that initiates contact with another person. In it, the writer should aim to establish an effective working relationship with the recipient. Request messages should observe the Seven Cs listed in Chapter Two: completeness, conciseness, clarity, coherence, correctness, courtesy, and credibility. As you plan your message, keep in mind that a major part of courtesy is making reasonable requests—don't ask the person to whom you are writing to do your work for you. Here are some questions to ask yourself before you write.

1. Am I being specific about what I want to know or what I'd like done?

2. Am I asking someone else to find information or perform a task that I could easily accomplish myself?

3. Is this request going to inconvenience the person in any unreasonable way?

4. Would I find a similar kind of request from someone else an imposition?

If you can easily find out the information you need from a library, the Internet, or other source, or if you are vague about what exactly you want, you should reconsider your request carefully. Do not compromise your ethos by asking others to do for you what you could easily do for yourself. Of course, you should also be sure to say please and thank you.

Writing a Response

A *response* is any message written in reply to a request or advertisement, or in reaction to a situation. These might include letters of recommendation, information, congratulations or condolence, adjustment, refusal, or complaint.

In addition to observing the Seven Cs, an effective response is both prompt and helpful. Always give the reader a positive impression and make sure your answer is complete and understandable. Guidelines for specific situations appear below.

Positive and Negative Messages

Within the two general categories of request and response are many situations that require a letter, a memo, or an e-mail message, and these may be broadly categorized as positive (or neutral) and negative messages. Negative messages present special problems for a beginning writer. Letters, memos, and e-mail messages carrying positive news, such as congratulations or recommendations, or those that carry emotionally neutral messages, such as a notice of a meeting or a cover letter for the company's staff directory, make it easy to establish a positive writer-reader relationship. However, messages that must convey negative, disappointing, or contentious information letters of complaint or refusal, for example—demand more careful handling.

All messages, whether they carry positive or negative information, require the same qualities of completeness, conciseness, coherence, correctness, clarity, and competence as other professional correspondence. Negative messages also require special attention to courtesy. In a difficult or challenging situation, especially one in which bad news is being presented, it is particularly important to be courteous to your correspondents in order to ensure cooperation and good will.

Be sure to moderate your tone when delivering bad news. Try to cushion your negative message with suitable language and an appropriately positive attitude. Be especially careful to avoid suggesting or implying that the other person is somehow responsible for the situation, even if you believe this to be the case. Always assume that you could be mistaken, and give your correspondent the benefit of the doubt if you can. What you really want is to resolve the problem, and creating unnecessary bad feelings will only lessen the chances of this happening. Be pleasant and take care to avoid a sarcastic tone.

Following are several common writing situations encompassing both positive and negative messages. Messages of congratulations, fundraising appeals, and public service announcements could be considered generally positive messages, while messages of complaint or refusal are considered negative. The models below show these types of messages in letter, memo, and e-mail formats. All of these types of messages may be communicated in any of the three forms.

Acceptance, Congratulations, or Acknowledgement

Of the four types of positive, or "good-news" letter discussed here, the letter of acceptance, congratulations, or acknowledgement is the most pleasant to write. Such a letter, memo, or e-mail message may be written to an employee, a colleague, or a client on the occasion of achievement or accomplishment. Perhaps someone you know has been accepted to a specialized training program, received an award, published a book or article, earned a promotion, made a significant contribution to an important project, or simply provided long or effective service. Whatever the occasion, the letter of acknowledgement or congratulation should be positive in tone and to the point. Such letters should be:

- specific (identify the achievement or occasion);
- positive (make sure your tone is warm and your language complimentary);
- sincere (nothing is more insulting than congratulations that sound insincere; avoid being too effusive or ironic); and
- appropriately brief (as in any business communication, say what you have to say, then stop—many people ruin effective acknowledgement or congratulations messages by not knowing when to quit).

Of course, your letter should also observe the Seven Cs; a sloppy, error-filled letter undercuts the sincerity of the congratulations and compromises your credibility.

Figure 4.1 contains a memo of acknowledgement from a supervisor to a staff member who has contributed an extraordinary year's work to the company. Figure 4.2 is an e-mail message expressing congratulations to a colleague who has received the company's Employee of the Year award. Compare the two. Based on your knowledge of professional communication, and the guidelines provided above, which is more effective? Why is this so?

Letters of Complaint

Although not as pleasant to write, negative messages are unfortunately much more common than messages of acknowledgement, partly because none of us is immune to error. Most companies do their best to handle correspondence and requests in a professional and efficient manner; however, occasionally mix-ups do occur. Documents may be misplaced, cheques lost, computer files erased, wrong parts sent, or mailings waylaid.

Most companies will do everything in their power to keep such mistakes from happening, since any organization functions more smoothly when good will is maintained through effective client or staff relations. Still,

FIGURE 4.1   Note how Allan Dolovich's warm tone and use of specific details make his letter of congratulations effective.

Dolovich and Cowan Technical Services
Internal Correspondence

DATE: April 14, 2004

TO: Gwynne Nishikawa, Management Team

FROM: Allan Dolovich, President

RE: Appreciation of your contributions during the 2003/2004 fiscal year

Just a note to say thank you for all the work you have done on Project Management this past year. I have especially appreciated your willingness to aid in the vacation period when I needed someone to fill in for ill personnel, and to take on an extra project when one of our engineers left the company.

Considering that you also managed to design a new training program for Staff Development, your contribution has been truly commendable and beyond any requirements of your job description.

Thank you for all that you have done and for your valuable assistance. Few, if any, members of the management team have done more to make this a successful year.

Please accept my heartfelt thanks for a job well done.

Allan

cc.   Personnel File

AD:jm

FIGURE 4.2 In what ways is this message of congratulations flawed? (Consider both content and relation.) What does it suggest about its author?

Date: January 25, 2005

To: Peter_Holowaczok@dolovicow.ca

From: Randy_Alexander@dolovicow.ca

Re: Employee of the Year Award

Congratulations on winning the Employee of the Year Award. It doesn't usually go to somebody who has been in the company only two years—you must really be something special. I have been here for four years and though my work is really deserving, nobody seems to notice. I haven't even been nominated. I suppose I haven't made friends with the right people; I never was much good at lobbying.

I have heard it said that the work you've been doing in Public Relations is really outstanding, especially that slick brochure you produced for the new Project Management Strategies campaign. Most everybody believes that it's really well designed. I saw it—it is pretty good, but could I give you a little advice? I didn't think that the photo of Dolovich and Cowan cutting the ribbon on the plant project was appropriate for the cover. I have a talent for graphic design, and it's not what I would have chosen.

Please let me know if you'd like to get together sometime to talk about layout; I took a course in graphic art in college and could probably pass on a few useful hints. Also, I'd like to have a chance to talk with the new golden boy from PR.

Congratulations again on winning the award.

Sincerely,

Randy

despite all our good intentions, such problems can take place. However, you can prevent them from becoming a major inconvenience if you always make it a rule to assume, at least initially, that the mix-ups you encounter in your professional dealings are honest errors. If you treat such incidents as unintentional and keep a positive attitude, your complaints will be dealt with more promptly and positively.

The first rule for letters or memos of complaint is to be especially courteous. No one wants to receive abusive, sarcastic, or threatening messages; phrase your bad news in as positive terms as possible. Problems will be more easily solved if you allow your correspondents room to correct the situation without making them look foolish. They will be more interested in helping you—and more anxious to maintain your good will—if you approach them in a friendly, non-threatening manner.

It is also important that in a complaint you specifically identify the nature of the problem and the action you want taken. For example, if you have ordered technical manuals and after waiting a reasonable length of time have not received your order, you will want to identify the missing items by name, catalogue or item number, catalogue issue, and page number. You should state the date of your order, the cheque number, if there is one, and the amount of the order. If you have a standing account with the firm, cite your account number. Be sure that all of this information is correct. Be sure also that you tell the reader exactly what you want done about the problem. The reader's idea of a satisfactory solution may differ from yours.

Finally, before you send any letter of complaint, let it sit for a day and then take another look at what you've said and how. Messages composed in the heat of the moment are rarely effective, and may unintentionally inflame the situation. Give yourself a chance to cool down, and always reconsider your tone before you send the message.

Below are the points to remember for letters of complaint.

- Phrase your comments positively.
- Be sure to identify the exact nature of the problem immediately.
- Provide all relevant details.
- Request specific action.
- Be courteous. Thank correspondents for their help.
- Always allow a "cool-down" period, and reconsider your complaint before you send it.

In the letter in Figure 4.3, Lam Huan is making a complaint regarding an error in the processing of a scholarship award. Note that though the awards officer is at fault, Huan wisely does not cause more difficulty by being sarcastic or abusive.

FIGURE 4.3 What features make Lam Huan's request for adjustment effective? Compare Huan's letter with the letter in Figure 4.4; in what ways does Ing Jang violate the principles of effective professional writing?

90 Victoria Crescent
Bruce Mines, ON N2E 2R4

November 20, 2004

Talya Melniczok, Awards Officer
Student Awards and Services
Science Division, Northwoods College
Box 456
Bruce Mines, ON N0E 040

Dear Ms Melniczok:

Re: Scholarship for 2004-2005
          Student number: 987-451-03

I have just received word from the Office of Student Accounts that my entrance scholarship, which came into effect in 2003-2004, has been discontinued. The clerk I spoke to told me that their records indicate that the award was for one year only, and that I am liable for full tuition payments for this academic year.

I am certain there has been an error in communication between your office and the Accounts office. I have attached a copy of the original letter I received from your office, which clearly states that the scholarship will automatically be renewed for subsequent years if my grade average is above 80% and if no individual grade is below 75%. I believe you will find that I have met these requirements; I have attached a photocopy of my grade report showing that my average for last year stands at 86%, and that the lowest grade I received in any course was 79%.

I hope you can assist me in straightening out this error. Would it be possible for you to contact the Accounts office to confirm that the scholarship is still in effect? The person who is handling my case is Vasjli Gajic.

Thanks very much for your assistance. If you need any further information from me, I can be reached at lph123@std.northwoods.cc.ca or by phone at 546 7089.

Sincerely,

Lam Huan

Lam Huan

FIGURE 4.4 How effective is this letter of complaint? Can you think of any ways it could be improved?

22 Ulethe Crescent
Calgary, AB T3R 5Q8

November 14, 2003

Registrar
Western Plains University
PO Box 666
Calgary, AB T2K 3M4

Dear Registrar:

The other day I was in the office to pick up a transcript that I need for an internship job I'm applying for. If I don't send it right away I can't be considered. So anyway I talked to this clerk of yours who should be fired for her rudeness to me. Who do you people think pays for your salary?

She told me that the transcript costs eight dollars. Eight dollars! For what? And anyway she said I couldn't have it because of this so-called library fine that she said was outstanding from last term. I needed to have that book to study for my final exam and then I accidentally left it home when I came back here in September. It's not like anybody was reading it over the summer anyway, and now she tells me that the fine is seventy-five dollars. So this means it's going to cost me eighty three dollars to get a stupid transcript.

Well, you can imagine I wasn't very happy about that. But she had no right to call the security people. It was only a joke when I said I know where she parks her car. She embarrassed me half to death. Now how am I going to apply for my internship? If I don't have enough money for next term it will be your fault, and anyway this business of having to wait for a week to get a lousy transcript is stupid.

I think you should fire the woman I spoke to and anybody else who is rude to students. After all, this is a public institution, so as a student you people work for me. I hope I don't see her anymore.

Sincerely,

Ing Jang

Ing Jang

Letters of Conciliation or Apology

Letters of apology or conciliation are particularly challenging to write because the readers to whom they are directed are upset, disappointed, angry, or hurt. Normally, these letters are written in circumstances in which hard feelings have already been created and expressed, and these feelings of hurt or frustration must be dealt with explicitly before the rest of the issue can be attended to. If they are ignored, the apology will be rejected as insincere, and the situation may be further inflamed.

A message of apology or conciliation must recognize the violation that has been done to the relationship of professional trust between the writer and the reader; if it does not do so, it will not only fail to correct the situation, it may even escalate it by hardening attitudes even further. An apology thus requires careful attention to the relational element of the message—that is, it must deal forthrightly and sincerely with the hurt and disappointment of the reader. A sincere and effective apology recognizes these feelings and supports the reader's expression of them.

Most important to an effective letter of conciliation is that the writer must also be prepared to take responsibility, either personally or on behalf of the company or organization, for whatever part was played in causing the upset. The responsibility may be diffused across the company or organization, or may rightly belong to some junior member of the staff, but for the apology to be effective, the writer must personally express regret, and must recognize the actual or potential pain, damage, or inconvenience that the action has caused to the reader. If possible, the writer should offer a gesture of conciliation—perhaps the original damage cannot be withdrawn or corrected, but some compensatory measure can be made. Without these gestures, sincerely given, rifts may not be healed and divisions may be increased and irrevocably entrenched.

A message of conciliation or apology, therefore, must do the following:

- Acknowledge the wrong done to the injured party.
- Recognize the legitimacy of feelings of hurt, anger, frustration, disappointment, or betrayal.
- Take responsibility for any damage inflicted.
- Sincerely apologize, on a personal level.
- Offer a gesture of support and compensation, if possible.

Below are two letters of apology, one ineffective and the other effective. The ineffective letter (Figure 4.5) responds to the situation outlined in Problem 8 in Section A, Sharpening Your Skills, which can be found on pages 116–117 of this chapter. The effective letter (Figure 4.6) responds to the letter of complaint shown in Figure 3.4 of this book; note how it has been crafted to support and acknowledge the reader's concerns, and how it carefully attends to the relational element of the message.

Engineering Students Society

Western Plains University PO Box 666 Calgary, AB T2K 3M4

September 13, 2003

Valerie Price, Vice President
Student Affairs and Services
Western Plains University
PO Box 666
Calgary, AB T2K 3M4

Dear Mrs Price:

It has come to my attention that you complained about the recent prank by two engineering students. The dean of Engineering said I had to write to you about it.

I am unable to reinmburse you because WPESS policy doesnt require us to take responsibility for your problem, and anyway we have'nt got sufficient funds available to cover the incident. And besides, the students were'nt representing WPESS, so it's not really our problem.

If you want to recover the money, I suggest that you return the new binders to Office World. If you want I can give you the names of the two students involved in the prank but other then that I do not see what else you expect.

Thank you for your time and attention to this matter.

Sincerly,

Davin Larbour

Davin Larbour, Esq.
President

Waskasoo College

Office of the Dean of Student Services, Division of Science and Technology

PO Box 1977, Forestville ON P6R 2T7

www.waskasooc.on.ca/sci-tech.htm

July 3, 2003

U.V. Ray
513 - 2 Avro Place,
Erin Mills ON M4D 5Y7

Dear Mr. Ray:

I have received your letter of May 12, regarding the "Communication for the Future" solicitation letter that was mailed to you in the spring. I would like to thank you for taking the time to write, and assure you that I am as concerned as you about the quality of correspondence that goes out from this office. I am more than a little embarrassed to say that the letter you received was printed and sent in error. It was in fact a rough draft prepared by a junior member of my staff and sent out prematurely with my "electronic" signature, before it received final editing. Needless to say, it was not the letter I would like to have sent, and I accept full responsibility for this error. At the same time, I want to assure you that such an oversight will not be repeated.

As Dean of Student Services for the Division of Science and Technology, I recognize the importance of strong skills in communication, and I promise you that we are committed to providing a high level of education in this area. I am happy to be able to draw your attention to the development of new communication curricula in the division, including advanced course options in the field. These changes are being led by Dr. Laura Patterson, who recently joined us as Distinguished Professor of Technical and Professional Communication. Among her responsibilities will be curriculum and programme development and preparation of workshops and seminars on communication; in future she will also be doing editorial consultation for my office.

Waskasoo College has always valued a strong relationship with its alumni. I appreciate your taking the time to raise these concerns, and I especially appreciate the constructive manner in which you have done so. Please accept my apology for the quality of the letter you received, and my sincere promise that I will do all that I can to prevent such an oversight in the future.

Sincerely,

Mikhail Bruin, Dean of Student Services

Refusals

These negative messages are often the most difficult to write. Great delicacy is required whenever you must turn down someone's request, whether that request has come from a candidate for employment or from a client who wants an adjustment of a fee or a resolution to a conflict.

As in all negative messages, tone is important to a refusal. You must be tactful in refusing to provide the service, and you must preserve the client's good will if possible. You should use as positive terms as possible under the circumstances to cushion the refusal, and be sure especially to avoid sarcasm or accusation. While you do not want to trivialize or underplay a difficult situation, you also don't want to exaggerate it through unnecessarily negative language. State the message briefly and then politely explain your reasons for the refusal.

At least initially, it is best to avoid placing responsibility for the refusal on the reader, even in cases where that person shares part of the responsibility. It rarely helps to point fingers at anyone; it is preferable instead to stress your own inability to comply with the request. Keep in mind that your most important goal is maintaining good will. For example, if you are writing a letter of refusal to someone who has been turned down for a job, stress that the position was offered to another candidate because that individual more closely suited your needs, not because the person you are writing to is inadequate. Likewise, if you are refusing a reference, it is better to stress your inability to supply it, rather than suggest that the person is flawed. The following are some points to keep in mind when writing a letter of refusal.

- Identify the topic in a subject or "re" line.
- Indicate briefly your inability to comply with the reader's request, using as positive terms as possible.
- State your reasons simply, taking responsibility for your refusal.
- Avoid sarcasm or accusation.
- Be polite and sincere.
- Suggest someone else your correspondent might approach for assistance, if possible.
- Offer the person your best wishes for better success elsewhere if it's appropriate to do so.

In the letter of refusal in Figure 4.7, the bad news has been cushioned with positive comments, and the reader has been invited to reapply for a future training session. As well, the writer has recommended additional steps that might enlarge the candidate's skill base in preparation for future training intakes. Contrast it with Figure 4.8; both letters have a similar message to relate. Which of the two would you rather receive?

Remember, no matter what kind of letter or memo you are writing, be sure to identify your main point first and communicate your message as clearly and concisely as possible, maintaining a polite tone throughout.

FIGURE 4.7 What strategies did Lazar LaFleur use to soften the news of rejection for Molly Crump? Contrast this example with the e-mail message in Figure 4.2. Which would you prefer to receive?

Testamek Engineering Ltd.
64 Weniam Street • Vancouver, British Columbia • V1R 9V1
www.testamek.com

February 15, 2004

Ms. Molly Crump
55 Elwood Close
Calgary, Alberta
T5W 3F6

Dear Ms. Crump

Thank you for participating in our recent screening session for our Technical Management Trainee Programme. Although the interviewers felt your qualifications and aptitude tests showed solid potential, I am afraid we will not be able to offer you a spot in the programme for this year's intake.

This decision is not meant to reflect on your management potential; rather, it is an indication of the large number of very fine applicants who participated in this year's screening interviews. We simply could not accommodate all who applied.

We would like to encourage you to reapply for next year's intake. In the meantime, we suggest further upgrading your communication skills, either through additional courses or through one of the many commercial training programmes available. We believe such an experience would enhance your readiness for the demands of management training.

We extend our best wishes and look forward to seeing you again next year.

Sincerely

Lazar LaFleur
Training Director

The Collegian

The Voice of Waskasoo College

Student Union Building, C-870

April 1, 2005

Gaetan Jamberneau
125 Coulee Court
Forestville, ON P2W 0W3

Dear Gaetan:

We have received your latest "experimental" close-up photos of the gadgets in the engineering building. As you know, the *Collegian's* editorial staff usually welcome submissions from students and members of the community at large.

We would prefer, however, not to receive any more strange photos from you. Although I admit that we published one of your photos last year, you should know that "art" photography is not really our "thing" and we used the other shot only because we were short on submissions and had some extra space to fill.

We are more interested in other kinds of submissions—cartoons, normal photos of stuff on campus, and articles written by the college community; I am pretty sure that our readers don't know much about experimental photography and don't really want to see these weird extreme closeups in the *Collegian*. If they did want to look at stuff like that, they could take an art course! After all, we are interested in things that are relevant to real life and most people agree that technical devices are boring. Besides, I'm not even sure how to judge the quality of your work, since I don't know what you mean by "chiaroscuro" or whatever you called it.

I hope you won't be too upset by my letter, but I thought you would want to hear the truth. If you decide to take up a more interesting kind of photography, please let us know. We might like to have something else from you.

Sincerely,

David Kaminski

David Kaminski

PERSUASIVE MESSAGES

Throughout this book, I emphasize the necessity in any professional writing of identifying the reader's needs, expectations, concerns, and interests. Such an understanding of your reader is always necessary if you are to write effective professional messages, but is especially important when you move from merely informative to persuasive writing. All persuasive communication must engage the attention and interest of its intended audience and motivate that audience to act as requested. As well, an effective persuasive message will help to guide the reader to the appropriate action through an enabling strategy, a device that makes it easy and convenient for the reader to take the requested or recommended action.

Grant applications, research proposals, tenders, and design reports are among the forms of persuasive writing in the technical professions, although they are not the only ones you may have observed. Though persuasive messages of these kinds can be considered good news letters and should, like the previous types, be positive in tone, they are more challenging to write because you are not only asking the reader to accept your message, but inviting the person to respond with action. A persuasive message must influence your readers, motivating them to adopt a proposal, accept a tender, donate to a cause, volunteer time or resources, or provide some authorization or sanction. In order to persuade effectively, you need to establish a human connection with the reader through the relational aspect of your message; to do this, you must demonstrate credibility through a positive, rhetorically balanced approach that is honest and sincere. Outline to the reader the advantages of the course of action you are advocating, and don't bully, patronize, or pander to the reader. Finally, to demonstrate credibility and courtesy, you must avoid obvious gimmicks.

Persuasive messages may be sent to colleagues, clients, or managers who are already positively disposed to your project, or to those who are not "in the loop." Your approach may differ slightly depending on which is the case for you. The following are some of the things your persuasive message should do if it is sent to someone outside your sphere of influence:

- Engage the reader's attention quickly. You may do this with a question, an unusual statement, a brief case study, or a quotation. If you can, make a connection with the reader's interests, expectations, immediate challenges, or needs.
- Motivate the reader to respond by demonstrating how your idea will help to solve a problem the reader is currently facing, fulfill a pressing need, remove a difficulty, or satisfy expectations. Link your proposal to the reader's concerns by showing how it will help to fulfill, alleviate, or resolve them; show her how she will benefit.
- Make it easy for the reader to respond to your suggestion or implement your plan. Be sure to provide a description of what specific action should

be taken to get the plan moving. If your audience cannot clearly see exactly what they must do, they may be unmoved to action, and even if they are convinced by your argument their commitment may be lost. The audience will expect a proposal to hold out a reachable goal and show a plausible first step.

- Keep a positive, enthusiastic tone. Aim to resolve difficulties, not to intensify them through a negative approach. Your recommendation should clearly and explicitly offer advantages to the reader, either by solving an existing problem, or by enhancing a current method or system.
- Communicate your sincerity. No matter how earnest you feel, your reader will reject your message if you don't sound as though you mean what you say. Be sure that your sincerity is reflected in what you write. Especially avoid the standard clichés of professional writing, which will make you sound insincere or even phony.

The persuasive message to those already familiar with and supportive of your work is similar:

- As someone already familiar with and positively disposed to your ideas, your reader will be more inclined to read your message through, so engaging attention and interest will be simpler.
- The established contacts have already shown commitment to your idea; briefly remind them of its advantages. If relevant, outline achievements to date. Remind your supporters of the benefits of the proposal and the ways it will help to serve their needs.
- Be enthusiastic. Outline advantages in a positive tone.
- Communicate your sincerity by addressing the reader directly and genuinely.

All persuasive correspondence must appeal to the reader strongly enough to make that person respond actively. It should be positive, warm, and persuasive, appealing to the needs, interests, and expectations of the reader.

A Note on "Hard Sell"

Persuasion, as you know, requires careful identification of reader interests and concerns. Accommodating your reader is a perfectly legitimate, basic principle that underlies all effective persuasive communication. However, there are other strategies used to persuade that are less respectable: instead of identifying and responding to an existing need, some promotional material seeks to create a false sense of urgency or fear. This kind of advertising appears to be designed to manipulate readers by appealing to their greed or pandering to their fears and doubts. Some writers also overstate their case for persuasion, compromising both their trustworthiness and their audience's commitment. Contemporary readers, who are overwhelmed with persuasive messages from television, radio, the Internet, and print media, are quite sophisticated about pushy "hard sell" approaches to persuasion. They are likely to be wary and suspicious of such methods, and will reject your message as insincere if they detect such an attitude. Remember to keep your tone positive and warm, not crass or offensive, and to keep your audience's concerns genuinely in focus.

The letter below was circulated in a downtown neighbourhood to advertise a new security service that has just opened. Contrast it with the persuasive appeal in Figure 4.10, which encourages donations for a public cause.

FIGURE 4.9 A promotional letter like this one is a good idea for a new service, but these writers could use a little help. What improvements to their letter would you suggest?

**GET THE FEELING
SOMEBODY'S WATCHING YOU?**
555 Broadway, Whitney Pier NS
(902) 345 2473

Dear Neighbour:

We are recent graduates of the Electronic Technology Program at East Coast College, and we have just opened our very first home security service to offer you the best in up-to-date remote-observation electronic household security. In school, we learned just about every technique imaginable for electronic and computer monitoring, and we are anxious to try them out on your home. Take a chance with your security: Try us!

As the new kids on the block, we are anxious to make a go of this venture, and we think it will be good for you to try somebody new for a change. We need to establish a clientele or we won't be able to stay in business for long, so we hope you will try us out.

To make our new service more attractive to you, bring this letter with you when you come in to register your home for a security check. We will perform a free "casing" service of your belongings, and will give you 5% off your first month's rates. Also, we guarantee to satisfy your security needs, so if there is a theft while we are monitoring your property, we promise we will provide an additional three months' outstanding service for no cost. So instead of going with the old-fashioned security companies, come in and try us. We're anxious to please you and to have a chance to practice our various computer-based techniques. We think you'll be really surprised at the quality of our service.

Sincerely,

Ali Mahoud

Ali Mahoud

Kevin Heppner

Kevin Heppner
Co-Owners

FIGURE 4.10 What are the qualities that make this fundraising letter effective? What methods do Maniel and Mfu-Mansa use to engage readers' interest and gain their support?

West Coast College
PO Box 4646, Burnaby BC V6J 2T7

April 29, 2004

Dear Friend:

Like many of us who live in the city of Burnaby, you may have fond memories of playing at the old City Recreation Park. Unfortunately, in the last ten years the park has fallen into disrepair and is no longer a place where children and families can go for recreation and fun. With your help, that can change. This spring students from the Civil Technology Program at West Coast College have been at work on a special project: with permission of the mayor and city councillors, they have begun to refurbish the old recreation park. They have already embarked on a clean-up of the site, where they plan to establish a playground, picnic park, and games field.

As you may know, what is left of the old playground equipment is in dire need of repair. The students have donated their time to the clean-up and have organized teams to help with the building of new facilities and the repair of any of the old equipment that can be salvaged. Once the park is rebuilt, the mayor has promised to fund supervised recreational activities for children throughout the summer.

Unfortunately, all this hard work leads nowhere without financial support. That's where you come in: the Civil Technology Student Society of West Coast College invites you to participate in this worthwhile project with a donation of equipment, materials, or money to assist the students in rebuilding the park into a site we can all enjoy and be proud of. Our goal is to raise $5000 in corporate and private donations to support the work the students are doing. Donations of any amount will be welcomed.

We hope you will come out to the site to see the progress we have already made, and that you will be able to assist us in our efforts to complete this worthwhile project. Cash donations may be sent directly to the CTSS or dropped off at any Benjamin's Drug Store location. To donate materials or equipment, please contact Louis Mfu-Mansa, Chair of the Civil Technology Department.

Sincerely,

Jerry

Jerry Maniel, President

Louis

Louis Mfu-Mansa, Faculty Advisor

THE PARTS OF A LETTER

Whether the letter is sent by ordinary post or via fax, all letters consist of the same standard parts and may be written in any of three standard formats. All are also single spaced.

1. *Return Address*

   This is the mailing address of the writer of the letter; it does not normally include the writer's name, which customarily appears below the signature. In a personal letter, the return address is your home address. If you are writing on behalf of your organization, it is the organization's name and address. If you are using company letterhead, you do not need to include a separate return address.

2. *Date*

   Though there is a move toward dating letters numerically, it is still better to write out the date; styles of numerical dating vary, and this inconsistency can cause confusion, especially during the first twelve years of the new milennium. There are currently three forms in use:

Canadian	12/06/02	Day/Month/Year
American	06/12/02	Month/Day/Year
"Metric"	02/06/12	Year/Month/Day

   Looking at these examples, you can immediately see how confusing numerical dating can be. Without some further cues for interpretation, it's very difficult to ascertain whether this date is December 6, 2002, June 12, 2002, or February 6, 2012. If your employer prefers numerical dating, you should use the format the company uses, but otherwise, avoid this unnecessary confusion by writing the date out in full—in this case, June 12, 2002.

3. *Inside Address*

   This is an important part of the business letter; it provides filing information for the company to which it is sent. It should include the following, in this order.

Name and title	Mr. Louis Mfu-Mansa, Chair
Name of organization	Civil Technology Program, West Coast College
Address of organization	PO Box 4646 Burnaby, BC V6J 2T7

   If the person's title is very long, it might be placed on a separate line, but the order of the parts remains the same.

Name	Ms. Soo Liang Chan
Title	Assistant Manager, Human Resources

Name of organization	Waskasoo College
Address of organization	PO Box 1977
	Forestville, ON
	P6R 2T7

4. *Salutation*

This is the opening of your letter; traditionally, it's "Dear. . ." Use the name of the person to whom you are writing, if you know it. If you don't know it, and it is important that you have it (for instance, in a job application letter), check the Internet or telephone the organization and ask the person's name. If you are unable to obtain the name by these means, you may wish to delete the salutation altogether, an option that has become more acceptable in contemporary correspondence. Do not use "Dear Sir" if you don't know the name; women in professional positions might reasonably object to the assumption that all such positions are held by men.

Note that in Canada, unless you are on very friendly terms with the person you are writing to, it is never proper to address someone by their first name in formal correspondence. "Dear Barbara" or "Dear Mai Li" is considered an inappropriate usage for a person you do not know well. Instead, use their title and surname, as in "Dear Dr. Warnick" or "Dear Ms Chiu." If you are corresponding with a person whose given name is not readily identifiable as male or female—for example, K. Barnett, Terry Ferguson, Inderjit Bhanot, Saran Narang, or Pat Androgue—you may wish to write the full name in the salutation: "Dear K. Barnett" or "Dear Inderjit Bhanot."

5. *Subject or "Re" Line*

This useful device has been borrowed from the memorandum. It has become an important part of the business letter, since it forces the writer to observe the rule of putting the main information first and allows the reader to identify the main point immediately. This line should be brief and to the point, and should indicate clearly what the letter is about. It usually consists of a single phrase or two.

6. *Body, or Discussion*

This portion of your correspondence contains the main information you wish to communicate, in as clear a form as possible. Identify the subject matter at the beginning, giving a brief outline of the situation or problem. Follow with some pertinent details, carefully selected and organized so that the reader may easily understand your message. Finish with a specific statement that outlines what you expect of the reader. The body of the letter is single-spaced and divided into brief paragraphs for ease of reading. The number of paragraphs can vary, as can the length of the letter, depending on the complexity of the subject matter. Some letters are two or three pages long, but most are one page. Whatever the length of the letter, its message should be easily grasped in one reading. If your reader must reread the letter several times simply to understand it, it is poorly composed and ineffective.

7. *Complimentary Closing*

This can be any one of a variety of forms, including "Best Regards" or just "Regards"; the most common nowadays is "Sincerely" or even "Yours sincerely." "Yours truly," although not incorrect, is used less frequently than it once was. Whichever you use, be sure to note its correct spelling. If you are on friendly terms with your correspondent, you may even wish to use a more familiar closing, such as "Cordially" or "Best wishes" or even (if you are very well acquainted) "Cheers." Do not use these latter forms for formal correspondence, however. If you are in doubt, simply use "Sincerely."

8. *Organization Name*

Occasionally, a writer using organization letterhead will signify that correspondence is written on behalf of his or her employer by placing the organization name, in block letters, immediately below the complimentary closing, above the writer's signature. This precaution is observed to clarify responsibility for legal purposes. At one time, it was safe to assume that anything written on the company's letterhead was written on behalf of the firm, but frequent use of letterhead for personal correspondence has made this assumption impractical. Nowadays writers may take this extra measure to emphasize that the contents of the letter are indeed a matter of company business. It is not necessary to include this line in your business letters, but if it is common practice in your organization, by all means do it.

Yours sincerely,

DOLOVICH AND COWAN

Aidan Terry

9. *Signature*

This is the name of the writer only; it should not include any nicknames, titles, or degrees. A professional should also use a consistent signature, not Jennifer in one letter and Jen or Jenny in the next. Choose one form of your name (preferably not a diminutive, and certainly not a nickname) and use it consistently for your professional correspondence.

10. *Typed Name and Titles*

Since signatures can be difficult to decipher, a courteous writer types the name in full beneath the signature. If you wish to list any degrees or titles, type them along with your name. (These, you will remember, are never part of the signature.) If you typically sign your name with initials, or if it is otherwise hard to distinguish, you may wish to indicate the honorific you prefer. A woman who signs herself as J. MacLennan, for example, should not be surprised or offended if some correspondents make the incorrect assumption and address her as "Mr." You may also wish to include your position in the organization you represent.

Terry Lansdown (Mrs.)   Nancy Black (Dr.)
Client Services   Chief of Staff

Saran Narang (Ph.D.)   Soo Liang Chan (Ms.)
Editor-in-Chief   Assistant Manager, Human Resources

11. *Secretary's Notations*

As with the inside address, the typist's notation is useful for record purposes; it is sometimes important to know who typed the correspondence, whether any enclosures or attachments were included, and whether any other people received carbon copies or photocopies. The notations appear in the lower left of the page and are as follows.

/jml	Typist's initials—some organizations use this form to indicate that the typist actually composed the correspondence on behalf of the writer, though this is not the case in every office.
DFC/jm	Writer's initials/typist's initials—may be written this way to indicate that the typist typed what was written by someone else.
encl:2	Enclosure notation—indicates that two items were enclosed with the package.
attach:3	Attachment notation—similar to enclosure notation, but items were appended to the letter with a clip or staple. In this instance, there were three.
cc. R. Burton **R. Vandeven**	Correspondence notation — these people received copies of the correspondence. The notation "cc" stands for carbon copy, the method used for creating copies of documents before photocopiers became indispensable office equipment. Although no one now uses carbon copies, the notation "cc" continues to be used whenever copies, however generated, are sent to individuals other than the addressee. This notation occasionally appears as "pc" for photocopy. In either case, its meaning is the same.

12. *File Number*

A code consisting of a combination of letters and numbers may appear at the lower left or, more commonly, at the upper right of the page. This is a file number that assists the business in filing copies of correspondence in the appropriate location.

FIGURE 4.11 This sample letter shows the parts of a letter in full block format; the numbers correspond to those in the list of parts, above.

[1]   980 Main Street
      Saint John, NB E1G 2M3

[2]   January 21, 2005

[3]   David Cowan, President of Operations
      Dolovich and Cowan Technical Services
      403 Reindeer Street
      Vancouver, BC V3A 4D4

[4]   Dear Dr. Cowan

[5]   Re: Full Block Letter Format

[6]   This is an example of the full block style; note how all the parts begin at the left margin. Paragraphs are not indented, and lines are skipped between them.

      Also, you will notice that this letter is single spaced, as all business letters should be. Note, too, the optional use of open punctuation in this letter—this means no punctuation at the end of the salutation or the complimentary closing. Of course, if you wish, you may use a colon after the salutation and a comma following the complimentary closing.

[7]   Sincerely

[8]   MEDIACORP LTD

[9]   *Stan Cherniowsky* (signature)

[10] Stan Cherniowsky
      Communication Manager

[11] cc. Ian Hauen, Ryker Sheepskin Products Ltd.
      SC/jm

[12] 05-01-SD

The Parts of a Memo or E-mail Message

Although the letter and memo serve similar purposes and may carry similar kinds of messages, a memo is intended to remain inside the company or organization, and this fact affects its structure. It has no need of an inside address, return address, or salutation; instead, it has a relatively standard heading made up of four parts. The labels "To," "From," "Date," and "Re" (or "Subject") are usually arranged vertically at the top left-hand side of the memo, although the layout of these parts may vary slightly, as we will shortly see.

1. *To*

   The *To* line takes the place of the salutation and identifies, by name and title, the person or persons to whom the memo is directed. In an e-mail message, the *To* line consists of the e-mail address rather than the name of the recipient. The e-mail address of any additional recipients may also be added here.

2. *From*

   The *From* line identifies, again by name and title, the person who wrote the memo. In the case of a message sent by e-mail, your program will automatically fill in your e-mail address on any message sent from your account.

3. *Date*

   The *Date* line states the date on which the memo was written; as we discussed above, you should probably avoid numerical dating unless your firm recommends it, since it can cause confusion. You do not need to supply the date in an e-mail message, since the program itself will automatically record the date and the time of the message, which will be displayed when the recipient reads the message.

4. *Re* or *Subject*

   The *Re* or *Subject* line identifies for the reader exactly what issue the memo or e-mail message addresses and what you wish to say about that issue. This crucial line of your memo should contain the main point—the one that you identified in your rough draft with the key words, "The main thing I want to tell you is that...." Busy people should not have to scan the entire memo to find the gist of it. Often they decide whether to even take the time to read the memo by glancing at the subject line, so make sure yours is specific.

   This line is especially important in an e-mail message; regular users of e-mail can receive as many as twenty or thirty messages every day. Most people don't have time to read all of their messages carefully, and some even delete those that look unimportant without reading them at all. If you want to make sure that your message gets read, supply a subject line that will clearly show the reader what the message involves.

5. *Message*

The body of the memo, without a salutation, follows the headings. It may be separated from them by a solid line if you wish. Like the body of the letter, it contains the main information you wish to communicate, in as clear a form as possible. It is single-spaced and deals with the situation or problem as specifically as possible. The body of the e-mail message is similar to that of the memo, although many people like to personalize this impersonal form of communication by adding a salutation. If your e-mail message is to someone you know well and with whom you're on friendly terms, this salutation may be as familiar as "Hi, Gail." If you are corresponding via e-mail to a professional colleague or client, you may wish to use the same form of salutation that is used in a letter: "Dear Franco," or the more formal "Dear Dr. Berruti."

Like a letter, a memo or e-mail may deal with issues of varying complexity. Memos used for simple issues are usually less than a page long, but the memo format may also be used for a short report dealing with more complex situations. (Reports are covered in Chapters Five and Six.) Whatever its purpose, the memo or e-mail message, like the letter and report, provides specifics to support the main point given in the subject or "re" line. However, because these messages are often very short and to the point, they risk being curt or abrupt in tone. Be especially careful to observe courtesy in a memo or e-mail message in order to avoid offending your reader.

Use a memo for information that might be considered "official"; a casual message may be better delivered by telephone, by e-mail, or by a brief personal note. As a gesture of courtesy and respect to your correspondents, you should avoid contributing to the flurry of unnecessary memos or e-mails and refrain from cluttering others' e-mail queues with mass e-mailings of trashy messages; if others send these to you, avoid passing them on.

6. *Initials* or *Signature*

No complimentary closing is required, but a memo may be initialled or signed if you wish or if the practice at your company dictates. Though a signature is not essential on all memos, it is becoming more common to sign them, particularly if they are intended to confirm official arrangements or if they must record authorization for a project or procedure. Your initials may be placed adjacent to your name at the top, or they may be placed at the bottom following the message. There is no need to type your name again under your signature.

Though you may wish to type your name at the end of an e-mail message, it cannot, of course, as yet be signed. For this reason, e-mail is not yet used for very formal correspondence or for reports and letters that are legally binding.

7. *Notations*

Since memos serve the same purposes as letters, they make use of the same notations, especially secretary's initials and carbon copy designations. If it is appropriate, copies may be sent to superiors or other interested parties within the company. For example, a supervisor who writes a memo commending an employee for work well done might direct a copy to the personnel file; the head of a departmental committee might direct a copy of a meeting announcement to the department head to let that individual know that the committee is getting on with its work. Memos, like letters, may also contain file numbers for easy reference. E-mail messages may be sent to several people simultaneously; if so, their e-mail addresses appear in the *To* portion of the message rather than in a copy notation at the bottom of the page.

LETTER AND MEMO FORMAT

Letters may employ one of three formats widely used in Canada: semiblock, full block, and modified semiblock. As the Reference Guide in Figure 4.12 indicates, the main difference in letter format is in indentation: in *full block style*, everything, including paragraphs, begins at the left margin of the page. It is the preferred contemporary form of letter and the simplest to set up. Paragraph divisions are indicated by skipped lines. The fullblock format is the type illustrated in Figure 4.11, on page 106, and in most of the examples in this book.

In *semiblock format*, the return address, date, complimentary closing, signature, and typed name are indented so that they begin at approximately the centre of the page. Each paragraph is also indented one tab stop (which is usually preset at $\frac{1}{2}$" in most word processing programs); paragraph divisions may also be (but do not have to be) indicated by a skipped line if the writer prefers. This format has a more balanced appearance, but is also more difficult to type. The semiblock style is the oldest pattern and although it is still used, it has largely been replaced by the full block style.

Modified semiblock is a transitional form, a blending of the other two styles. It is laid out exactly like the semiblock style, except that the paragraphs are not indented. The return address, date, complimentary closing, signature, and typed name are lined up to the right of centre, just as in the semiblock style. All of the text in the body of the letter begins at the left margin, and paragraph divisions are always indicated by a skipped line, as they are in the full block style. Though the full block style is now the most

FIGURE 4.12 A handy Reference Guide for memo and letter formats.

MEMORANDUM	MEMORANDUM	MEMORANDUM
DATE:	DATE: TO:	TO: DATE:
TO:	FROM:	FROM:
FROM:	SUBJECT:	SUBJECT:
SUBJECT:		

Full Block Letter   Semiblock Letter   Modified Semiblock Letter

commonly used letter form, you should use whichever format your employer prefers.

The layout of the *memo*, with clear-cut headings at the top of the page, is simpler than that of the letter and so is generally easier to construct. All memos contain the standard four-part heading of To, From, Date, Subject, but the arrangement of these parts may vary, depending on the preference of the writer. The memo's heading section is normally arranged at the left margin, usually in the order *To, From, Date,* and *Subject,* or *Re,* though there is some variation to this standard order. Several common variations are displayed in Figure 4.12. The paragraphs in the body of the memo may be indented, as in the semiblock letter style, or not, according to the writer's preference. Paragraphs that are not indented should be separated with a skipped line.

You may also have noticed that the examples in this book use different punctuation styles. Some use an "open" style, with no punctuation marks at the end of the salutation or complimentary closing; others use "closed" punctuation, with a colon following the salutation and a comma after the complimentary closing. Either punctuation style is acceptable in any of the formats, but do not mix open and closed styles within a single letter.

DOCUMENT LAYOUT

Whichever letter format you use, give careful attention to layout. In any piece of professional correspondence, you want clarity, and the impression of clarity is enhanced by an attractive arrangement of the letter or memo on the page. Leave generous margins—at least an inch on all sides—and place the printed material as near as possible to the vertical centre of the page. Try not to crowd your letter or memo too close to the top of the page; unless it is very brief, you should space your letter so that approximately half of the print falls within the lower half of the page.

Another important factor in creating an attractive and readable layout is the font you select from those available in your word processor program. In general, a serif font (such as Times or Roman) is easier on the eye in a lengthy document than a sans serif font (such as Helvetica). Script fonts, especially in smaller sizes, are difficult to read, as is gothic lettering. Most computer fonts are proportionally spaced, which means that thin letters (such as "i" or "t") take up less space on the page than do wide letters (such as "w" or "m"), just as they would if the document were typeset. These fonts also automatically place an extra space after a period, so that sentences are separated by a slightly larger space than words within a sentence. However, a few fonts such as Courier, which is designed to resemble a typewritten font, are monospaced, which means that every letter—no matter its width—takes up exactly the same amount of space on the page. If you have selected such a font, you must add an additional space after the period, as used to be the case for documents typed on a typewriter. Finally, always select a point size large enough for the font to be read comfortably. Most are pre-set at twelve; you may go slightly larger than this for headings, but do not reduce the point size of the font in the body of your document below twelve. Compare the following examples for visual appeal:

Times	The quick brown fox jumps over the lazy dog. The lazy dog does not respond. The fox disappears through a hole in the hedge.
Helvetica	The quick brown fox jumps over the lazy dog. The lazy dog does not respond. The fox disappears through a hole in the hedge.
Courier	The quick brown fox jumps over the lazy dog. The lazy dog does not respond. The fox disappears through a hole in the hedge.

POINTS TO REMEMBER

1. Use a letter or memo if the information you are communicating must be retained on file, or if the information is of formal importance in the organization.

2. Do not use e-mail for messages that are confidential or for important information that must carry a signature.

3. Apply the Seven Cs of professional writing to all letters, memos, and e-mail messages.

4. Letters, memos, and e-mail messages can either initiate or respond to contact, and can contain positive or negative messages.

5. Make sure your letter contains all the appropriate parts, whether you are sending it by fax or by conventional mail: return address, date, inside address, salutation, subject line, body, complimentary closing, signature, typed name, and notations.

6. In a memo, the *To, From, Date*, and *Re* or *Subject* information appears first.

7. Be sure to provide a clear and relevant "re" line on all e-mails.

8. Recognize reader needs, expectations, knowledge, concerns, and relationships, and always avoid sarcasm.

9. Always be positive in tone and attitude.

10. Avoid contributing to e-mail clutter.

FIGURE 4.13 This letter violates the rules for a letter of congratulations. What impression does it make on the reader?

2100 Allendon Crescent
Peachvale ON L7B 3Z9

March 14, 2004

Ms. Kara Exner
90 Main Street
Oshawa ON M6K 3D4

My dearest Kara

Our class just heard the news from Professor Bolton that students from your little communication class are going to present papers at the Maritime Communication Association Conference in April.

I know just how you feel—I did a presentation for one of my classes last week too, and I'm sure it's good enough to present at a conference if I wanted to submit it (conferences can be such a mind-numbing experience, can't they?) And Halifax, what a place, it almost makes you think you should find somewhere interesting to go!

However, it is an accomplishment, dear—and who knows? Maybe someday you'll be shaking my hand as a fellow conference presenter!

Yours very sincerely

Donna Skruwyew

Donna Skruwyew

FIGURE 4.14 How easy will it be for Dr. Calvin to respond to this request? Why?

75 Snayall Drive
Couchgrass, Manitoba
R1A 3C6

October 30, 2003

Professor L.M. Calvin
Science Division
Northwoods College
Box 456
Bruce Mines, Ontario
N0E 0A0

Dear Professor Calvin:

I am writting you this letter to please ask if you would kindly act as a reference for me?

I took your class when I attended the college about 6 years ago. Now I want to go back to school in a similar program at Pineridge College here in Couchgrass. Could you send your letter directly to them? It would be very much appreciated.

Yours truly,

M. Jones

M. Jones

SHARPENING YOUR SKILLS

Section A

The following situations require letters; some of them will carry positive messages, and some will carry negative messages. In writing them, observe all the elements of style and format discussed above; as well, be sure you study the situation to ensure that you accurately determine what the communication challenge is. Add any details necessary to make your letters more convincing.

1. A recent fire in your home destroyed a number of your personal effects. Among the items lost are your educational records, particularly your college diploma. You are currently employed with Dolovich and Cowan, but are considering a career move, and you will need the certificate as proof that you graduated from college. Write to the office of the registrar at your college, giving all of the details they will need to locate your file, and inquire whether they will be able to replace your diploma.

2. You have heard that a former professor of yours has just been nominated for the Distinguished Teaching Award at your college. Write a letter of support for her or him (identify a professor of your choice) that may be added to the professor's dossier for the awards committee's consideration.

3. Write to a former instructor or employer, requesting permission to name that person as a reference in a job application. Be sure to include all the information the person will need in order to comply with your request.

4. You are nearing graduation and have been conducting a job search. You are lucky enough to have been offered two positions. The first is with the small firm where you completed an internship last year; the other is a more promising position with a multinational. The small company offers better pay initially, but the multinational firm offers has better prospects for promotion and benefits over the long term, and after carefully considering the two, you have decided to accept the position with the multinational. Write the letter to your internship supervisor thanking him for the offer, but politely refusing the job. He is: Gene Kirk, Testamek Engineering Ltd., 64 Weniam Street, Vancouver, British Columbia, V1R 9V1.

5. You are the former instructor of Pat Yorgason, who has just written to you asking for a letter of reference. It has been three years since Pat was in your class, and you have taught three hundred students per year since then. You never knew Pat well and you can't picture a face to go with the

name; all you can recall is the impression of an indifferent student. A check of your records confirms that impression: Pat's grade in your communications class was C-, but otherwise nothing much jogs your memory. In fact, you have taught lots of "Pats," and you can't even recall whether this one was male or female! You don't feel that you are the best choice to write a recommendation for this person. Write a letter to Pat politely explaining why another referee might be a better choice.

Pat Yorgason, PO Box 75, Okotoks, Alberta T0K 0K0

6. Your friend Ing Jang is a student at Western Plains University. He has recently had an encounter with the staff of the registrar's office that turned out very wrong, as you can see from Ing's letter (Figure 4.4). Ing wants to register his complaint, but before sending his letter on to the Registrar's office, he has forwarded it to you for your comments. After reading the letter, analyse Ing's communication situation in order to answer the following questions: Should Ing write a letter at all? If so, what should its purpose be? Describe the likely expectations, needs, and attitudes of the intended audience. Given what took place, what challenges does Ing face in repairing his credibility? What strategies can he adopt to manage the situation effectively? What elements of the current letter might actually inflame the situation further? Write your analysis as a letter of advice to Ing Jang.

7. When you filed your income tax return this past year, you found that you had mislaid one of your T-4 slips from a part-time job, showing the amount of $1150. You declared and paid tax on the income anyway, and attached a note to your return stating that you would forward the T-4 slip when you were able to locate it. Your subsequent letter of assessment from the government showed that they did indeed calculate tax on the amount when they processed your return. Several weeks later, you located the missing T-4 slip and submitted it, along with a letter of explanation. Today, you received a re-assessment of your income tax from the Canada Customs and Revenue Agency showing that they mistakenly added an additional $1150 to your total income. As a result, they have assumed you earned $1150 more than you actually did, and have billed you an additional $632.50 in tax. This is clearly an error, as you have already paid tax on the amount. Write to the Canada Customs and Revenue Agency in Winnipeg MB, R3C 4T4, explaining the situation and requesting an adjustment. You feel some urgency because they have already charged you $80 interest on the amount they claim you owe them, and will charge additional interest if you do not pay the full amount of the additional assessment within twenty days of receiving their notice.

8. You are on the executive of the Western Plains Engineering Students Society, and your organization has been involved in the university's orientation programme for new students. This event is organized through Student Affairs and Services, under the direction of the Director of Student Retention, Joanna Davidson. In addition to coordinating all the events, Davidson's office also prepares for each student who participates in orientation an information package in a presentation binder featuring the university's logo. These binders are fairly costly even at the bulk rate ($2.50 per student, plus $2 for photocopying), and the contents are inserted by hand. Orientation took place this past Saturday, and Davidson's staff were busy with preparations until well into Friday evening. Right before she was ready to leave the campus at about 9:30 p.m., Davidson decided to double-check that everything was in order. To her horror, she discovered that she was short three cartons of binders—thirty-six in all. She and one assistant had to stay until nearly 2:00 a.m. reprinting materials to be put into replacement binders; early in the morning, before orientation began at 10:00 a.m., her assistant had to purchase thirty-six binders at full price from an office supply store. By working frantically, they were able to get the extra binders assembled for the 10:00 a.m. start-up, but the last-minute replacements cost them $10 per student instead of the $4.50 they had planned on, driving up their costs by over $360. However, when the entire orientation staff assembled on Saturday morning, the two volunteers from engineering came forward with the three missing cartons, which they said they had hidden "for a prank." Davidson and her office staff were understandably annoyed: they spent several extra hours replacing the missing materials and were now stuck with three dozen binders of material that they couldn't use and that cost them over twice the budgeted price. Davidson's supervisor, Dr. Valerie Price, Vice President, Student Affairs and Services, has contacted the Dean of Engineering, Dr. R. Billinton, to complain and to ask for reimbursement of this significant cost. The dean has assigned you, as a representative of WPESS, the task of writing a letter of apology to Dr. Price and Ms. Davidson. Be sure to clearly define the purpose of the letter, and the needs of the audience, before you begin to write, and to provide a solution that will satisfy both Price and Billinton, and direct a copy of your letter to Dr. R. Billinton, Dean, Engineering.

9. The letters on pages 114 and 118 (Figures 4.14 and 4.15) contain weaknesses. Evaluate them according to the criteria you have learned and be prepared to rewrite them more effectively if your instructor directs you to do so.

DOLOVICH AND COWAN
403 Reindeer Street, Vancouver, BC V3A 4D4

To: All Staff

From: Ivana Petrovic, Social Committee Chairperson

Date: July 29, 2003

Re: Arrangements for Annual Centre Employee Picnic

This is to inform you that the arrangements for this year's annual company picnic are final and complete at long last. After lots of hard work and planning by this committee, it was decided that it will be on Saturday, August 6.

As you know, we needed to ask for volunteers to lend us various types of equipment for playing sports and games, and we also had to arrange for barbecues to be brought to the site. Luckily we have lots of willing volunteers who can help us out with these requests and they have agreed to bring their equipment for us all to use.

If you are one of those generous people who have agreed to volunteer to us any sports or games supplies or a barbecue or any other kind of item we will be needing, those who have done so are asked by your dedicated committee to arrive one half hour early. The entertainment subcommittee and the food committee, including myself among many other dedicated individuals will be on hand by 10 am to get things rolling right along.

The picnic begins properly at 11, though you can plan to arrive with your family anytime between 10 and 11, unless you are one of our volunteers mentioned above. As has been the case with our many previous successful annual company picnics, this one is to be held as usual at Ellsworth Conservation Park.

If you need directions how to get to the park, just contact me or anyone else on the social committee. Plan to bring everyone in your whole family for a super fun-filled day.

See you all there.

Message Composition – E-mail

To: students@cenmar.u.ca, faculty@cenmar.u.ca, staff@cenmar.u.ca

From: Phyllis Golemiec Phyllis_Golemiec@cenmar.u.ca

Date: October 1, 2004

Re: Town Hall Meeting on Student Computing

—————————————————————

On Wednesday, October 6, 2004, there will be a "Town Hall" meeting on Student Computing sponsored by the Vice President of Information and Communications Technology. This event will be held in room 107 Billinton Hall. All students, staff, and faculty are welcome to attend.

The purpose of this meeting is to provide an opportunity for students to ask questions and discuss issues relating to computing and technology at Central Maritime University. Computer facility managers and members of the Information Technology Services and other technology-related departments will be in attendance and will participate in the discussion.

Please plan to attend this event.

Section B

Write the following memos, bearing in mind all you have learned about professional writing style and format. Add whatever details are required to make your memos convincing. If your instructor directs you to do so, submit your assignment via e-mail.

1. As you have been discovering, the way you communicate has a profound impact on your career success; many studies have shown that employers in all fields—including engineering—rank communication high among the skills they seek in new recruits. Suppose you have applied for a job with such an employer. In the course of the interview process, you have been asked to identify three important principles of communication and show how they have contributed to your own effectiveness as a communicator. Present your discussion in standard memo format and address it to your communication instructor.

2. You have been in your summer job as a research assistant for one of the senior professors in your programme, Dr. David Stover, for three weeks. Your duties include data collection and analysis for crystal diffraction and immunomagnetic separation. You have fallen into the habit of taking lunch at your desk, and today you accidentally knocked over a full super-sized container of cola onto a pile of irreplaceable printouts from the diffractometer. Unfortunately, you didn't notice the spill right away, and most of the data have been obliterated. These data have not been stored anywhere else, and the loss means that the experiments that produced them will have to be re-run. This mishap has caused a loss of all the work you have so far accomplished. Write a memo to Dr. Stover explaining what has happened and outlining what you will do to compensate for the damage.

3. It's unbelievable, but your instructor, Maury Huknows, slept in and was late for the final exam in your economics course. As a result, the start of your three-hour exam was delayed for 45 minutes, and you were unable to finish. You have approached your professor, but he seems unwilling to make any allowances for your difficulties: he insists the exam should only have taken two hours anyway, and points out that several members of the class managed to finish all the questions in the time they had. You believe you should have had the whole three hours, and you are certain you would have finished the examination if you had had the full time allotment. Write a memo to the dean of your college, carefully outlining your situation and recommending a course of action: you can request either that you be allowed to re-take your final exam in a full three-hour period, or that you be graded for the full percentage of the exam on the questions you were able to finish in the time period. Be sure to watch your tone carefully and to justify your case in terms that will satisfy both the dean and your instructor.

4. You and a friend spent last summer operating a small manufacturing and retailing business to market a series of topological puzzles that you designed. The designs were your idea, but you needed a partner and your friend, a business major, seemed eager to be involved. The puzzles proved popular and the business was a success. You did not bother to draw up a formal business agreement, but the partnership seemed to work well, and the plan is to repeat the experience next year. At the end of the summer you returned to the city to find an apartment while your friend stayed behind in Banff to wrap up the business over the Labour Day weekend.

Your understanding was that you would split the remaining profits of the business in three ways: one-third to each of you, and one-third as seed capital for next summer. You haven't seen your friend since school began, and you still don't have a cheque for your portion, which should be $4500. You also haven't been advised about whether your friend has banked the seed capital. It's early in the term yet, and you know that the delay could be simply the result of the hectic pace of back-to-school, but you're getting a little nervous and you really need the money. You don't want to undermine the partnership or sour the friendship, especially if it's just a misunderstanding, but you do want to know where you stand. Write an e-mail message to your friend requesting that he provide this information, and your cheque, as soon as possible.

5. Recently, in class, a professor you like very much made what you believe is a serious error of fact in one of her presentations: she said that the date of the Riel Rebellion was 1875 (it was 1885). You had heard her make the same error in an earlier class, but at the time you thought it was just a slip of the tongue, and so said nothing. Your class is not actually about the Riel Rebellion, but you think this detail important enough to want to set her straight. You want to save yourself, and the professor, the embarrassment of pointing out the mistake face to face, so you decide to write an e-mail message to her explaining the mistake and correcting her error. You really like this professor and you want to retain her goodwill. Compose the message in which you politely point out Professor Yvette Bruneau's error. Submit it to your communication instructor via e-mail.

6. The memos shown in Figures 4.15 and 4.16 are weak for one reason or another. Read them through critically to see what improvements you might be able to suggest. Be prepared to rewrite them more effectively if your instructor requires you to do so.

CHAPTER 5:

Informal and Semiformal Reports

LEARNING OBJECTIVES

1. To understand the purposes for which reports are written.
2. To learn the standard parts of all reports: Summary, Introduction, Discussion, Conclusion(s), Recommendation(s), and Appendices.
3. To recognize common report formats and to know how to select the format appropriate for the report you are writing.
4. To know how to use the Report Writing Planner to plan your report and select appropriate information.

Of all the tasks performed by a technical professional, writing reports on the job is one of the most important. Technical occupations are typically notorious for the amount of paperwork they involve: regular project reports, incident reports, memos, minutes, funding proposals, tenders, log book entries, lab reports, and other forms of correspondence form a large part of the routine responsibilities of these positions. Reports provide a record of project work, of interactions with clients, and of strategic planning at the department level; they are essential for continuity should personnel or circumstances change. The reports you write as a routine part of your job may also be significant if you hope to be promoted, since, especially in a large organization, your superiors may know you only through the reports you write. If you wish to advance in your chosen career, you will need to report accurately, carefully, and clearly.

Furthermore, if you are employed as a technical professional in a government position, your reports will also be a matter of public record; in other words, they are legal documents. As such, they may be used as evidence in a court case or be required in an investigation, inquiry, or inquest. For all these reasons, your reports require careful attention to detail and clarity.

WHY WRITE REPORTS?

Reports are written for a variety of reasons, but their primary function is to pass on information to supervisors and co-workers. Routine incident reports provide an account of what occurred and a record of the action that was taken at the time of the incident. They also provide a means for staff to communicate with others in the organization about incidents or situations that might require further information.

Why Write Reports?
- Pass on information
- Provide a permanent record
- Keep track of trends and developments
- Allow effective decisions

Reports also provide a permanent record of events; as such, they function as a kind of diary of the company, of various departments or divisions within the company (for instance, personnel or payroll), of interactions with individual clients or employees, or of a chain of events leading to a particular action or result. Often, however, reports may constitute an important foundation for future decisions or actions, recording the details of project decisions or interactions with staff, clients, or the public. They can also form a record of how the company has carried out its business in accordance with statutes or government policy.

Apart from their role in record-keeping, reports are an important part of the total function of an organization or company, since upper levels of the bureaucracy get a comprehensive view of things from the information supplied in the routine reports of various employees. The larger the organizational structure, the more significant are individual reports to the day-to-day functioning of the whole company. Reports from those who deal with the day-to-day personnel, business, managerial, or technical issues can help upper management to assemble an overview of the operation of the company, can flag small problems before they turn into big ones, and can alert officials to the need for policy or procedure changes.

Finally, reports provide the information necessary for managers to make effective decisions by reporting actions and details as objectively as possible. A well-written report distinguishes between observed facts and inferences or opinion; analysis and judgements are clearly labelled as such. By reading such objective reports from a variety of sources in the institution over a period of time, a manager can be assured of acting in the best interests of clients and personnel.

As a technical professional, you will be required to write numerous reports and to read the reports of others; when you are faced with piles of paperwork on the job, you will appreciate the skills of conciseness, completeness, and accuracy, both as a writer who produces reports for others to read, and as a reader who must plough through the written work of others. Since writing reports will be such an important part of your professional life, you should cultivate these skills while you are still in training for your future career.

THE PARTS OF A REPORT

The Standard Parts of a Report

- Summary
- Introduction
- Discussion
- Conclusion(s)
- Recommendation(s)
- Appendices

You will recall from Chapter Two that all professional writing can be divided into six parts, which you may remember by the acronym "SIDCRA." The six parts, in their conventional order, are summary, introduction, discussion, conclusion, recommendations, and appendices. You may also recall that the last two of these six parts are optional, especially in short informal reports.

While in a letter or memo these parts may not be visually separated or explicitly labelled, a report is generally more highly structured and will frequently use section headings to assist the reader in following the direction and purpose of the message. The more formal the report, and the longer it is, the more likely it will be to feature formal recommendations, to include appendices, and to employ formal section headings and subheadings. Before we move on to discuss the formatting of the report, however, we should review the parts of the report in detail.

Summary

Since reports are longer than most other forms of professional communication, and since their intended readers are typically very busy, the report writer includes a summary or abstract, a brief statement that gives an overview of the situation or problem dealt with in the report, the general findings, and the specific action recommended. After reading the summary, the reader should know what to expect from the introduction, discussion, conclusion, and recommendations. The substance and the direction of your findings should be clear from the report summary. Further, the language of the summary (along with that of the introduction, conclusion[s], and recommendation[s]), should be straightforward and clear enough to be understood by the least expert of the intended readers. The length of the summary varies with the length of the report, and though there is no set length, you can think of the summary as being approximately one-tenth as long as the report itself. For example, a ten-page report may have a summary of approximately one page, while the summary of a formal report fifty pages long may be five pages. A short informal report using memo or letter format may have a summary consisting of a subject line and a brief initial paragraph.

Although the summary is the first part of a report that the audience reads, it is normally the last part to be written. Because it provides an overview of the findings, it usually cannot be written until after the report is finalized. The summary of a report provides a general sense of the purpose, constraints, outcomes, and recommendations that will follow.

Introduction

The introduction to a report sets out, as clearly as possible, the problem or situation being examined and provides any necessary background information; it may also set out the writer's approach and assumptions, and the limits of the report. In short, it prepares the reader for the discussion of the possible outcomes or solutions offered in the report.

Discussion

The main body, or discussion, of the report details the writer's method (including the criteria used to evaluate possible solutions) and the steps that led to the recommendations and conclusions offered in the report. It may describe those other possible solutions and show why, according to the writer's criteria, they were judged to be unacceptable. Exactly what you include in this section of the report will depend on what situation you are dealing with. However, if you have detailed technical or specialized data, you would place it in the discussion section. In general, the discussion is aimed at the most knowledgeable of your expected readers.

The discussion is the longest and most detailed part of your report, and is normally written first. It is typically made up of a number of shorter sections, each with its own heading. The word "discussion" itself rarely appears as a heading; instead, it is a broad term that is used to denote everything in the report between the introduction and the conclusion. The headings that are used in the discussion section are specific to the contents of the report. For example, a report that evaluates a training program offered by a local consulting firm might use these section headings within the discussion:

- Program Description
- Prerequisites
- Advantages of the Program
- Costs

- Limitations of the Program
- Resources

Conclusion(s)

Depending on the situation you are writing about, there may be several possible outcomes, or only one. Your conclusions represent the logical results of the investigation or presentation you have dealt with in your discussion. The conclusion lays out any judgements that can be made based on the facts presented in the discussion. Your conclusion should present no surprises for your reader, who has been led by your discussion to expect what appears there. For example, if your report evaluates three different computer systems that the company is considering for purchase, your conclusion should indicate which system best meets the company's needs.

Recommendation(s)

In this section you will recommend what you consider to be, according to your evaluation criteria, appropriate action in response to the conclusions you have reached. You may have one or several recommendations to make. Recommendations normally outline the action that the reader of the report should take, and occasionally your recommendation section will even list the actions you intend to perform yourself. For instance, your report evaluating the computer systems should clearly recommend the system, or combination of components, that the company should purchase.

Appendices

An appendix is any additional material or information that is attached to or included with a report. It is not considered a part of the report itself, but it provides additional support or explanation for points in the discussion. Any relevant supporting information that, for reasons of space or complexity, has not been dealt with in the discussion section of the report may be attached to the report as an appendix. The purpose of the appendix is to assist your reader in fully understanding your information. There may be one appendix or several appendices, and while not all reports have them, any report may have them. Formal reports are more likely than informal ones to include appendices.

No matter what the subject of your report, from a government white paper to a technical report, it should conform to this general structure. Once you are on the job, your employer may require a specific structure or organization pattern that differs in detail from the advice presented in this book, which is necessarily general. If this is the case, you should follow the requirements of your employer. However, as a general guideline, you can rely on the SIDCRA structure to help you organize your materials for presentation.

PLANNING AND DRAFTING YOUR REPORT

Reports, like other effective written communication, should be carefully planned and drafted, following the process and strategies presented in Chapter Three. The communication problem-solving structure is especially useful for writing longer documents such as reports, and you should refer to that chapter as you begin to plan and organize your materials. As well, the Report Writing Planner (Figure 5.1) will help you to identify the important elements of your message and the probable needs and expectations of your reader.

Following the strategies outlined in Chapter Three, assemble the main information to be covered in the discussion segment of your report, keeping your reader's needs, attitudes, expectations, and purposes clearly in mind. Be sure to follow the problem-solving approach to drafting and editing your report and to pay attention to all three of the keys to effective communication: effective audience appeal, a coherent and sound message, and personal credibility.

Following the steps in planning, preparing, and polishing your report, develop the relevant details, giving enough information to enable your reader to make an informed decision. Consider carefully how best to present the points in your discussion, choosing your words with care as you work, and selecting an appropriate organizing principle. Then write the rough draft of your discussion, beginning with the summary statement. The clause "The main thing I want to tell you is that... " may help you to focus your rough work, but remember to delete it in the final version of your report.

Revise your discussion, using the guidelines in Chapter Two, and observing the principles of coherence and the Seven Cs of professional writing. Once you have developed a reasonable draft of the discussion, plan and write the conclusion, recommendations, and introduction. Let the document sit for a while, then read it from beginning to end, revising as you go. When you have a nearly complete draft, write the summary.

As in all your professional communication, you will find the writing process easier if you define the purpose before you begin, keeping in mind the goals of the communication, the probable needs, concerns, and expectations of the intended audience, the relationship between you, the reader, and the issue under discussion, and the most likely use to which the report will be put. The Report Writing Planner (Figure 5.1) will help with these tasks.

FIGURE 5.1   Use this planner to identify your main message and your reader's needs and expectations.

REPORT WRITING PLANNER

Before beginning your report, answer these questions as fully as possible.

1. What is the topic of this report?

2. What is going to be done with this report? Why is it needed? Who asked for it?

3. Who are my readers? What are their interests in this subject? What background information is already known to them and what will I have to fill in so that my report can be understood and acted on? What will be the readers' likely attitude to this information?

4. What is my main message? What will my Summary Statement be?

5. How will I briefly outline my introduction, providing the appropriate background information?

6. How will I outline my discussion, providing any relevant main points and details?

7. How will I clearly state my conclusion and recommendation(s)?

REPORT SITUATIONS

In the initial chapters of this book, you learned the importance of the Seven Cs of effective professional writing: completeness, conciseness, clarity, coherence, correctness, courtesy, and credibility. These qualities of effective style remain as important in a report as in any other job-related correspondence. However, in work settings involving the application of government standards or regulations, routine report writing takes on an added element of legal responsibility. In such a case, providing correct and complete information is not only desirable, but crucial, since your report forms part of the legal record of the company.

Reports may be written in a variety of situations and for a variety of purposes. However, the most frequent types written as a matter of daily routine can be classified into one of the following categories.

Common Report Purposes

- Log books
- Laboratory
- Incident or Occurrence
- Accident or Injury
- Assessment or Evaluation
- Progress, Status, or Project Completion
- Investigative or Analytical
- Research
- Proposals

Log Books

You may not think of the log book as a report, since it is routinely seen only by the individual who writes it. Nevertheless, it must be readable by other engineers and specialists, because it forms a record of research being conducted or of the development of a project. Because it performs many of the functions of a report, it fits into this category.

The purpose of a log book is to keep a record of the daily research and design activities of a practising engineer. It is the raw data from which a formal report may be written at the conclusion of the project, or upon which regular progress reports will be based. It is normally updated every day. This practice gives the engineer or technical support staff a record of the activity that went into a particular design, but it also does much more. Because the log book is a legal record, it can become a crucial piece of evidence in legal disputes over patents or design rights. In such cases, the log book entries establish a date, or even in some cases a time, at which a critical insight or idea was first noted. Such evidence can be critical in establishing ownership of an idea, technique, or device in patent decisions or other legal proceedings.

In preparing your log book, you must observe the following critical requirements. The log book must be a bound notebook from which pages cannot be easily detached. (A coil-bound notebook from which pages can be torn will not serve this purpose, nor will any stapled or temporarily bound notebook.) Every page must be numbered consecutively, and no pages may be removed from the book. Every entry must be dated, and there may be no significant blank spaces between entries where information could be inserted at a later time. Any such blank spaces should be blocked off with a stroke or an "X" through them to show that they were intentionally left blank. Entries may—and in some cases should—include calculations, thumbnail design sketches, diagrams, or charts. These, too, should be dated. Finally, the log book should be written in indelible ink and should not contain any obliterations or adjustments in correction fluid. Any corrections that are made should be crossed out neatly with a single pen stroke so that they remain legible. In many firms, the practice is to sign and date every page as it is completed.

Though there is no set layout or format for the log book, it should be kept as ordered and readable as possible so it can serve as a clear future reference or legal evidence if the need should arise. The dates and details should be consistently presented and easy to find. The book is useless as a legal submission if there are obliterations, if pages are torn out, if it cannot be understood by anyone other than the writer, or if there are gaps into which information could be inserted at a later time. The records may be kept on both sides of the page, or on one side only. If you choose to write on one side of the page only, you must be consistent; never enter any information onto the back side of a page, for all the reasons already given.

The log book should also be carried to technical meetings where the design project will be discussed, so that the discussion may be recorded. Most engineers keep their log books handy at work and make time each day to record what has been thought, said, and done. The log book should record all your ideas, calculations, diagrams, experiments, and tests. It should be clear and detailed enough to allow another engineer to understand and recreate your work. When a book has been filled, it should be stored in a safe place.

Lab Reports

Lab reports differ from log books in that they are clearly intended for a reader other than the writer, in that they report the findings of a single experiment or a connected series of experiments. These are used in technical environments such as engineering firms, hospitals, and computer companies to record findings from laboratory work. They may be forwarded as part of a longer research report or kept on file locally. The lab report write-up is normally a more formal presentation of material already recorded in the log book.

The format used for lab reports can vary significantly from one employer to another or even from one college course or instructor to another. Most laboratories have developed a set format that is used for their lab reports, and you should follow the format that your employer or instructor prefers. In general, however, despite variations in structure, the lab report is fairly consistent in content, and if you have not been given any other guidelines, the format described here is the one you should use.

The lab report describes why the experiment was conducted, how the work was carried out (the equipment and method used), the findings, any problems that occurred, the results that were produced, and the implications of these findings for future experimentation. The headings used in a lab report normally reflect these steps, so that the general structure of the lab report follows some variation of the following headings.

- Purpose
- Procedures
- Equipment
- Problems Encountered
- Results or Findings
- Implications for Future Work

You should note that these headings, though they use different terminology from the SIDCRA outline we have been discussing, present the same kinds of information in the same general order. As in the SIDCRA structure, introductory information precedes the fuller discussion, which is followed by conclusions and recommendations.

Incident or Occurrence Reports

The incident or occurrence report is probably the single most common type of report after lab reports and logs, since it is used in nearly every profession and field. Written in response to an unexpected (usually problematic) incident, this report is primarily an informative one that outlines details of an unusual event. It may suggest ways in which the event has influenced work currently in progress and outline the steps being taken to correct the setback. Incident reports cover any atypical occurrence, even if it seems minor at the time it occurred; over a period of time, a series of minor incidents may point to a significant pattern or reveal a failure of a particular procedure or practice. This type of report provides details of the event as completely and fully as possible.

In Chapter Two, we discussed the elements of a complete message. You may recall the questions—when, where, what, who, why, and how—that were recommended as a guideline to ensure that you have included all the information your reader needs. Answering such questions is even more important in a routine incident report than in other situations; complete information will make future decisions and actions easier and more effective. Keeping these prompts in mind will help ensure that nothing is left out of your report.

- *When?* Record the exact day, date, and time of the incident.
- *Where?* Describe the location in as much detail as necessary for officials to reconstruct the incident should it be necessary to do so.

- *What?* Explain exactly what occurred from the moment you became aware of the situation until its completion. What did you hear? What did you actually see? What action did you take?
- *Who?* Identify those who were directly involved in the incident and those who were present at the scene.
- *Why?* It may not be possible in every circumstance to answer this question, but if you have access to such information, you should include it in your report. If the statement of motive or cause is the result of your own inference, you should clearly label it as such in the report.
- *How?* You may not be able to answer this question with certainty, but in some situations (for example, an equipment failure or security breach that is the result of faulty maintenance practices) such information may be crucial if future occurrences are to be prevented.

In addition to the standards of completeness that apply to all of your professional correspondence, the incident reports that you write should adhere to a further standard of objectivity because they are part of the legal record. When reporting an accident, injury, or unusual occurrence in the workplace, you should refrain from making unwarranted inferences, recording only what you know for certain happened and avoiding speculation about who was at fault or what caused the mishap. A useful guideline for writing such reports is the formula "I was, I saw, I did."

"I was" records where you were and what you were doing at the time of the incident. It clearly sets the scene for the reader, allowing a clear understanding of the circumstances surrounding the occurrence.

"I saw" records exactly what you personally observed; it does not involve judgements or inferences you have made about the circumstances. For example, you may be driving in a convoy of vehicles heading for a work site. It is winter, and the roads are icy. As you round a corner, you observe one of the trucks that had been ahead of you lying in the ditch. When you stop your vehicle and go to investigate, you discover that the truck has collided with another vehicle that is also in the ditch. The driver of the car is injured and your own driver is unconscious. You are not an accident specialist, but it looks to you as though your driver ploughed into a car that had already been in the ditch when he rounded the corner. You guess that he was momentarily distracted by the sight of the first vehicle and lost control on the ice. However, even if it seems clear to you that this is what occurred, you should realize that this is a conclusion you have drawn from the circumstances. Write in your report only those events that you observed for yourself; if you must include your conclusions, be sure to clearly label the additional information as an inference or judgement.

"I did" tells of the actions you performed in response to the incident: did you handle the situation yourself? If so, what exactly did you do? Did you call in your supervisors, medical personnel, or the police? Record every detail of your actions with respect to the incident so that others will be able to clearly understand the sequence of events that occurred.

An effective incident report records only information that the writer knows for certain is accurate. Keeping to the "I was, I saw, I did" formula will help you maintain objectivity in your incident reports and will ensure that your report records as fact only what you actually observed.

Accident or Injury Reports

An accident or injury report is a special kind of incident report written when an accident or injury has occurred in the workplace. Like the incident report, it records the details of the accident, the action that followed, and the treatment that was provided. It follows the same pattern of information as the incident report: it identifies the day, date, time, and location of the accident, the series of events that took place, the observations of the report writer, and the resulting action. Because an accident or injury in the workplace is a serious occurrence, it may involve police investigation, a workers' compensation claim, or an insurance claim or investigation. The writer of an accident or injury report must be sure to include all known details as accurately and as clearly as possible. This report, like the occurrence report, should answer all pertinent questions—when, where, what, who, why, and how—and follow the formula "I was, I saw, I did."

Assessments or Evaluations

An assessment or evaluation report, unlike the various types of incident report, does more than record information. In contrast to the incident report, in which the writer refrains from offering inferences or judgements, the assessment report calls upon the writer to make a judgement or offer a professional opinion of a situation, an action, or a client. Assessment reports draw upon the experience and training of the technical professional in order to evaluate what action, response, or intervention is needed.

Evaluation reports may be specially commissioned as part of the investigation of a proposed course of action or an existing situation. In this case, the writer is asked to assess the likely outcome of a proposed policy change or to investigate the nature of a problem and suggest possible solutions. The evaluation report may be used whenever more information is needed, either on its own to evaluate an existing situation or proposed action, or as a follow-up report to evaluate the recommendations made by another report. Rather than analysing causes and effects, it usually measures a solution or a situation against a set of explicitly-stated criteria in order to determine the suitability or unsuitability of that solution or situation. Its conclusions are based on a careful comparison between the initial criteria (usually identified by the person who commissioned the report) and the suggested action or solution.

Unlike an incident report, which outlines the facts of the case and describes the action already taken, an assessment report calls for the writer to apply professional judgement and usually involves recommendations for future action.

Progress, Status, or Project Completion Reports

A progress report, also known as a status report, is a kind of assessment report that details the movement or development of a long-term project or program over a period of time. The project completion report is the final progress report for a given project; as its title indicates, it is written after the project has been completed.

Progress reports are of two types. The first, the *periodic report*, is delivered at regular time intervals—every two weeks, for example, or for very long-term projects, every few months. In a college or school, students receive periodic reports of grades at the completion of each term. A company's annual report is another example of a periodic progress report.

The second type, the *occasional progress report*, is delivered whenever some significant stage in a project is completed, and the time interval between reports may vary. For instance, if I am overseeing the construction of a new facility, I may write reports only when significant stages are completed; since each part of the project takes a different amount of time to complete, my reports will be delivered at irregular time intervals.

Progress reports are part of the record-keeping function of a business or organization. Progress reports are used for purposes besides tracking engineering projects; they may also be used to assess a new program, method, or system. As well, the regular assessments you receive from your supervisor on the job are also progress reports.

Investigative or Analytical Reports

The investigative report, which is usually commissioned or requested by someone other than the writer, examines and analyzes a particular problem or question that has been identified by someone in the company. It is usually called for by someone higher up in the bureaucracy, and may even be the result of a government inquiry; a recent example would be the Romanow Report on Health Care. An investigative report usually evaluates causes and effects, frequently offers solutions to the difficulty, and may even apportion blame. Its conclusions are based on careful research, such as results from a controlled scientific experiment or data collected from testimony, interviews, surveys or questionnaires. Such reports are generally beyond the scope of the daily routine for most employees; nevertheless, their recommendations may have implications for the way your company carries out its activities.

Research Reports

The research report, like the investigative report, examines a particular question of scientific or other importance. It may take one of two forms: it may report the results of your own experimental research, or it may survey and synthesize existing work in the field and offer a summary of current thought, practices, or ideas. Academic writing in the technical professions frequently combines these two functions.

Even when it is used for the second of these two purposes, the research report differs from the investigative report in that it does not necessarily seek to resolve a specific problem within the company. Instead, a research report may be used to gather information helpful to the company in developing a new product or line, procedure, policy, or organizational structure. The major function of this kind of report is to present the findings from your research, whether it is experimental or qualitative research; the report may or may not offer recommendations based on its results. As part of your college program, you may be assigned a research report of either type. You may have to report on the results of your experimental research, or you may be required to consult research sources in order to determine the importance of a particular issue to your field of study. The credibility of such a report may depend to a large degree on the thoroughness and authority of your research sources.

If your report draws on material from outside sources—others' research, government documents, books and professional journals, articles in the popular press, Internet sources—this material should be properly documented, according to standards appropriate to your field of study. You will find more information on documentation in Chapter 6, on page 163.

Proposals

The proposal is a kind of report intended specifically to persuade the reader to follow a course of action the writer believes is appropriate, necessary, or beneficial. A proposal may be initiated by the writer or written in response to events or circumstances that present an opportunity or a difficulty for the company. A proposal may be requested by upper management, or may also be developed in response to a client's request for a service. For example, I might respond to a client's request for a computer training seminar by submitting a proposal for a seminar designed to fit the client's needs. The goal of a proposal is to stimulate action by its intended reader; it may suggest a positive change or innovation that the writer thinks his or her company should adopt. For instance, I might write a proposal any time I wish to create an innovative program in my department. A proposal may offer a course of action as a solution to an existing or potential difficulty; or it may request approval or funding for a new or existing project.

Proposals may be directed to readers within the company—to managers who have the power to enact the policies advocated by the writer—or they may be directed to agencies or readers outside the company (for example, a funding proposal to external agencies or government ministries). Because a proposal is intended to solicit authorization or funding for the writer's project, it must be sufficiently detailed and convincing to gain the reader's acceptance and approval; like other reports, it must answer the significant questions of when, where, what, who, why, and how, and it must attend carefully to all Seven Cs of professional writing: completeness, conciseness, clarity, coherence, correctness, courtesy, and credibility. Proposals are treated in more detail in Chapter Six.

STANDARDIZED REPORT FORMS

Although some reports are lengthy and comprehensive book-length formal documents, many common types of short and highly standardized day-to-day reports are submitted on prepared forms. Standardized forms make the job of managing routine information easier for both reader and writer. Instead of having to invent a report structure for each client interview or incident that occurs, you can organize your responses using the appropriate form provided by your agency or organization.

Report forms ensure consistency in cases where large numbers of reports containing similar information must be kept by many people as a matter or routine. Nearly any kind of report filled out on a regular basis can be organized into a standard form. In some cases, the information may be entered directly into a computerized report form on your screen. Because the kind of information required is always the same, a form helps to ensure that each person applies the same standards and collects the same details. In this way, too, much repetitious work is eliminated. Insurance claim forms, student grade reports, workers' compensation forms, income tax forms, registration forms, hospital charts, requisition forms, travel claims, some project status reports, and even job application forms are some common examples of standardized report forms. Other types that are often standardized in large organizations include occurrence or incident reports, accident or injury reports, and performance reviews.

All of these forms are designed to prompt the writer to provide the information pertinent to each type of report. Nevertheless, to ensure that your information is complete, you should keep in mind the questions outlined above—when? where? what? who? how? why?—so as to be sure that all the necessary details are included. Figure 5.2 shows a sample project completion report form of the sort that might be used in an engineering firm; Figure 5.3 shows a performance evaluation report form.

FIGURE 5.2   A sample project completion report form, such as might be used in any engineering firm. Note that the form standardizes the information that is invited, and provides a checklist for attachments. It also carries the signature of the person who submits it, and, like all important documents, is dated.

DOLOVICH AND COWAN
Project Completion Report

Department: _____   Date: _____

Project title: _____

Project number(as per inventory): _____   Location: _____

Project manager(s): _____

_____

Affiliation: _____

Date of approval: _____   Initiation date: _____

Scheduled completion date:_____   Completion date: _____

Funding source: _____

Completion report done by: _____

Parent Projects: _____

_____

Related Projects: _____

_____

Abstract: attached ☐   Rationale: attached ☐

Methodology: attached ☐   Benefits/Accomplishments: attached ☐

Publications/Presentations: attached ☐

Key words: _____

_____

Signature of Reporting Personnel _____

FIGURE 5.3 A career progress report is a periodic assessment report which may be completed annually, semi-annually, biennially, or quarterly, depending on the job in question.

CAREER PROGRESS REPORT

Evaluation for the period: _____ to: _____

EMPLOYEE'S NAME: _____

Department: _____

Position Duties: _____

_____

_____

_____

Additional Responsibilities Since Last Assessment: _____

_____

_____

_____

Achievements: _____

_____

_____

_____

EVALUATION SUMMARY	Superior	Competent	Developmental
Overall performance	[ ]	[ ]	[ ]
Job-related goals	[ ]	[ ]	[ ]
Development of others' goals	[ ]	[ ]	[ ]
Relationship goals	[ ]	[ ]	[ ]
Potential for advancement	[ ]	[ ]	[ ]

MERIT INCREASE RECOMMENDED   [ ] yes   [ ] no

NARRATIVE STATEMENT OF ASSESSMENT: _____

_____

_____

_____

_____

_____

Suggestions for Professional Development: _____

_____

_____

_____

_____

_____

SUPERVISOR:_____ Date:_____

Signature: _____

Employee's Comments:_____

_____

_____

_____

_____

I have read this summary and enclosed comments and discussed them with my
supervisor.

Employee's Signature: _____ Date _____

INFORMAL AND SEMIFORMAL REPORTS

Report Formats
- Informal
- Semiformal
- Formal

Although many routine reporting tasks can be standardized on forms such as those shown in Figures 5.2 and 5.3, not all reports can be presented in this way. In some situations, you may be required to generate your own structure, as well as to select the information required to "flesh out" the body of the report. Reports may be *informal* or *formal* in their structure, or they may strike a balance between the two, as is the case in the *semiformal* report. An informal report is structured in memo or letter format, though it may be up to five pages long. A formal report, by contrast, looks more like a book manuscript or a long formal thesis, complete with its own cover and table of contents. Length and layout are the most visible differences between formal and informal reports: the informal report is usually short, with an average length of three to five pages; it is also less detailed, has fewer distinct parts, and a less elaborate layout. An informal report may make use of headings to assist the reader in locating information, but because it is often less than three pages long, such headings are not always necessary.

In writing your reports, remember that format is meant to serve function, just as it does in the standardized report forms used for repetitive tasks. Choose the format that delivers your message most effectively. Generally, the longer and more complex the report, the more formal it will be, and the more likely it will be to use headings and other organizational devices to assist the reader in understanding the material presented. The formal report format allows clearer organization of large amounts of material, while short reports may be better presented simply. We will deal with the structure of formal reports in the next chapter; for the remainder of this chapter, we will discuss informal and semiformal reports.

As you may already have guessed, reports are not strictly divided into informal or formal types; occasionally you will find that you must write a report that, while important enough to warrant more formal treatment than an informal report, is not really long enough to require all of the elaborate formatting of the formal report. Since the organizational structure of the formal report, with its table of contents, special headings, and fancy cover, can overwhelm a short report of under ten pages, and a memo or letter format may come across as an overly casual treatment of your information, you may wish to use a third style of report, called the semiformal report, which combines aspects of the informal and formal report styles. Rather than opening with a memo or letter-like format, the semiformal report usually has the title, author's name, and date at the top of the first page; it is also more likely than the informal report to use headings to separate report sections. The semiformal report is not really a distinct type, but is a variation on the informal report. It is used in place of the memo or letter format when you want a more formal appearance for your short reports. In this book we will use the term "informal" to refer to letter or memo reports, and the word "semiformal" for short reports that use a slightly more formal style on their first page and that incorporate some other formal features in a report of under ten pages.

Although your employer may sometimes require a specific format for reports, often you will have to select your own format. How do you know whether to write a formal, an informal, or a semiformal report? Asking yourself the following questions may help you decide.

1. ***What is your purpose?*** If you are addressing a relatively minor issue, your report will most likely be informal; if the situation is important, your report will be semiformal or even formal.

2. ***Who is your audience?*** The more distinguished or the wider your audience, the more formal your presentation should be. A brief document to your immediate supervisor that no one else is likely to read will likely be informal; a detailed proposal being sent to the company president and advisory board, or sent outside the company, is likely to be formal.

3. ***How detailed is your analysis?*** The more complex the problem or issue and the more detailed and thorough your presentation, the more carefully you will have to organize your information, and the more you will use the titles, headings, table of contents, and support materials of the formal report.

In choosing a report format, you should be guided by the complexity of the problem or issue—that is, how much detail or research is required—and the intended audience or readers of the report. Most of the routine reports you write on the job will be informal or semiformal. Figure 5.4, the Report Format Decision Scale, provides an at-a-glance guideline to help you decide whether your report should be informal, semiformal, or formal.

Informal Reports

If you need to write a brief report about something for which there is no standardized form, you may choose to present your information in an informal report. Compared to the formal report, which will be addressed in the next chapter, this kind of report has a fairly casual format. For instance, it does not have a title page or table of contents. It may be written without enumerated sections, references, or appendices, although any of these could be included if they were needed. Unless your employer provides a standardized form, the informal report format is commonly used for regular progress reports, incident reports, evaluative reports, and informal proposals. If you have been commissioned by your boss to write a short report for his or her eyes alone, chances are you will be writing an informal report.

FIGURE 5.4   To use the report formal decision scale, put an "x" on each line of the scale to indicate the probable characteristics of your report. If most of your marks are to the far left of centre, your report should likely be informal in structure; if most of your marks fall to the far right, the report should be formal. If the marks tend to cluster in the middle, you should probably write your report in a semiformal format.

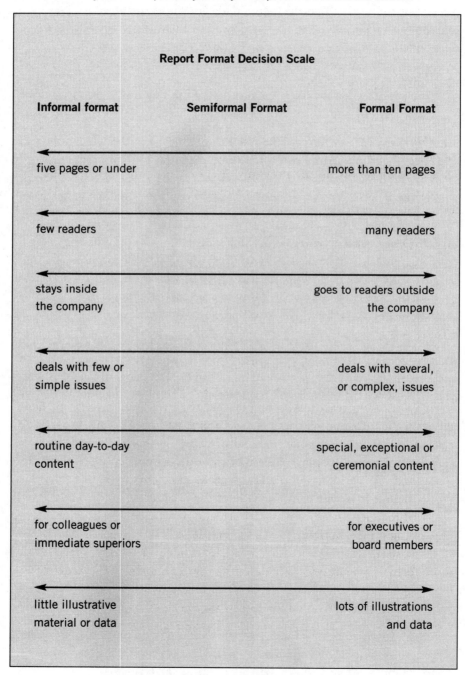

Report Format Decision Scale

Informal format	Semiformal Format	Formal Format

five pages or under — more than ten pages

few readers — many readers

stays inside the company — goes to readers outside the company

deals with few or simple issues — deals with several, or complex, issues

routine day-to-day content — special, exceptional or ceremonial content

for colleagues or immediate superiors — for executives or board members

little illustrative material or data — lots of illustrations and data

Informal reports are typed on one side of the page only, observing standard margins: 1" at top, right, and bottom, and $1\frac{1}{2}$" at left. The informal report begins like a memo or a business letter. The standard format of the memo (To, From, Date, Subject) or letter (return address or letterhead, date, and inside address) identifies the primary reader, the writer, and date. The subject or "re" line normally states the report's primary recommendation. The first page of the report may be typed on company letterhead or the internal correspondence form used for memos. If the report is longer than one page, subsequent pages are typed on plain paper rather than on letterhead. Remember that informal reports, like all other types, may be double or single spaced, according to the practice of your employer.

An informal report, depending on how long it is, may contain section headings. If so, then each new section begins immediately following the previous one, on the same page. Compare the following list of report parts with the sample informal report that follows.

Contents of the Informal Report

- Memo or letter opening
- Statement of recommendation(s) contained in subject or "re" line
- Brief summary statement
- Introduction
- Discussion: Background to issue or situation Outline of important facts and details Possible outcomes, results, or solutions
- Conclusion
- Recommendation(s)
- Appendices: (optional): Charts Supporting data Diagrams

Sample Informal Report

Informal reports can be presented in either memo or letter format. The class observation report in Figure 5.5, a variation of the assessment report, is written as a letter. Had the writer chosen to do so, it could also have been formatted as a memo and sent along with a cover letter to the recipient.

Semiformal Reports

Regular progress reports, incident reports, evaluation reports, and proposals may also be presented as semiformal reports. The primary difference between the informal report and the semiformal report is the more formal appearance of its first page: instead of using memo or letter format, the semiformal report displays the organization's name, report title, author's name, and date at the top of the first page, as shown in the sample semiformal report below. It is typed on one side of the page only, observing the same standard margins as the informal report, and may be double or single spaced, depending on the preference of your employer.

The semiformal report format is used whenever your reports require a more formal appearance than a memo or letter allows. It is also generally a bit longer than the informal report, running usually between five and ten pages, though it can be used for reports up to twenty pages long. Its contents are usually divided into short sections headed with appropriate titles.

FIGURE 5.5   This informal report offers a brief assessment of the seminar conducted by a visiting professional engineer as part of the Department of Civil Engineering's Professional Practice course. Though the report is presented using letter format, it could equally well have been presented in memo form.

Department of Civil Engineering
Central Maritime University
Box 5005 Truro, Nova Scotia B2S 4M3
<www.cenmar.u.ca>

26 February 2003

Veronica Scott
Dolovich and Cowan Technical Services
Eastern Office: 403 Frederick Street,
Halifax NS B3A 4D4

Dear Ms Scott:

Re: Seminar Observation Report

Thank you for contributing a session to our third year Professional Practice course. I found the session both interesting and informative, and am happy to contribute this observation report for your files.

The session I saw was conducted with a group of a dozen or so students, along with yourself, seated around a large central table in a seminar room. The principal method you employed was organized discussion. Though you had a clear agenda for the session, the students also seemed to feel free to ask questions and contribute to the table talk—and they did so freely. Your written outline on the board behind you served to both structure the session and keep the discussion on track.

I found your interaction with the group to be very professional, striking exactly the right balance of cooperation and respect. I noted that you established a warm and supportive atmosphere by sitting at the table with the students, a move that signalled your respect for them as future colleagues, and encouraged their engagement and active participation. As well, your willingness to share your personal experiences as a student, an intern, and a partner confirmed the friendly professionalism of your approach, and I'm sure the students will benefit from your generous advice regarding the selection of technical electives, the identification of firms offering professional internships, and other details of their remaining education.

As a practising engineer yourself, you were able to draw extensively on direct knowledge of the kinds of preparation and experience they've had to date, and on what lies ahead for them as they undertake their first jobs. You offered a very helpful framework to assist them in assessing their own interests and selecting a position that will be right for them. Your extensive experience, your level-headed advice, and your genuine enthusiasm helped to make the session a valuable and helpful one for all the students.

Overall, I came away from your session impressed with your ability to reach the students at their own level of concern and need, a factor that in my view indicates your strong commitment to the welfare of these young professionals. The presentation was clear, organized, thoughtful, and genuinely useful. We are delighted that practising professionals like yourself are willing to take the time to share their insights and advice. I know that the students join me in expressing my thanks, and in extending an invitation to return next year. Thank you very much for your contribution to making CivEng 351 a success.

Sincerely,

Alphonso Malle

Alphonso Malle
Associate Professor of Civil Engineering and
Coordinator, CivEng 351

Unlike the informal report, where the summary of recommendations is presented in a "re" or subject line followed by a brief summary statement at the beginning, the semiformal report has a title rather than a subject line and presents its summary in a short paragraph at the beginning of the report.

Because the semiformal report is a variation on the informal report, the distinctions between them are not entirely clear-cut; in some cases, the same material can be presented in either format, depending on the circumstances. For a very short report, your choice of format will be influenced primarily by your purpose and your audience's needs. The more important these are, the more likely it is that you will choose a semiformal style over the informal one. If the report is over ten pages long, but the issue presented is fairly straightforward and direct, the semiformal format is more appropriate. If your report is likely to be more than ten pages and is divided into many complex sections, you should consider using a formal format.

Sample Semiformal Report

Contents of the Semiformal Report

- Report title, author's name and title, and date at top of page one
- Summary
- Introduction
- Discussion: Background to issue or situation Outline of important facts and details Possible outcomes, results, or solutions
- Conclusion
- Recommendation(s)
- References/Bibliography (optional)
- Appendices (optional): Charts Supporting data Diagrams

The development of most major projects is preceded by a formal proposal, which may be presented in an informal, semiformal, or formal format. The semiformal report shown below is the project proposal for the book you are now reading. Remember that a proposal, like any other type of report, can be presented in formal, semiformal, or even informal format.

Note that the headings used in the body of the report are those indicated in the publisher's outline. Most publishers have a standard formatting guide for such submissions, and the section headings are designed to tell them quickly what they need to know to gauge the viability of the book, the potential market, and the credibility of the author proposing the manuscript. The summary offers information about when the entire manuscript will be available, to allow the publishers to schedule their editorial commitments.

Although this report could also be arranged in informal format, semiformal presentation is more appropriate for my purpose and intended audience.

FIGURE 5.6 A semiformal report offers a more formal appearance than an informal memo report, and is therefore more suitable for treating complex issues or analyses. A proposal may be presented in any of the three formats—informal, semiformal, or formal.

Manuscript Proposal
EFFECTIVE COMMUNICATION FOR THE TECHNICAL PROFESSIONS
Jennifer MacLennan
D.K. Seaman Chair, Technical and Professional Communication
College of Engineering, University of Saskatchewan
19 October 1999

SUMMARY

The proposed project is to be a textbook and handbook in technical and professional communication, suitable for a one-term course at the post-secondary level. The manuscript is currently in development, and will be ready for submission in fall 2001.

INTRODUCTION

While the proposed book will in some respects resemble my *Effective Business Communication* (3e, 1998) and *Effective Communication for the Helping Professions* (at press), it is not simply an adaptation of the existing texts. Like them, it is intended for a one-semester university or college course in writing in the workplace, and its approach will be similarly grounded in a theoretical model. However, it differs from my previous books in that it will focus upon the specific communicative demands of the *technical* workplace, and will incorporate formats and constraints unique to this context. It will also be more detailed, and will include elements of critical analysis and careful reading.

STATEMENT OF PURPOSE
A. The Book
1. Description

I propose a book of approximately 400 pages, which will treat common writing scenarios faced by students in the technical professions, using an approach that emphasizes an understanding of audience, purpose, and communicative competence.

2. Organization and Structure

The book will be designed to compete favourably in the current market. As such, it will deal with writing situations commonly faced by those in the technical professions: memos, short reports, log books, proposals, and formal reports. It will also present a comprehensive treatment of the job search process. However, rather than being simply a collection of models and precepts, this book will be designed to encourage students to think about audience and strategy and to understand the role these elements play in shaping content and format of communication. A preliminary chapter outline is attached as Appendix A.

3. Outstanding Features

My book will be different from others for the same reasons that my previous texts have distinguished themselves: it will be designed to build the kind of discretion and effective judgement students need to respond to any communication situation. The specific features which will distinguish this book from its competition are:

- it will feature Canadian contexts and examples;
- it will be supported by my experience as a communication specialist housed in an engineering college;
- it will incorporate successful methods and strategies used in my previous books on communication;
- it will provide a solid theoretical foundation;
- it will feature assignments designed to develop the skills of *reading* carefully and analytically;
- it will include numerous samples of both effective and ineffective communication, and will help students to develop the discrimination they need to edit their own work for clarity and effectiveness;
- it will offer a comprehensive treatment of the job application process, which is aimed squarely at the target audience (engineering students) and is based on years of practical research;
- it will offer assignments and situations for writing that, while addressed to a technical audience, are accessible and understandable to students who have not yet amassed years of professional experience;
- it will avoid cookie-cutter formats; instead, like my earlier communication texts, it will be designed to inculcate in the students the judgement they need to respond appropriately to a variety of communicative demands in the workplace.

4. Apparatus/Pedagogical Aids

In addition to its discussion of communication principles and its treatment of the various contexts and formats of technical writing, the book will incorporate a variety of assignments designed to encourage mastery of the concepts and formats being learned. Most technical communication texts offer writing practice only; this book, by contrast, will offer examples for critical reading and analysis as well as models for writing. The variety of writing samples, some of which violate the principles of good writing that the students are learning, can be used to help the students develop skill in assessing the appropriateness and effectiveness of different communicative strategies. By comparing effective examples with ineffective or flawed examples, students will develop greater discernment and critical skill that in turn will assist them in editing their own work. An instructor's resource manual which includes additional background material and suggestions for teaching the course will accompany the text.

5. Level

The book is intended for first-year or second-year post-secondary students who must develop skill in communicating effectively for the technical or professional workplace: it includes attention to critical reading, clear writing, and effective presentation skills, and will offer an approach to communication that is theoretically grounded in the discipline of rhetoric. It will not, however, be top-heavy with theory; instead, its aim will be to develop skill and judgement, ability and understanding, in combination. I have used this approach in a variety of contexts and situations, for students in both technical and non-technical fields, and it has worked consistently well across disciplines.

B. Market

1. The Primary Course

The book is designed for a one-semester course in communication for technical professions, at both college and university level. Our college offers approximately 25 sections per year, with a current enrollment of between 15 and 20 students per section. As the course is redesigned (re-engineered!) with the implementation of our new curriculum, the enrollment may increase to between 20 and 24. All engineering students are required to take this course at some point in their programme.

2. Other Courses

Any one-semester course in communication designed for technical students. Most Canadian universities and colleges, if they offer programmes in engineering disciplines, teach such a course.

CONCLUSION

The book described in this proposal is currently in development. Many of the readings and all of the teaching methods have already been extensively classroom-tested. I will naturally continue to use the approaches outlined here in my own teaching of technical communication, and in training others to teach the course at the University of Saskatchewan.

PROJECTED COMPLETION DATE

Should the proposal be accepted, I anticipate being able to submit a complete manuscript of approximately 500 pages by November 2001.

APPENDIX A

Projected Table of Contents for *Effective Communication for the Technical Professions*

CHAPTER FIVE Informal and Semiformal Reports

No matter what kind of report you are writing, you must prepare thoroughly and organize carefully.

1. Understand the context in which the report will be received, and what is to be done with the information it contains.

2. Identify your purpose or the action you wish the audience to take.

3. Identify your reader by needs, expectations, knowledge, goals, concerns, and relationship.

4. Use the problem-solving approach to define the communication challenge and develop your report content.

5. Develop your points fully and observe the Seven Cs.

6. Answer the questions *when? where? what? who? why?* and *how?* in each report.

7. Follow the "I was, I saw, I did" formula for writing incident, accident, or misconduct reports.

8. Wherever appropriate, use the standardized forms supplied by your employer or instructor.

9. In choosing a report format, you should be guided by the complexity of the problem or issue—that is, how much detail or research is required—and the intended audience or readers of the report. Most of the reports you will be writing will be informal or semiformal.

SHARPENING YOUR SKILLS

The following report situations vary in complexity and requirements. Whatever report you are writing, observe all of the rules we have discussed and add any specifications or details you need to make the report convincing. Before beginning to write, you should review the strategies for writing in Chapter Three. You may also wish to use the Report Writing Planner (Figure 5.1) to outline your two principal elements: reader and main message. Be sure to edit your work carefully.

1. You are in charge of a placement student at the company where you served your internship, and are required to provide feedback to the college regarding the student's performance. Your brief report should comment on the following:

   a) the student's duties and responsibilities;
   b) the number and titles or responsibilities of the staff;
   c) the student's expertise and command of business principles;
   d) the student's general attitude and behaviour;
   e) any special strengths the student has demonstrated;
   f) any areas of the student's performance that need improvement; and
   g) your overall evaluation of the student's competence.

   You may use as a model someone at your own placement (do not use the person's real name), or you may model your evaluation on yourself. You should provide concrete and specific details and examples, recommending at least one area for improvement.

2. As a technical professional, you will be required to write numerous reports on both short-term and long-term projects. Some of these will take the form of periodic or occasional progress reports, including reports on your own performance on the job. In most instances, you will have to evaluate your achievements and your failures along the way, indicating how you plan to overcome any obstacles you have met with. This term, you are enrolled in a professional writing course, and it is now mid-term. Your task for this assignment is to prepare and submit a short (informal) report to your instructor outlining and evaluating your own progress in the course. The report will include such topics as your initial objectives or expectations, your achievements thus far, any failings or obstacles you have encountered and what you have done (or plan to do) to overcome them, the work that has yet to be done, and your expected grade or performance. You will want to supplement your report with evidence such as mid-term or assignment grades, course projects, and topics covered. Keep in mind that you are not evaluating the course *per se*, but your own commitment to and progress in the course. Essentially what you are preparing is a self-evaluation report such as you might occasionally be expected to prepare for annual performance reviews on the job.

   Although the assignment is intended to focus on your professional communication course, you may wish to use another of your courses instead. If you want to do this, get the approval of your instructor.

3. You have learned that your college offers a course option known as "Directed Study," which allows a student to work closely with a faculty member to develop an individualized course in any area of the discipline. There are several guidelines for directed study courses: they are normally offered at the senior level (as course number 398 or 498 in any department); they usually build on or extend from existing courses; and they must not duplicate existing courses in the department. Normally a full course outline is drawn up and a final exam is required. You have identified an area of study in your discipline in which you would like to pursue further study. Professor Horace McAllister has agreed to supervise your work, but he has asked you to draw up a preliminary course outline and requirements. Your proposal must include the following information: the proposed department and number of the course; the course title as it will appear on your transcript; the name of the instructor; the term in which the course will be offered; a course description of not more than fifty words; a list of prerequisites, if applicable; lecture, seminar, and laboratory hours per week; methods of evaluation; required textbooks; and a brief outline of the topics to be covered in the course. Set up your report using appropriate headings and submit it to your communication instructor.

4. The dean of your college is conducting a study of the first-year experience with an eye to dealing with a number of concerns, including heavy workload, class sizes, and opportunities for co-op work placements. In the course of this investigation, she has called for contributions from students. You have decided to offer your dean some comments about areas of your programme that you feel should be improved and areas that should remain unchanged. Assemble your comments into a short informal report addressed to Dean Deborah Cottreau.

5. For this assignment, you are asked to analyse and assess the communication challenge inherent in a report-writing project from one of your technical classes, and to outline the strategies you will use to approach it. Present the results of your analysis in a short report of approximately 750 words (three pages) using either memo-report or semiformal format, as directed by your instructor. You may draw on class discussions and course readings to help frame your analysis. Your report should consider the problem as fully as possible and should use a standard SIDCRA structure. Remember that, in the discussion segment of the report, you will need to devise appropriate headings for the information you present, and organize the sections in a coherent manner. Below are some suggestions for issues you might consider in your analysis.

Summary: Identify the specific writing project that is the subject of your report and describe your goals as a writer or speaker. Briefly state the outcomes, or results, that *this* report will present.

Introduction: Describe the writing project you are working on in your other course (what is the assignment? what are the format and content requirements? what is its purpose?) and show how it can be understood as a communication challenge (what kind of report is it? how many are collaborating in the writing? will the material be presented orally? how many readers will receive the written work? what decisions will they have to make on the basis of your report?) Set the tone and expectations for the discussion that follows.

Discussion: Draw on what you've learned so far to position the project report as a communication challenge. Show how this project presents a challenge to you as a professional communicator. Analyze the communication requirements in terms of audience, purpose, your own credibility as a writer and researcher, the specific circumstances under which the report will be written (that is, is it a group project?), ethical considerations, and the constraints you are facing. Describe your communicative goals and identify the specific strategies you hope to use, or the steps you hope to follow, to achieve them. Why have you selected these strategies? Which do you think will be the most difficult of the challenges that will face you in preparing the project report? How do you plan to overcome these challenges? To what extent is a successful outcome within your control? In what respect do your goals coincide with, or differ from, those of your intended reader(s)? You may also wish to refer to the work plan you have drawn up for the pacing of this project; in what sense is it a communication tool?

Conclusion(s): What qualities of communication can be understood from doing a report project such as the one you're analysing? In what sense is the report a technical exercise? In what sense is it a communication exercise? How independent are these two goals? In what way can you use what you've learned in this course to help you with the report?

Recommendation(s): Use this section to lay out the specific steps that you plan to follow. These should clearly follow from the analysis presented in the discussion segment of the report.

Appendices: You should provide relevant materials that support the discussion segment of the report. These might include: the original assignment sheet; the definition of the technical problem you will be working on; a description of the procedure you are required to use to prepare the report; a copy of the work plan your group has drawn up. Other items may also be included if they are relevant. Be sure to refer to the appendices in the body of the report and to label the items appropriately.

CHAPTER 6:

Formal Reports and Proposals

LEARNING OBJECTIVES

1. To learn the basic parts and format of a formal report.
2. To learn about the use of visuals in a formal report.
3. To learn about writing an effective technical description.
4. To understand the purpose and focus of a formal proposal.
5. To study a model of a formal report.

The formal report, although similar in purpose and organization, is generally more complex and detailed than the informal report. It tackles more difficult problems, analyzes them in greater depth, and considers more extensive evidence in support of its recommendations. It is usually much longer than informal or semiformal reports, being anywhere from ten pages to several hundred.

You should use a formal report format if your subject matter is of considerable significance to your company, if your readership is likely to be large or prominent, or if your findings are extensive. Usually a project resulting in a formal report will involve several or all of the above considerations. For example, you would use a formal report structure for a lengthy report from your department to the president of the company that makes important recommendations for major department changes.

Chapter Three outlines the strategies for drafting and writing a formal report, while Chapter Five discusses the various situations for and purposes of reports. Any of the report types discussed in Chapter Five may be presented as formal reports. The formal report, since it is longer and deals with more extensive and complex issues than informal or semiformal reports, presents a greater challenge simply because there is so much more information to manage. For this reason, the problem-solving strategies and the "Three Ps" of planning, preparing, and polishing are especially valuable for the formal report writer. In this chapter, we will consider the details in structure that distinguish the formal report from the semiformal and informal types.

When to Use Formal Report Format

- The topic is significant.
- The report is long.
- The readers are numerous or prominent.
- The findings are extensive.
- The report contains lots of illustrations and data.
- The report is destined for outside readers.

THE PARTS OF A FORMAL REPORT

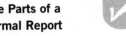

The Parts of a Formal Report

- Cover
- Letter of Transmittal
- Title Page
- Summary
- Table of Contents
- List of Tables and Figures
- Acknowledgements
- Introduction
- Discussion
- Conclusion
- Recommendations
- References
- Appendices

A formal report, especially if it is to be sent outside the company, is a reflection of the company's professionalism and corporate image. As such, it must conform to standard conventions of layout and structure. It should be error-free and printed on one side of the page only, with standard margins of 1" at top, right, and bottom of the page and $1\frac{1}{2}$" at the left side. The text may be single-spaced or double-spaced, according to your company's practice, though double-spacing is more common. The formal report is usually bound in a cover that bears the company name, report title, author's name, and date, and it is usually bound. The formal report employs the same six standard parts that make up shorter reports. However, the physical structure is more elaborate, employing a title page, table of contents, lists of figures and tables, and section headings. The parts of a formal report, in order of presentation, are as follows.

The Cover encloses a formal report. It may be a plain purchased cover or one that has been specially designed. If you are buying a cover, don't choose one with a gaudy picture or design. A plain-coloured, good-quality cover is preferable. The cover, like the work inside, should make as professional an impression as possible, and one in gray, black, or navy makes a more dignified impression than a wildly coloured one. Avoid cheap, poor-quality report covers. Spending a little more for a good cover will make a better overall impression. If you have specially designed a cover that displays the title of your report, choose the title carefully to reflect the content. It should be neither too long nor too brief. A subtitle may help to describe the material presented within the report.

A Letter of Transfer or Transmittal may be attached to the outside of the report cover or bound inside the cover just ahead of the title page. This placement is a matter of preference, and in writing your report you should follow the practice used in your company. The letter is a formal business letter from the writer (you) to the person or persons to whom the report is addressed. It should briefly outline the reason for the report and point out some of its important findings or features. Like all business letters, the letter of transmittal is normally single spaced. If your report is to stay inside your company, you may wish to use a memo instead of a letter; however, if it goes to readers outside the firm, write a proper business letter on company letterhead.

The Title Page, containing the name of the company or institution, the title and subtitle of the report, the name(s) and title(s) of the person(s) who commissioned the report, the name(s) of the author(s) and their title(s), and the date, comes next. If you are provided with a cover page format by your employer, follow that; otherwise use the format shown in the sample title page on page 188 of this chapter (Figure 6.11B). The formal report should always contain a title page, whether or not the title appears on the cover of the report. The title page is never numbered.

The Summary of Recommendations, also known as the *Executive Summary* or *Abstract*, usually precedes the Table of Contents in a standard formal report, although some report formats place it immediately following. The summary, as discussed in Chapter Five, is a brief overview of all the important parts of the report and should include condensed versions of the introduction, discussion, conclusion, and recommendation(s). Although there is no strict guideline about the length of the executive summary, it should normally be no longer than one-tenth the length of the report as a whole; that is, a ten-page report would likely feature a summary of one page, while the summary of a 100-page report could be as long as ten pages. After reading your summary, even your least expert reader should have a clear idea of your findings and your approach. The summary is not considered part of the report proper, and so is not numbered as part of the main text. However, if desired, front matter (Summary, Table of Contents, List of Illustrations, Glossary, and Acknowledgements) may be numbered using lower-case Roman numerals at the bottom centre of the page.

The Table of Contents, a detailed listing of numbered sections and the pages on which these are to be found, comes next. It is designed to help the reader locate information in the body of the report. Sections are listed in order and may or may not be numbered. The sections of a very lengthy report may be further divided into subsections; if so, these should also be listed in the Table of Contents. Like the summary, the Table of Contents page is front matter and is not considered part of the body of the report.

The List of Tables and Figures, or *List of Illustrations*, may follow the Table of Contents, especially in a technical report. Like the Table of Contents, it is an ordered list of items, in this case the visual materials, that appear in the body of the report. It lists, by number and title, the figures and tables presented in the report, along with the numbers of the pages on which they appear. Like the Table of Contents, this list is an aid to the reader who wishes to quickly locate information in the body of the report.

A List of Technical Terms, also known as *Nomenclature* or *Glossary*, may follow the List of Illustrations if the report is highly technical or extremely complex. Such a list provides an at-a-glance reference for specialized terminology, making a complex report more reader-friendly and usable. Terms should be defined clearly and in a manner that will allow the reader to quickly grasp the meaning. If the technical terms are relatively few, but essential, this segment may be placed in the body of the report rather than being set up as a separate section; otherwise, place them in their own list as part of the front matter. In general, this section is used for technical or specialized terminology that is essential to the report; unnecessary jargon should be minimized or avoided completely.

Acknowledgements, if they are appropriate to your report, are placed on a separate page following the various lists of contents. On the Acknowledgements page, the authors of the report should recognize and thank all individuals or groups who assisted in the preparation of the report, or contributed to the project it discusses. The Acknowledgements page is considered front matter, and like the Table of Contents page, the List of Illustrations, and the Glossary, this page is not numbered as part of the main text. It may, like other front matter, be numbered with lower-case Roman numerals at the bottom centre of the page.

The Introduction begins the report proper, and its first page is considered the first page of the report. It is numbered "1," usually at the bottom centre of the page. As you will recall from the discussion in Chapter Five, the introduction not only introduces the subject matter of the report, but prepares the reader for the report's particular focus and its findings. It also briefly outlines any necessary background information, states the problem or issue, describes the situation, and sets out any limitations that may have been imposed on the investigation or analysis, as well as giving specifics about the direction that the analysis has taken. It is not meant to do any of these things in depth, but to prepare the reader for what will come in the body of the report. The introduction may occupy between one-tenth and one-fifth of the report proper; that is, the introduction to a hundred-page report will likely be between ten and twenty pages.

The Discussion, or main body, of the report follows the introduction to the formal report, just as it does in the informal and semiformal reports discussed in the previous chapter. It sets out the writer's method (including the criteria used to evaluate possible results, solutions, or outcomes) and presents a detailed analysis of the problem, issue, or situation that led to the conclusions and recommendations offered in the report. It should discuss the important facts of the situation, including relevant history, details, formulae, calculations, and examples. As well, it should itemize any possible outcomes or courses of action, indicating the one that has been recommended and detailing the reasons for rejecting the others.

The Conclusion outlines any inferences that can logically be drawn from the material presented in the report; it shows the outcome of the analysis. As you will recall from the last chapter, the conclusion summarizes the findings of the report and should be a natural result or extension of the point of view presented in the discussion. It should not contain any unexpected revelations or outcomes. Instead, it should satisfy the expectations created by the rest of the report, and draw together the arguments that have been presented in the discussion. Like the introduction, the conclusion of the report generally occupies between one-tenth and one-fifth of the report, so that the conclusion of a hundred-page report would generally be between ten and twenty pages long.

The Recommendations serve the same purpose outlined in the discussion of the informal and semiformal reports; that is, they spell out the action that the report writer expects will be taken on the conclusions presented. If a conclusion says "this is what I think about this situation," the recommendations say "here's what we should do about it." The recommendations may include several steps that the reader is expected to take; if so, these should be listed and numbered individually so that they are easy to identify and follow.

References or *Bibliography* (a listing of sources referred to in the discussion) may also be included in a formal report, since many such reports make use of research sources. While many of the reports you will write as part of your routine will rely chiefly on evidence drawn from the workplace itself, reports that are intended to influence policy, corporate vision, workplace practices, or product design strategies frequently draw on material from outside sources—others' research, government documents, books and professional journals, articles in the popular press, Internet sources—in order to assist managers and administrative officers in making informed decisions about the future of the company or organization.

Some style guides use the terms "References" and "Bibliography" interchangeably; however, this usage is not universal. Others insist that "References" should include only those sources directly quoted in the discussion of the report, while "Bibliography" is usually understood to also include sources consulted but not directly quoted. You may wish to resolve this difficulty by listing as "References" only those sources from which you have quoted, which is the path many researchers take.

However, you may find that you have made use of research sources that, though not cited, helped to shape your understanding of the issue or problem, and many believe these also should be acknowledged. In such a case, you may wish to divide your references into two lists, one showing "Sources Cited" and the other indicating "Sources Consulted." Either way, your list of references is intended to provide the reader with the information needed either to do further reading on the subject of the report or to check the accuracy of the writer's interpretations of previous research.

Appendices are often employed in a formal report, since the information presented there is sometimes quite complex. As in the informal report, these attachments may include any supporting data that are either too cumbersome or too complicated to be included in the body of the report. Some examples might include detailed charts or diagrams, calculations and derivations, or extensive supporting data.

Each appendix should be clearly labelled and given its own title, as, for example, in "Appendix B: Grammar Review" that begins on page 359 of this book. Appendices should be inserted into the report in the order in which they were mentioned in the discussion. The first to be mentioned is labelled "Appendix A," the second is labelled "Appendix B," and so on. Although it is an attachment to your formal report, an appendix should also function as a self-contained document, with its own brief introductory paragraph and its own conclusion.

Documenting Your Sources

The credibility of a research report can depend to a large degree on the thoroughness and authority of your sources. For this reason, all sources you consult, including Internet sources, should be properly documented, according to standards appropriate to your field of study. Once you have determined that the information you have collected is reliable and authoritative (an issue of particular importance in dealing with Web sites), you should be able to record all the information your readers would require to locate the cited sources for themselves if they should need to do so.

When you refer to your research sources in the body of your report, you may use one of two methods. The first of these is a format known as author-date, which places the last name of the author and the year in parentheses immediately following the relevant statement, as in (Moffatt 1999). The reader may then refer to an alphabetical list of references at the end of the report to find the complete documentation information for Moffatt's work. This format is currently the most widely used in nearly all fields.

The second method, commonly used in mathematics and the natural sciences, is to simply insert a number in parentheses immediately following the relevant statement, as in (1). The reader may then refer to a numbered list of references at the end of your report. The sequence in which the sources are numbered depends on the order in which they were cited in the body of the report, and it is in this order that they appear at the end of the report.

Documentation information for research sources in your list of references should include information about author, title, place of publication, publisher, and date. For Internet sources, you should identify the author, the name of the posting or Web page, and the date of the posting, along with the electronic address (URL). A style manual will provide you with clear guidelines for recording this information, along with samples of layout. Here are four examples of source citations using the author-date arrangement recommended for the natural and social sciences, which can be found, for example, in the *Chicago Manual of Style*. Even if you number your references, the information they contain can still be presented using this format.

Print Sources

A book:

> MacLennan, J.M. 2002. *Technical Communication: Engineering Messages*. Toronto, ON: Pearson Education Canada, Inc.

An article or chapter in a book:

> MacLennan, J.M., and G.J. Moffatt. 2000. Language and personal identity. *Inside language: A Canadian language reader*. Scarborough, ON: Prentice Hall Canada.

An article in a book edited by someone else:

> Irwin, L. 1997. The real and the imagined: students as teachers in the rhetoric classroom. In D. Zarefsky and J.MacLennan, *Public speaking: Strategies for success* (Canadian ed.). Scarborough, ON: Prentice Hall Allyn and Bacon Canada.

Electronic Sources

A page or entry on an Internet site:

> MacLennan, J.M. 1999. What students say about the study of rhetoric. On line. URL: http://www.engr.usask.ca/dept/techcomm/studntsay.htm

An e-mail message:

> Moffatt, G.J. 1999. Life under deadlines. E-mail to Jennifer MacLennan (20 January).

There are many acceptable styles for reference entries in addition to the ones shown, and, though no one style is more "correct" than another, you should use the form preferred by your employer, instructor, or editor. This book uses the author-title format widely preferred in the humanities.

ASSEMBLING THE REPORT

The parts of the formal report are usually arranged in the order listed, beginning with the front matter (Summary, Table of Contents, List of Illustrations, Acknowledgements). Frontal matter is not numbered as part of the body of the report, but may be numbered using lower case Roman numerals placed in the bottom centre of the page. After you have assembled the front matter, you should assemble the main report itself. Each part in the main report (the Introduction, the sections of the Discussion, the Conclusion, and the Recommendations) normally begins with its own title on a new page, almost like a chapter of a book. Starting with the Introduction, the pages of the report are numbered using Arabic numerals. The first page of each section may be numbered or left without a number.

Although the usual placement of page numbers in the bulk of the text is in the upper right corner of the page, your company's format may centre them at the top or bottom of each page. Use the format recommended or required by your college or your employer. Page numbers in a report consist of numbers alone; they are not accompanied by the word "page."

Remember that part of the effectiveness of a formal report, as with any professional communication, depends on its visual appeal, so it is important that your report be printed in a professional manner. Follow an accepted format carefully, and take great care that no spelling, grammatical, or typing errors mar the quality of your report.

Formal reports also make use of frequent paragraphing and employ headings and subheadings to assist the reader in following the reasoning of the writer. As in all professional communication, layout is part of your organization, so you should use white space effectively and choose a readable font, preferably in a 12-point serif font. Do not make corrections to a formal report in ink or pencil; instead, reprint the page. Always make corrections using the same font you used for the rest of the report.

USING VISUALS IN A FORMAL REPORT

A well-written formal report should contain straightforward, readily understandable explanations, but your words need not work alone to communicate your message. Instead, they are frequently augmented by visuals—graphic or pictorial material used to simplify or clarify the verbal material. You should try to translate your written information into visuals wherever it will help you to clarify and enhance your discussion. Well-chosen visual aids can help to communicate your message more effectively, and they should be carefully selected to fulfill this purpose.

Types of Visuals Used in a Formal Report
- Tables
- Graphs
- Charts
- Diagrams
- Line drawings
- Photographs

Visuals can include tables, graphs, charts, diagrams, line drawings, and photographs that are designed to help the reader understand what is being discussed. In many cases, they serve not only as handy presentation devices, but as aids to analysis. However, you should never use visuals simply to decorate your reports. Tables that are overly cluttered, diagrams or charts that present an incomprehensible jumble of acronyms and detail, photographs that are muddy or indistinct, or drawings that do not isolate the information they are meant to communicate will dtract from the clarity and effectiveness of the report. If your visuals are not immediately understandable to the intended audience, consider redesigning them or replacing them with something else. Never allow your visuals to unduly complicate your information, confuse your reader, or distract the reader from your message. If the visuals do not communicate clearly, they are not fulfilling their purpose.

Depending on their size, comprehensiveness, and immediate relevance to the written material, visuals may be placed either within the body of the report or in an appendix at the end. If they are necessary to the reader's immediate understanding and if they are simple enough, visuals should be positioned close to the appropriate paragraph in the report, preferably on the same page or the one immediately following. Large visuals may be rotated 90 degrees, with the top toward the left side of the page so that it is caught in the bound edge. Some experts advise that visuals may even be placed on the back of the preceding page so that they face the text where they are discussed.

Number visuals sequentially (Figure 1, Figure 2, and so on) and identify them by an appropriate title and brief caption, positioned beneath the visual. The text of the report should refer to the visual aid by figure number and title when discussing the material shown in the visual, and the placement of the visual should occur close to the discussion in the text.

If the visual is very complex or if it is not necessary to the reader's immediate grasp of the situation, it can be placed in an appendix. If it is very complex but necessary to the reader's immediate understanding, the complex version can be placed in the appendix and a simplified version placed in the body of the report. Below is a discussion of when and how to use different types of visuals.

Tables

Tables are widely used by writers of reports to help consolidate, present, and analyze comparative information. They are especially useful for comparing large amounts of material in a small space. While tables cannot take the place of your discussion, they do allow you to organize and effectively present comparative information in a format that is readily grasped by your readers. In fact, you should expect that some readers will turn first to the table for an overview of the facts, before they read the discussion of the details. For this reason, the tables must be clear and easily understandable on their own, presenting complex information according to some clear comparative principle.

Most word-processing programs allow you to design and insert tables right into the text of your report. If your program does not allow you to do this, you can create tabular columns using the tab key. Here are some guidelines for creating an effective table:

- Provide a heading for your table that includes a number and a suitable caption.
- Space items so that they do not look cramped.
- Design and place your table so that it fits on one page.
- If the table becomes too wide for the page, turn it 90 degrees, with the top of the table toward the bound side of the page.
- Label all parts of the table clearly so that readers will know what they are looking at.
- Keep your table simple enough that readers can take in the information quickly. A table that takes ten minutes to decipher is not fulfilling its purpose.
- Number your tables in the order in which they are presented in the text.
- Numbers and captions for a table are usually placed above the table; for other illustrative material, they are typically placed below.
- If possible, compare information vertically (in columns) rather than horizontally (in rows). People can scan vertical information more easily than horizontal information.

Tables 6.1 and 6.2 both deal with educational attainment in Canada, but communicate different information. Notice how the arrangement of each table is adapted to the table's purpose.

TABLE 6.1 Tables allow us to compare large amounts of information conveniently and clearly. Comparisons may be made along only one variable (for instance, by year), but more commonly, tables compare information along two variables, as shown in this example, which compares educational attainment by both year and level of schooling.

Canadian Population by Highest Level of Schooling (percentage)						
	1976	1981	1986	1991	1996	1999
Less than grade 9	25.4	20.7	17.7	14.3	12.4	11.0
Grades 9 to 13	44.1	43.7	42.5	42.6	40.4	37.7
Some post secondary education	24.1	27.6	30.2	31.7	33.9	36.6
University degree	6.4	8.0	9.6	11.4	13.3	14.8

Source: © Census of Canada, Statistics Canada. Population over the age of 15; "some secondary" and "graduated from high school" categories combined; "Some post secondary" and "Post secondary certificate or diploma" categories combined.

TABLE 6.2 Tables are effective for communicating complex comparisons in a relatively compact space, and for making statistical information easier to grasp.

University Graduates by Field of Study, 1997					
Field of Study	Bachelor's/ First Prof. Degree	Diploma and Certificate	Men	Women	Total Undergrad
Education	20 319	2 937	6 670	16 586	23 256
Fine/applied arts	4 047	514	1 458	3 103	4 561
Humanities	14 869	2 911	6 431	11 349	17 780
Social sciences	47 054	8 684	22 744	32 994	55 738
Agricultural/ biological sciences	9 538	586	3 951	6 173	10 124
Engineering/ applied sciences	9 030	722	7 691	2 061	9 752
Health professions	8 701	1 752	2 529	7 924	10 453
Mathematics/ physical sciences	6 992	559	5 124	2 427	7 551
Other	3 474	1 836	1 663	3 647	5 310
TOTAL	**124 024**	**20 501**	**58 261**	**86 264**	**144 525**

Source: © Statistics Canada

Graphs

Graphs provide a visual means of displaying relationships between variables. Line graphs plot one variable as a function of the other, with the vertical axis representing one of the variables and the horizontal axis representing the other. Since line graphs are frequently used to display successive change or growth over time, a convention has developed for orienting them. The notches along the horizontal axis (bottom) of the graph typically represent time periods (days, weeks, or months), while the notches indicated on the vertical axis (up the left side of the graph) normally represent growth units (pounds, number of items, or profits). Growth is represented by a line that slopes either upward (for an increase) or downward (for a decrease) inside the graph.

For example, you could use a graph to track weight gain or loss, showing the time interval along the bottom (horizontal axis) and the weights along the left side (vertical axis); you might use it to track temperature change in a wet-lab chemistry experiment, showing the time interval along the horizontal axis and the temperature range along the vertical axis; you might use it to compare sales figures for a product, indicating time periods along the horizontal axis and the number of units sold along the vertical axis. In each of these examples, the time intervals recorded along the horizontal axis of the graph would differ significantly: for weight gain or loss, relevant intervals might be measured in days or weeks; for the chemistry experiment, meaningful intervals would be minutes or even seconds; for sales figures, significant intervals might be measured in months, quarters, or even years.

Graphs may also show a comparison of two or even three growth lines, but any more than three or four is confusing. For instance, when you keep track of your own weight loss or gain, if you also record a friend's progress on the same graph, you are using it comparatively.

Like line drawings and diagrams, graphs can be produced on a computer. In fact, for those without training in drafting or commercial art, computer-generated graphs are often preferable to hand-drawn ones, since it is usually easier to achieve a professional-looking result on a computer than by hand. Some programs allow for elaborate shading, colouring, and three-dimensional effects, all of which can enhance the appearance and clarity of your report, provided they can be clearly reproduced using the method you intend to use for copying the report.

However, though fancy graphs and diagrams are attractive and can be fun to create by computer, it's easy to get carried away with visual effects and lose sight of their primary purpose. It is tempting at times to include fancy visuals just because they look attractive rather than considering whether they really serve to communicate information clearly. Keep in mind that a graph is meant to provide an easy way for the reader to visualize significant information by providing a quick overview. If you decide to create and include graphs in your report, consider carefully whether they will do the job you want them to do. Some graphs, although attractive to the eye, can actually complicate rather than clarify information that the reader needs. Graphs that force the reader to puzzle through complicated information ultimately detract from the impact of the report.

Don't try to make a graph do too much. If you use a graph to show comparisons in weight loss or gain, drug dosages, sales figures, absenteeism, or costs of training, for example, you could use a different colour to represent each of the items compared (absenteeism in four different departments, costs of two different training programs, weight loss for four individuals). Although coloured lines can be used effectively to differentiate tracks on a single comparative graph, you should avoid using colour unless your report is being printed and copied in colour, or unless you are making only a few copies and can draw all the lines by hand after the report is

FIGURE 6.1 A graph enables the reader to compare easily the relative success of the three house models.

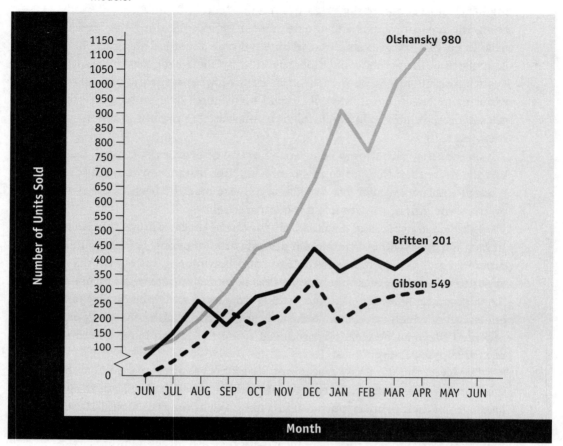

copied and assembled. If your report is to be photocopied on a black and white copier, show comparisons by using lines of varying thicknesses or a combination of broken, dotted, and solid lines. The graph in Figure 6.1 compares the sales of three home models by the same builder, while the one shown in Figure 6.2 tracks weight gain and loss among members of the hockey team.

Charts

Like graphs, charts are used to communicate comparative information quickly and clearly. They simplify complex information into an immediately understandable visual form. Charts come in different types, the most common of which are bar charts, pie charts, and flow charts. Bar charts are used to compare a single significant aspect of two or more items; each bar on the chart represents one of the items being compared. The length of

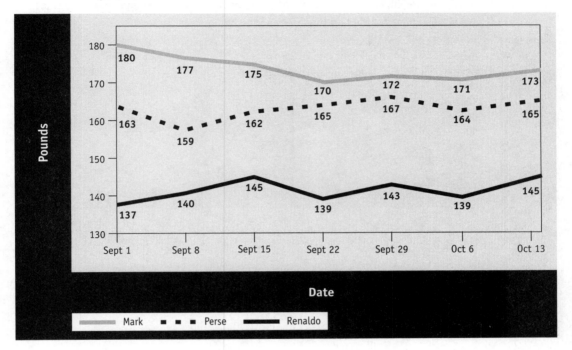

the bars may be easily compared to give the reader a quick impression of the difference among items. Bar charts may be drawn vertically or horizontally. The first vertical bar chart (Figure 6.3) is a visual representation of data from the second column in Table 6.2, showing undergraduate degrees, by field, granted in Canada in 1997 (the most recent year for which data are available). The second bar chart (Figure 6.4) compares total sales, during the first three months of the year, of the three home models shown in the graph in Figure 6.1.

Pie charts are used to show percentages or parts of a whole: how a budget is spent, the percentage of employees who have college diplomas, the breakdown of total business expenditures, or the portion of total sales made up by sales of one item. This kind of chart is best suited to emphasizing proportional relationships, as is suggested by Figure 6.5. The pie chart shown presents data from Table 6.2, showing undergraduate degrees awarded in each category as percentages of the total number of degrees granted in Canada during 1997 (the most recent year for which data are available). Notice how the labelling starts at the top of the pie and follows the segments in order clockwise—this helps the reader grasp the information quickly.

FIGURE 6.3 A bar chart can be used to highlight dramatic comparisons in a highly visual way. Because it simplifies the information it presents, a bar chart is best for illustrating large differences in scale. This chart compares numbers of undergraduate degrees granted across disciplines, based on the information contained in Table 6.2.

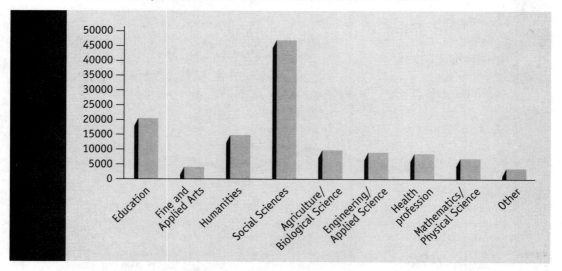

FIGURE 6.4 A bar chart can be used to compare sales of three house models over a particular time period.

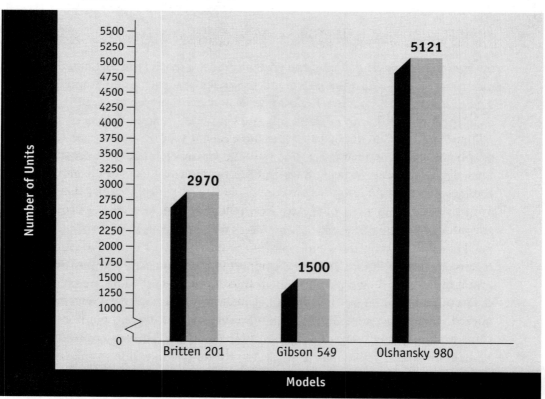

FIGURE 6.5 A pie chart shows the relationships of the parts to the whole. The reader can see at a glance the percentage of total degrees granted in each area or field.

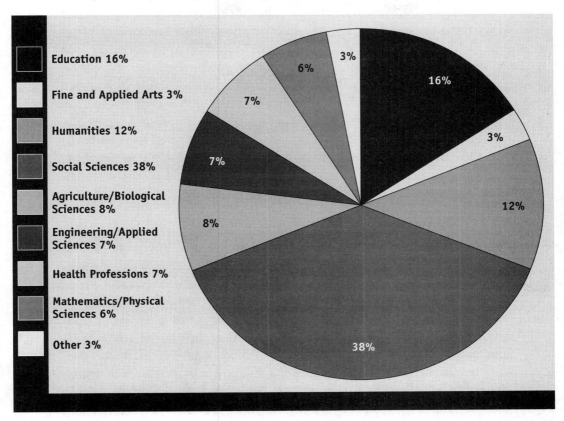

Education 16%

Fine and Applied Arts 3%

Humanities 12%

Social Sciences 38%

Agriculture/Biological Sciences 8%

Engineering/Applied Sciences 7%

Health Professions 7%

Mathematics/Physical Sciences 6%

Other 3%

Photographs

Whenever you must describe a site, a scene, or a product, a good photograph can aid your reader's understanding. Colour photocopying, scanners, digital cameras, and colour laser printers can make it easy and relatively inexpensive to include both colour and black-and-white photographs in your day-to-day reports, provided the results are clear enough to display the information you require.

However, obtaining a good photograph is harder than it sounds, since it must be sharp, with no fuzziness or unnecessary clutter in its composition, and it must be focussed tightly enough to communicate the information it is intended to communicate. Finally, it must reproduce clearly when it is copied. Remember that only the sharpest and clearest photos will reproduce well; others will be rendered muddy or unclear. Make sure that the photograph you select is detailed enough to show the information you intend it to display. If the shot is cluttered or indistinct, or taken from too far away, you should choose a different method of illustration.

Always try making a copy of the photo using the method you will be using to reproduce the report. When they are reproduced, even clear photographs may become too blurry for use, losing the very detail that you want them to communicate. As well, you should be aware that ordinary black and white photocopying will not produce sufficiently clear prints even of black and white photos. If the photograph you have chosen will not reproduce clearly enough using the printing options available to you, you might consider having the exact number of copies of the actual photograph made and pasting them into position in each copy of the finished report. You must decide for yourself how many reports you are prepared to assemble by hand like this. An advantage of this hand assembly is that it may allow you to use colour photographs of high quality in a report that is otherwise simply photocopied.

Line Drawings

When the information a reader needs from the visual is likely to be unclear in a photograph, or if a photograph would be unsuitable, a line drawing may be a better option. Line drawings, or technical drawings, can be used to illustrate any item, in any scale, from any angle. They can be as simple or as complex as required. For this reason, they are extremely valuable for communicating complex visual detail. A line drawing may be used, for example, to illustrate items in a parts list, to show detail in a particular construction, to demonstrate the layout of a building or work site, or to display the package design for a new product. Such drawings are often used in technical manuals or reports to illustrate parts, products, or design details; they may also be used in assembly instructions for furniture or equipment, in film-making to outline the sequence of scenes and action, and in architecture to show the proposed appearance of a finished building.

Line drawings, or technical drawings, are used when photographic illustration would be too expensive or would be insufficient to communicate the kind of information you require. You need not be a professional artist or technical draftsperson to do a simple line drawing, especially with the help of computer-aided design packages. Whether produced on your printer or by hand, your work must be neat and easy to read. It should be drawn in black ink and clearly labelled, and should be as uncluttered as possible. Figure 6.4 illustrates the parts and lock-washer assembly for attaching eyes to children's toys; Figure 6.5 shows one stage in an assembly diagram for a desk-bookshelf combination; and Figure 6.6 shows a view of a staircase and windowed wall in a proposed house addition.

Diagrams

Unlike line drawings, which show realistic detail, diagrams are usually simplified illustrations designed to highlight one or two aspects of a site, object, or process. Diagrams are particularly useful if your report explains how

FIGURE 6.6   A line drawing can be used to illustrate detail in situations where photographs may be insufficient to capture certain types of detail, as in the assembly of these child-safe plastic lock-in eyes for children's toys.

Lock-in Eyes for Children's Toys

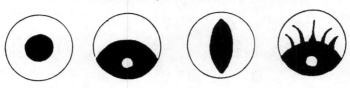

Some styles of lock-in eyes

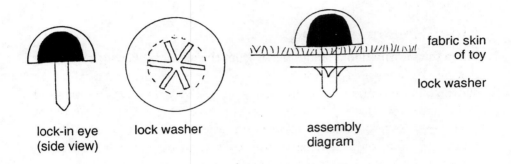

lock-in eye
(side view)

lock washer

assembly
diagram

fabric skin
of toy

lock washer

FIGURE 6.7   A line drawing can clearly show steps in a construction process that may be difficult to capture in a photograph.

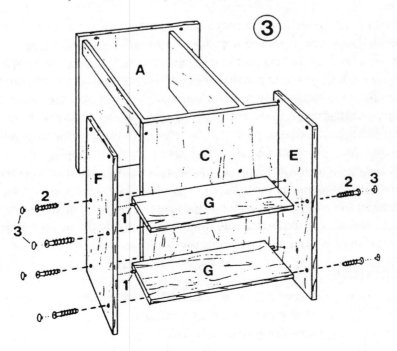

FIGURE 6.8   A drawing can also help us visualize what doesn't yet exist, as in this architect's rendering of one
view in a recreation room addition.

something works or is assembled. They are not meant to show what the object in question looks like in any realistic way; instead, they are a simplification of some particular feature of the item or process that is being illustrated. For example, a diagram may show the floor plan of the company's proposed new office space; it may lay out the circuitry in an electrical system; or it may illustrate the operation of a nuclear reactor or a distilling process.

Diagrams, like line drawings, should be clearly drawn in plain black ink. Though colour may be used effectively, you should avoid it if your report will be photocopied on a black and white copier. Shading, stippling, and cross-hatching can be used instead. Be sure to clearly label all elements of the diagram so that the reader can apprehend quickly the information that you want to communicate. The diagram in Figure 6.9 shows the floor plan for the house addition illustrated in Figure 6.8. Figure 6.10, a circuit diagram, demonstrates how a diagram can be used to present intricate detail.

FIGURE 6.9 This diagram of the floor plan for a house addition helps the householder visualize the architect's plans for the family room–sunroom addition.

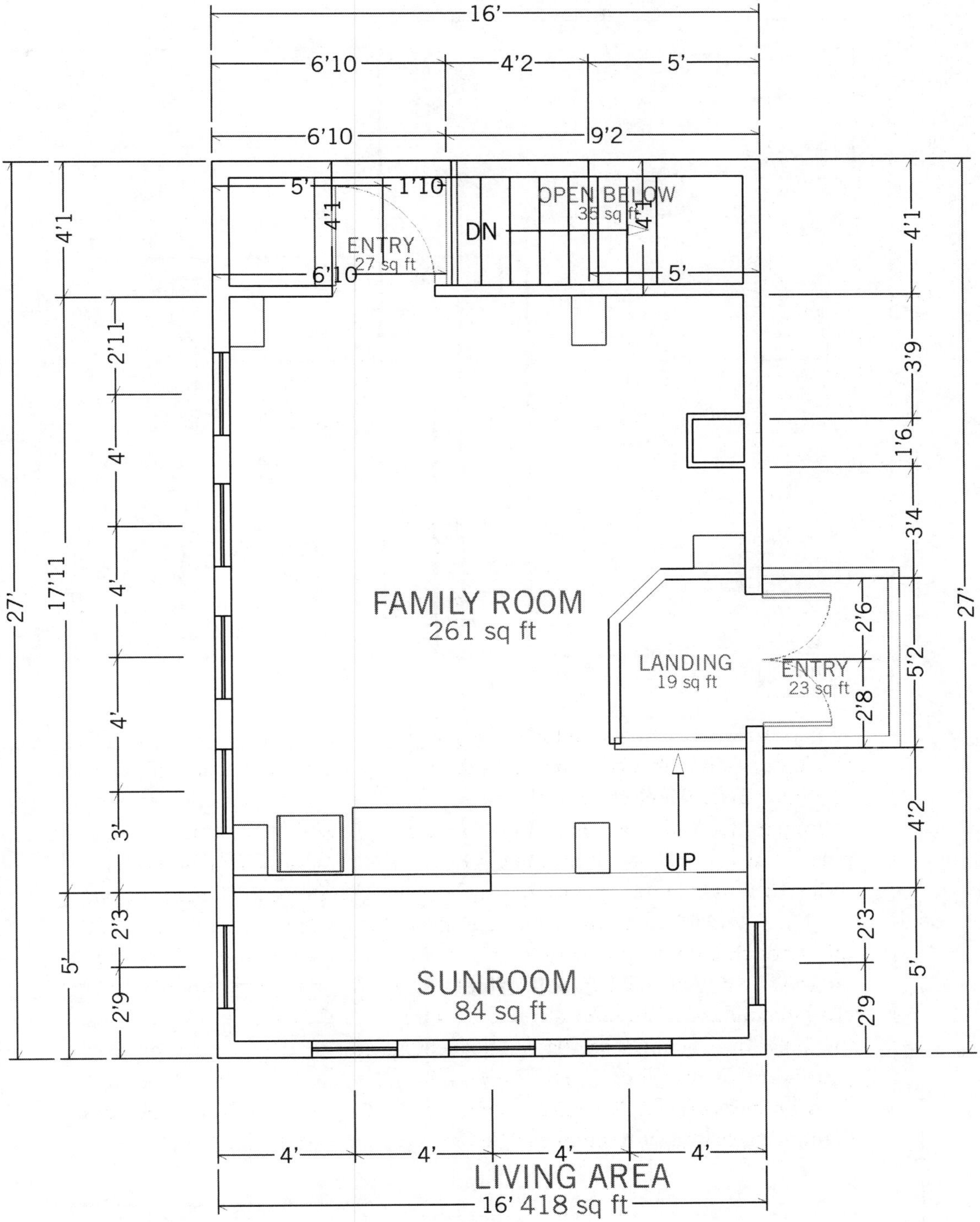

FIGURE 6.10 Reading a wiring diagram requires special expertise, but its symbols are clear to those who are trained to understand it. A simplified version of a highly complex circuit diagram would likely appear in the body of the report, while this complex version might be placed in an appendix.

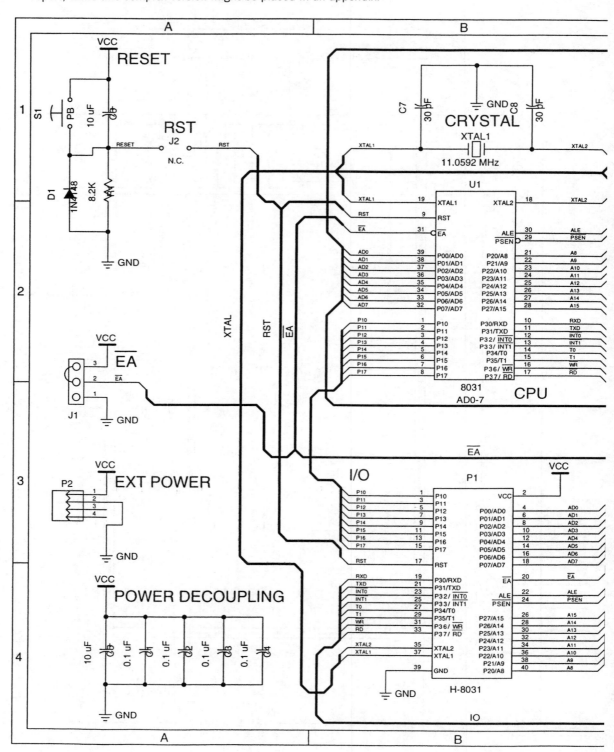

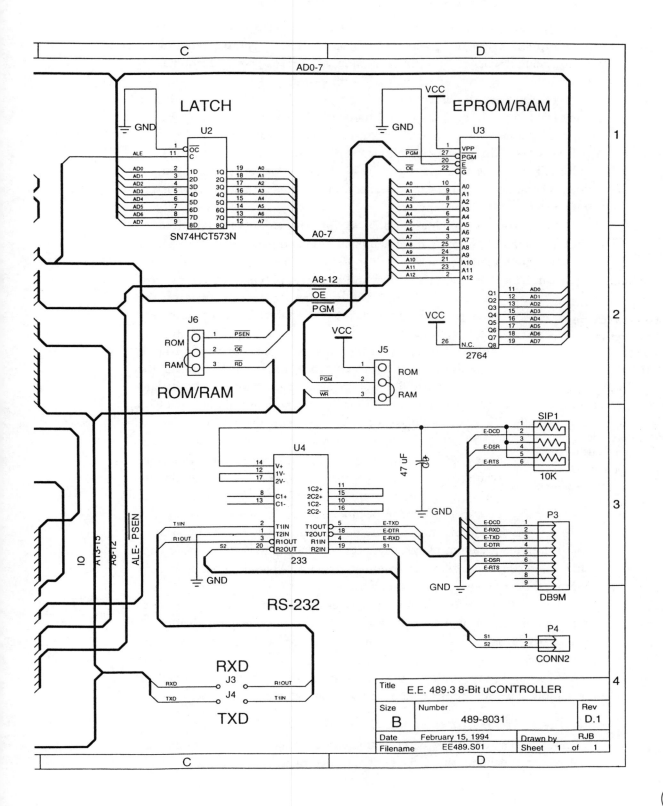

WRITING TECHNICAL DESCRIPTIONS

Description, or making pictures with words, is part of all good writing. Description is designed to create an image in the mind of the reader of the thing being discussed. In technical writing, description is often accompanied by labelled diagrams, line drawings, or photographs.

The purpose of description is to clarify for the reader the details of a physical layout, the size, shape, and function of a device or object, or the dynamics of a situation. Among other uses, description helps manufacturers and retailers to sell their products; contractors or businesses to secure a bank loan; filmmakers to get financial backing; medical personnel to monitor a patient's condition; or engineers, designers, and architects to outline the nature of a project under development.

Apart from its usefulness as a device for clarification of detail, description is also an important part of demonstration and persuasion. If you hope to motivate a reader to act on a proposal or recommendation, a description can help that person visualize and understand what you are proposing or asking for. A person who can readily visualize the outcome will be more likely to act on the recommendation. For instance, an architect or interior designer will want to help the client to envision the finished room, building, or landscape design in order to convince that person to engage the firm. A detailed description, accompanied by clear floor plans and different views of the finished room or building, will help the customer understand what is being proposed.

Finally, in cases where a reader must be able to carry out a task based on your instructions, description is an important strategy for enabling your reader to perform the action you are requesting or recommending. Many a convincing proposal fails in its purpose because the reader is not made clearly aware of what is to be done next or how it may be accomplished. In many circumstances, a description can help to make clear the steps that the reader must follow in order to carry out its recommendations, rendering the report or proposal much more effective and convincing.

Description provides the answer to questions about the nature and function of a device or object, or the general structure of a geographical or architectural layout. It answers several specific questions and follows a systematic organization. A description should communicate clearly to the reader the character and purpose of the item, its appearance, and (if appropriate) its assembly. The questions that a description should answer include:

What is it? What is it used for? What does it do?

Before you can effectively describe the item in detail, you must establish for the reader its general nature: the item to be described is a computer program,

Questions Answered by an Effective Technical Description

- What is it? What is it used for? What does it do?
- What does it look like? What is it made of?
- How does it work?
- How was it put together? How is it to be assembled?

an accounting system, an industrial sewing machine, a metal lathe, a can opener, an adhesive, a paint, an item of protective clothing, an anti-theft device for a vehicle, a high-speed cutting tool; or the layout shown is for a kitchen renovation, a medical office, a classroom, an artist's studio, a wood-working shop, a business office. Without this general orientation of under-standing, it will be impossible for the reader to figure out what is being presented.

All descriptions begin by letting the reader know exactly what is about to be described. Often, this task is performed by identifying the use to which the object or device or layout will be put, by emphasizing its function (a fastening device, a floor plan, or a solar cooker) or showing its uses (screws are used to firmly fasten objects to wood surfaces; a polymer stamp plate maker allows small-scale manufacture of rubber stamps; a personal organ-izer is a small computer that stores appointment, contact, and scheduling in-formation).

What does it look like? What is it made of?

Normally, a description provides information about the dimensions and fabrication of the object. For example, a real estate listing normally de-scribes the property in terms of its square footage before moving on to pro-vide other details about number of rooms, special features, location, or price. If the description is of an object or device, the user or reader will need to know precisely how large the item is, or its operational capacity, or both, in order to make a determination about its suitability for purchase. For instance, a coffee maker that is 15" high by 10" wide by 17" deep may fit the available counter space in my staff room, but if it makes only five cups of cof-fee at a time instead of the twelve I require, it does not meet my needs.

The client or customer will also need to know the number of parts, the material of which the device is constructed, and, if it is relevant, the design and colour of the object. The specific details of appearance and construction that are required in a description will depend upon the use to which the object is to be put. For example, a plastic housing may be preferable to cast iron for a portable household sewing machine, but in an industrial ma-chine, where durability is more of an issue, the cast iron housing may be preferable. The purpose for which the description is written—as instruc-tions for assembly, perhaps, or as a sales document—will also influence the kind of information it includes.

How does it work?

Depending on what the object is, and the purpose of the description, the reader may need to know whether the item is electronic or manual, whether it is digital or analogue, whether it runs on electricity or compressed air, whether it can be plugged in or must be supplied with batteries, whether it has a timer, a delay setting, auto-defrost, a water-saver function, a remote starter, a safety-lock operating switch, an instant-on feature, or an automatic power-surge protector. To operate the device effectively, the client will need to know exactly how it is used, the sequence of steps involved in operating it, how to troubleshoot problems, and how to refine its use to achieve more sophisticated effects. If the operation of the device is simple (for instance, a set of locking pliers or a vegetable peeler), the description may be brief. However, if the device is complex (for instance, a polymer stamp plate maker or a computerized engraving machine), the description may take up a substantial portion of a user's manual.

How was it put together? or How is it to be assembled?

Like the other elements of description, the answers to these questions depend on the nature of the device and the purpose of the description. If the description is part of a set of instructions for assembling an item of furniture, a bicycle, an appliance, or an instrument, for example, the description would certainly involve step-by-step set-up instructions. Similarly, if the description is part of the training of an operator who does not have to assemble the machine but who needs to know exactly how it functions, it should provide an analysis of the components of the machine, instructions on its disassembly for cleaning or maintenance, advice on routine replacement of worn parts (a furnace filter, guitar strings, printer cartridges, bandsaw blades) and a description of the function of each of its parts.

Like the incident and accident reports discussed in Chapter Five, an effective professional description should be as objective and impartial as possible. It should filter out personal impressions, judgements, and reactions, focussing instead on visible (or audible) details that will be evident to other observers and that will assist readers in visually reconstructing the object or scene.

Specifications

Many devices and structures we use every day must comply with particularly precise and exacting descriptions known as specifications, which set standards for safety, quality, and performance. Specifications provide requirements for the design, construction, use, and maintenance of a particular product. In construction and manufacturing industries, specifications may be set and controlled by professional standards and by law. Because those who use the product could suffer injury or death if it should fail, manufacturers or contractors who "cut corners" on meeting specification requirements are liable under the law if the equipment, device, or structure should malfunction.

Specifications for most products, from children's toys to lawn furniture, from electrical appliances to sports equipment, from office buildings to aircraft, from meat products to cereals, provide standards governing the following aspects of manufacture and installation:

- the methods used to build, install, or manufacture the product;
- the materials and equipment that may be used;
- the conditions under which the product may be manufactured;
- the size, weight, strength, and shape of the finished product.

Specifications set out in detail what is to be done and how it is to be done. Because specifications will be interpreted, assessed, and applied by diverse audiences, these detailed descriptions must be written clearly and unambiguously enough to ensure identical understanding and interpretation by a wide range of possible readers, who may include:

- the client or end user, who has a vision of what the product can do;
- the designer or architect, who develops the product within the limits, or constraints, of the specifications;
- the contractor, builder, or manufacturer whose company produces the product;
- the supplier of parts to the manufacturer, whose products must meet the particular requirements of the device or building in which they will be used (for example, building materials that are considered adequate and safe for residential use may not stand up to the heavier requirements of industrial or commercial buildings);
- the labourers who work on the actual construction or assembly;
- the inspectors who approve the final product;
- the advertisers who market the product to the public.

Each of these participants in the process of product development, manufacture, and marketing is required to adhere to the standards laid out in the specifications. The safety and welfare of the general public, and the solvency and success of the company, rely on these standards, which in many instances are set and protected by law.

The Elements of Effective Description

In a professional or technical document, description is meant to clarify and convince. It should reasonably reflect details observable to anyone who examines the site, product, object, or patient. A technical description may stand alone as an informal report, with its own introduction, discussion, and conclusion; it may be incorporated into the discussion of a longer formal report; or it may be attached as an appendix to a long report. The components of an effective description are fairly consistent, and include the following:

A clear and specific title. You should provide a title that will immediately communicate to your reader what you are describing. A title that mystifies unnecessarily, or that does not specify the item described, will render your description confusing and ineffective.

**Elements of
Effective
Description**

- Clear and specific title
- Overall appearance and component parts
- Illustrations or visuals
- Functions of all parts
- Appropriate details
- Summary and operating description

A description of overall appearance and component parts. Most descriptions employ a general-to-specific pattern, beginning with an overall depiction of the thing to be described. The reader needs to have a general sense of what the object or product is and how it functions, or of the size and layout of the site or building, before turning to the specific details of the description. Present your audience with a general overview before you get to the details.

Iillustrations or visuals. As you are already aware, descriptions of products, objects, or layouts are made much clearer if they are accompanied by illustrations. These diagrams, drawings, or photographs can be labelled to show the location of each significant component or part. Illustrations should be clear and easy to understand; they should correspond to the object in a way that is obvious and recognizable to the reader.

An explanation of the function of each part. The description should appropriately explain the purpose, dial settings, and operation of each part, clarifying what each does and showing how it relates to the whole. For instance, a polymer stamp plate maker uses a process similar to the development of black-and-white photographs. The ultraviolet lights inside the unit are used to expose a photo-sensitive liquid acrylic material through a photographic negative. The exposed photo-sensitive material hardens, forming a stamp that can be used for hand printing on paper and fabrics.

Appropriate details as required. Technical descriptions are as varied in detail as the objects they describe, so no two will include identical details. Your description should present the details that the reader will need in order to understand what he or she needs to know. As a user, I do not need to know how ultraviolet bulbs are manufactured for my polymer plate making machine; I don't even need to know what ultraviolet light is. However, I do need to know their function, position, and operation details with respect to the machine I own, and I need to know how to remove them and install replacements when they have burned out.

The details given in your description should be presented in the clearest sequence or organizational pattern. Though there are several organizational patterns to choose from, only three of these are generally useful for organizing a description: a spatial pattern, a functional sequence, or a chronological order. The first might be best suited to a description of a site or room layout; the second to a complex machine or device with several operations; and the third for describing steps that must be performed in sequence, as in the assembly of a device or object. A description of a complex mechanism might employ a combination of these. You can find more on organizational patterns in Chapter Three.

A summary and operating description. Your description should conclude by explaining, where appropriate, how the parts work together to make the whole item function. The summary and operating description correspond to the conclusion of a short report.

PROPOSALS

As you may recall from Chapter Five, the proposal is a persuasive report that is intended specifically to persuade the reader to follow a course of action recommended by the writer. A proposal is particularly challenging to write, because you must not only provide all necessary information, but also influence your reader to approve the project or implement the plan you are putting forth. In writing proposals, as in all persuasive writing, you will therefore need to develop skill in identifying and accommodating the reader's needs, expectations, concerns, and interests.

Proposals may be directed to readers within the company or to agencies or readers outside the company. They are written in a variety of situations, and may take the form of tenders, grant applications, project proposals, or marketing summaries. These may be initiated by you or invited by someone else, or they may respond to a general call for submissions.

For example, in preparing to write a book such as this, I must draft a project proposal to the publishing house. I may do this on my own, as a result of an original idea I have for a book, or I may be approached by a representative of the publisher who specifically invites me to consider writing a book on this topic. Finally, I may even respond to a general call for submissions. In each case, my submission follows a set of general guidelines laid out by the company or organization to whom I am submitting the proposal, with variations depending on whether I have initiated the process or the publisher has done so. You can find the proposal for this book, in informal format, in Chapter Five.

The main parts of a proposal—summary, introduction, discussion, conclusion, recommendations, and appendices—are the same as for any other report, and as in other reports, the headings in the discussion section reflect the specific subject matter of the proposal. Because a proposal asks the reader to act on what is recommended, to support the idea with approval, a contract, or funding, you must be sure to demonstrate the value of what is proposed and show how it fulfills the reader's interests and needs. You must establish your credibility by displaying an understanding of the challenges of the situation and the details of implementation, including costs and obstacles.

Above all, as the writer of a proposal, you must demonstrate that you really know what you're talking about and have done your homework thoroughly. You may believe that your proposal is sound, but unless you can convince your reader that you have thought the project through and have anticipated any problems, you will not gain that person's confidence, approval, or funding. Don't expect the reader to act on faith if your presentation is incomplete or unclear. Do your research.

Here are some elements to consider in the writing of a proposal:

- Following the guidelines for submission obtained from the granting agency, publisher, or company, clearly identify what is being proposed.
- Demonstrate the need for the action you are proposing, particularly if the proposal originated with you. If you are responding to a call for tenders or submissions, indicate as much in your proposal.
- Communicate the advantages of the proposed action or change. If it will solve an existing difficulty; if it will save the reader time, work, or money; or if it will increase sales, efficiency, or profits, show how.
- If your proposal responds to an invitation for submissions, be sure to show how it is the best means of reaching your reader's desired objectives. Because the reader has requested a proposal designed to meet specific needs, you must make sure that the details of your proposal match those requirements.
- Make it easy for the reader to implement the proposal. Give all pertinent details of the situation and point out any existing resources that can be put to use. Try to anticipate any questions your reader will have and answer them in advance.
- Present information positively, even when outlining disadvantages. Naturally you feel that the advantages of your proposal outweigh any disadvantages it may have. Your task is to make your reader share this view.
- Indicate the steps to be taken to bring about the proposal. Remember that to help your reader accept your project, you must show how it can be done.
- For tenders and grant applications, and for some project proposals, including a projected budget is mandatory. Be sure to think through all aspects of the project, and provide a budget that indicates as accurately as possible what implementation will cost. Remember as you draft your budget that people may be more willing to commit money, time, or effort after they have been convinced of the worth and viability of the project.

At times, you will be invited to submit a proposal for implementation of a project someone else has suggested. As with original proposals, you must still present all the information the reader will need to evaluate your suggestions, but in this case the proposal will differ slightly, since you will be responding to requirements that the reader has outlined for you.

Remember that a proposal, whether originated by you or requested by someone else, must be especially persuasive to convince a reader to implement the very good idea you've presented.

SAMPLE FORMAL REPORT

A formal report may be used for any number of purposes. Annual reports, research reports, progress reports, evaluation reports, proposals, and feasibility studies are some types. The following example shows a project status report, but the format is similar to that of any other type of formal report.

FIGURE 6.11A   The letter of transmittal. Note that this cover letter normally precedes the title page of the report.

Dolovich and Cowan Technical Services
Eastern Office: 403 Frederick Street, Halifax NS B3A 4D4 (902) 987 6543

February 14, 2003

Richard Hynes, Chair
Sherman Bay Watershed Cleanup Consortium
PO BOX 959
Sherman Bay, NS

Dear Mr. Hynes:

We have attached the project status report for the Murdoch Creek Watershed Cleanup Project.

The report details progress made so far in the jointly-funded hazardous waste cleanup and decontamination project. It sets out a description of the challenges inherent in such a project, outlines the strategies to be undertaken, and describes the tasks so far completed; it also provides an overview of budget expenditures to date.

In our recent meetings, you also asked that we consider the extent to which costs may exceed the current budget; we have attempted to estimate the potential overrun, but because the exact nature of the contaminants is still unknown, such an estimate is necessarily sketchy.

However, as this report indicates, there is little likelihood that the contamination can be contained or removed within the budget restrictions we currently face; when this project has come to a close, it is likely that the "Scum Dump" problem will remain a serious environmental and health issue for years to come.

We will be present at the next meeting of the Consortium Board and will be prepared to answer questions on this report or any other aspect of the Watershed Cleanup Project.

Sincerely,

Bruce James, Project Manager

Jennifer Neilands, Consulting Engineer

Murdoch Creek Watershed Cleanup Project

Status Report

Prepared for the Sherman Bay Watershed Cleanup Consortium
by
Bruce James, Project Manager
Jennifer Neilands, Consulting Engineer

Dolovich and Cowan Technical Services
Eastern Office
403 Frederick Street
Halifax NS B3A 4D4
(902) 987 6543

Summary

Sherman Bay, Nova Scotia's infamous "Scum Dump" area is Canada's worst hazardous waste site, containing over 700,000 tonnes of toxic sludge. Toxicity tests conducted in the nearby neighbourhood showed levels of at least a dozen toxic chemicals well above those permitted by the CCME (Canadian Council of Ministers of the Environment).

In May 2002, federal, provincial, and municipal governments announced a $62 million cost-share agreement to fund activities recommended by a Watershed Cleanup Consortium formed of members from the surrounding communities, government, technical specialists, and volunteers. The present report presents a summary of activities and achievements to date.

i

FIGURE 6.11D   The Table of Contents for the formal report.

Murdoch Creek Watershed Cleanup Project

TABLE OF CONTENTS

ii

FIGURE 6.11E The Introduction is the first page of the report proper.

INTRODUCTION

Sherman Bay, Nova Scotia's infamous "Scum Dump" area is Canada's worst hazardous waste site, containing over 700,000 tonnes of toxic sludge—over 20 times the amount of toxic waste in New York's infamous Love Canal. Among these are polychlorinated biphenyls (PCBs), arsenic, lead, and benzene, all at levels well above those permitted by the CCME (Canadian Council of Ministers of the Environment).

The entire area is bordered by residential and commercial sites. A large number of health problems, including elevated rates of cancer, asthma, and other life-threatening conditions, combined with the extent and nature of the contamination, adds to the urgency and challenge of cleanup at this site.

The process of cleanup is underway. This progress report details strategies that have been implemented to date and outlines work to be completed. It also recommends further study of and action to resolve this complex and dangerous contamination.

2

BACKGROUND: SITE DETAILS

For nearly a century, Sherman Bay, Nova Scotia, was the site of a large steel-making and coking operation situated near Murdoch Creek in what has become the centre of the city. The plants have been closed for nearly a decade, but the site they occupied, and its nearby "Scum Dump," today constitute Canada's worst hazardous waste site.

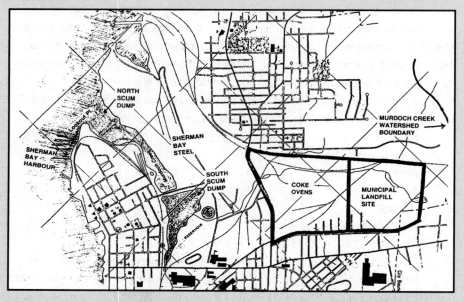

Figure 1: Map of the Sherman Bay "Scum Dump" and surrounding area

The area known to local residents as the "Scum Dump" is, in fact, an estuary opening into Sherman Bay Harbour at the mouth of Murdoch Creek. Although at one time it used to accommodate large sailing ships, today this estuary is narrow and shallow, filled with at least 700,000 accumulated tonnes of highly toxic by-products of coking and steel-making accumulated over 80 years of dumping from the nearby plants. The estuary daily receives additional tonnes of untreated sewage from 30 outlets that drain into Murdoch Creek. An unlined municipal dump sits at the head of Murdoch Creek and also leaches into it. In all, the 200-hectare site is estimated to contain over 20 times the amount of toxic waste of New York's infamous Love Canal.

FIGURE 6.11F (Continued)

3

It is known that at least forty-thousand tonnes of this waste is polychlorinated biphenyls (PCBs). It is also known that the coke plant's benzene tank leaked for years, saturating the surrounding ground. A remaining tank, open to the air, contains an unknown mix of these and other chemicals, and piles of coal, coke, and sulphur also remain at the site. Beneath the old coke plant site, 160 kilometres of underground pipes carry more deadly chemicals. The soil around the area regularly erupts in flames which cannot be extinguished, and authorities fear that attempts to dismantle the pipes may cause an explosion.

Toxicity tests conducted on the site and surrounding area showed high levels of arsenic, molybdenum, benzopyrene, antimony, naphthalene, lead, toluene, tar, benzene, kerosene, copper, and polyaromatic hydrocarbons (PAHs). All are at levels well above those permitted by the CCME (Canadian Council of Ministers of the Environment). For example, the concentration of arsenic has been shown to be 18.5 times as high as acceptable levels under federal/provincial guidelines; naphthalene levels are 8.9 times as high, and molybdenum and benzopyrene levels are six times the recommended limit.

Above the coal plant is a century-old dump, originally used for slag and other hazardous wastes from the steel mill, which continues to serve as a clandestine junkyard; every week unrecorded industrial discards are dropped amid the slag. An aging incinerator on the site emits mercury, and a brook that flows through it runs a bright rust shade. When it rains, puddles turn fluorescent green.

The entire area is bordered by residential and commercial development containing homes, ball fields, playgrounds, schools, supermarkets, and even restaurants. Recent (one year ago) tests have shown that the surrounding land, the groundwater, and a brook where children play have all become contaminated with a variety of toxins known to cause various cancers, birth defects, heart disease, kidney disease, brain damage, immune deficiencies and skin rashes.

FIGURE 6.11F (Continued)

4

CONTAMINANT TEST RESULTS

Over 1000 samples of soil, surface water, and groundwater have been collected at the "Scum Dump" and coke plant sites to determine the nature and extent of contamination. Samples were compared with provincial and Canadian Council of Ministers of the Environment (CCME) guidelines to determine if further study is needed.

Preliminary findings of soil and ground water analyses reveal:
- PAH levels in some surface soil samples above CCME guidelines;
- metals, including arsenic and lead, at levels above CCME guidelines;
- impact on groundwater, primarily in samples taken near the landfill site;
- discolouration and levels of iron and manganese in groundwater exceeding CCME guidelines.

These latter features—discolouration and high metal levels—are not considered a health issue because they are common to the area.

HEALTH IMPLICATIONS

Recent studies have confirmed the presence of polycyclic aromatic hydrocarbons (PAHs) at concentrations of up to 8000 ppm; concentrations of PCBs range from 50 to 1000 ppm. Exposure to both chemicals, and to the heavy metal contaminants at the site, has been linked in lab animals to cancer, liver damage, birth defects, heart disease, kidney and skin diseases, brain damage, immune deficiencies, and reproductive malfunctions.

Cancer rates in the area have never been adequately documented, but some estimates place them at almost double the national average. A study of the medical history of 117 deceased long-term steel plant employees found that an alarming 63 per cent were treated for some form of cancer prior to their death. Current residents complain of "sore throats, dry eyes, nausea, headaches and a long list" of other ailments (Shawn, 2000), including "watering eyes, scratchy throat, and shortness of breath" (Toxins cloud history, 2000) and "asthma, ear infections, eye infections... bronchitis." (Unknown dangers, 2001)

FIGURE 6.11F (Continued)

5

STRATEGIES IMPLEMENTED TO DATE

1. Preparations for Site Demolition

Before extensive decontamination can be carried out, stacks, buildings, and other structures remaining at the site must be dismantled. As well, piles of coal, coke, and sulfur must be carefully removed so as to prevent further distribution of pollutants. During demolition and removal, "separation zones," which have been determined based on the nature of the contaminants at the site, will be used to establish safe distances for the protection of those who live and work nearby. Work methods will be carefully controlled so as to minimize the possibility of additional contamination. A new access road, designed to divert traffic from residential areas, is currently under construction, as is a pad for washing vehicles and equipment. When these accommodations are complete, demolition will begin. The first of the demolition activities will be to remove the two remaining stacks and derelict buildings.

2. New Fencing

To ease residents' concerns about children and pets wandering onto the demolition site, construction of 3.5 km of new fencing around the site was completed in January of this year. The entire site is now enclosed and the fencing will remain in place throughout the cleanup.

3. Air Monitoring Program

An expanded air-monitoring program has been implemented to measure both "real-time" and long-term air quality during site demolition and beyond. Hand-held monitors will allow an immediate response to changes in air quality during on-site work. A mobile unit will be transported to sites around the area as an extra safety measure.

4. Landfill Closure

The landfill, which sits at the top of a hill overlooking the main site, contains nearly a century's accumulation of industrial and household waste. The landfill has

FIGURE 6.11F (Continued)

6

been closed and sealed to public access in preparation for site containment procedures, the goal of which is to stop the flow of contaminants from the landfill into the coke plant site. Plans include capping the site and diverting the brook that runs through the area. Construction of a clay landfill cap is underway. The landfill containment is expected to be complete within a year of the date of this report. A leachate management plan—to keep contaminants from leaving the site—will also be implemented.

5. Diversionary Sewer Project Phase I: Design

Complicating the cleanup operation is the fact that at least 30 sewer outfalls feed into Murdoch Creek. To deal with this large volume of waste water, which includes storm and sanitation sewers, a diversionary sewer will be constructed to redirect wastewater to an area off nearby Lookout Point where a water treatment plant development is planned. Design plans for the diversionary sewer are complete, with construction set to begin immediately, weather permitting. Construction is expected to take eight to ten months.

As an added precaution, soil removed to make way for the underground pipes will be tested for contaminants prior to disposal. Contaminated soil will be treated according to environmental regulations.

6. Heritage Resources Impact Statement

The new diversionary sewer will be routed through a portion of Lookout Point thought to be one of Sherman Bay's first settlements. A Heritage Resources Impact Assessment revealed that a military hospital was formerly located in the area, possibly with an on-site cemetery. An archeological dig will be required to ensure that no burial ground will be disturbed during construction.

FIGURE 6.11F (Continued)

7

SUMMARY OF BUDGET AND EXPENSES TO DATE

The $62-million cost-share agreement for the Murdoch Creek Watershed Cleanup Project is managed by Dolovich and Cowan, with the assistance of the provincial government. Approximately $16.4 million has been spent to date (Figure 2). This has covered extensive fieldwork and design plans for the landfill closure, the construction of the fence, purchase of an emergency response vehicle, and plans for the diversionary sewer. A complete budget can be found in Appendix A.

Budget Allocations to Date	(in $millions)
Environmental Studies and Assessments	6.0
Identifying Cleanup Technology and Strategies	1.2
Emergency Response Equipment and Training	0.3
Site Separation (includes past relocations)	1.1
Site Security Fencing and On-Site Patrols	1.9
Health Studies	0.4
Landfill Closure	1.5
Regulatory Environmental Health Assessments	0.5
Secretariat and Activities	1.0
Project Management and Public Communication	2.5
Total expended to date	16.4
Total budget allocation	62.0

Figure 2: Budget allocations to Date

FIGURE 6.11F (Continued)

8

WORK REMAINING

1. Diversionary Sewer Construction

An archeological dig will be conducted in the Lookout Point area to ensure that no burial ground will be disturbed during construction; a further assessment will be designed to confirm that no munitions were buried in the area. Construction of the diversionary sewer will begin later this year when these assessments have been completed.

2. Demolition of Remaining Structures

The first of the demolition activities will be to remove the two stacks and the by-products building which still remain on the site. This work will begin when the road and washing pad construction is complete. These are expected to be complete within four months of the date of this report.

3. Decontamination Proposals

Total cleanup of the Sherman Bay "Scum Dump" is still far from complete. The extent of contamination at the site will require further study beyond that budgeted for in this project, in order that strategies for decontamination may be identified and carefully assessed. Further environmental and health impact studies must be carried out, and appropriate technology for dealing with extensive soil and water contamination at the site will have to be identified and implemented.

FIGURE 6.11G The conclusion of the report.

9

CONCLUSION

The Murdoch Creek Watershed Cleanup remains on schedule, and several strategies identified in the original proposal have been implemented. Demolition, site closure, containment, and construction of the diversionary sewer can continue to move ahead as soon as preliminary adjustments to the site are carried out. Removal of structures and surface contaminants will follow. However, total cleanup of the Sherman Bay "Scum Dump" is still far from a reality. It's too soon to determine which technology will be found to be most appropriate for decontamination procedures, or to foresee what further complications may emerge as the site structures are dismantled. It is estimated that cleanup may take ten or more years and cost up to $1-billion.

FIGURE 6.11H Recommendations follow the Conclusion.

10

RECOMMENDATIONS

The recommendations arising from this project status report are as follows:

1. that the Murdoch Creek Watershed Cleanup Project, under its current consortium and budget, move ahead as planned;
2. that further plans be made to commit an additional budget of at least $40,000 for implementation of appropriate and extensive decontamination proposals;
3. that pyroengineering specialists be brought into the consortium to deal with the explosive and flammable potential of unknown combustibles at the site;
4. that testing of various decontamination methods begin as soon as possible, so that they may be implemented when preliminary demolition and removal of waste materials has been completed;
5. that a strategic plan be developed for handling unknown chemicals in the tank on the site.

FIGURE 6.11l   References follow Recommendations.

11

REFERENCES

Cancer study left in federal hands. *Sherman Bay Record,* 19 November 1999.

Environment Canada. Government support to Murdoch Creek watershed initiatives. On line. http://www.ec.gc.ca/murdo.htm 12 October 2001.

Fact sheet on the infamous Sherman Bay "scum dump." Enviro Club Canada. Press Release, 24 July 2000.

Hare, Brenda. Coke plant cleanup group awaits word on next step. Enviro Club Canada. On line. http://www.murdo.htm 16 January 1999.

Losing patience on "scum dump." *Nova Scotia Chronicle,* 23 November 2000.

Shawn, Gwen. Sherman Bay "scum dump": It's time to clean up Canada's 'national shame.' *Environmental Hazards Magazine.* April/May 2000.

Toxins cloud history of Sherman Bay street. *The World Post,* 20 July 2000.

Unknown dangers lurk in toxic wasteland. *The Toronto Comet,* 20 July 2001.

FIGURE 6.11J Appendices follow References.

APPENDIX A: BUDGET

Murdoch Creek Watershed Cleanup Project Current Budget			
Item/Category of Expenditures	Allocated	Expended	Remaining
Environmental Studies and Assessments	8.2	6.0	2.2
Demolition and Coke Plant Cleanup Activities	3.0	0.0	13.0
Identifying Cleanup Technology and Strategies	5.0	1.2	3.8
Emergency Response Equipment and Training	1.5	0.3	1.2
Site Separation (includes past relocations)	1.7	1.1	0.6
Site Security Fencing and Security Patrols	1.9	1.9	0.0
Health Studies	1.7	0.4	1.3
Landfill Closure	12.5	1.5	11.0
Regulatory Environmental Health Assessments	3.5	0.5	3.0
Secretariat and Activities	3.5	1.0	2.5
Project Management and Public Communication	9.5	2.5	7.0
Total Budget	62.0	16.4	45.6

FIGURE 6.11K Supplementary information belongs in the appendices.

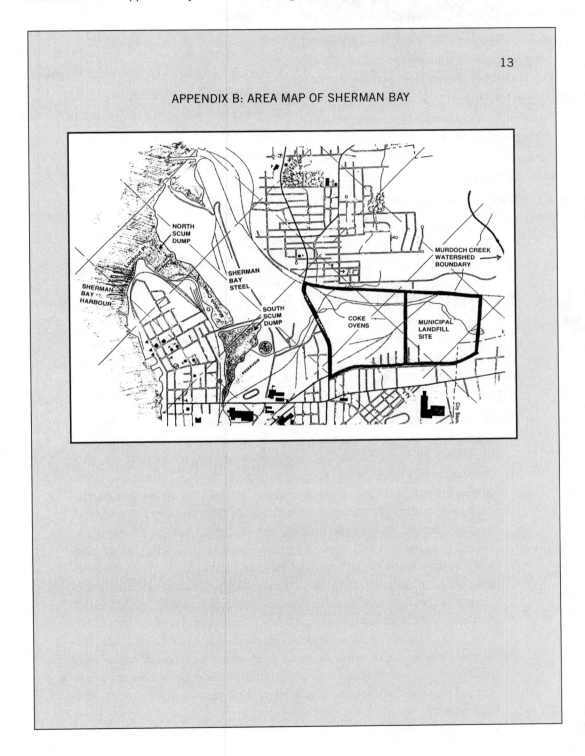

13

APPENDIX B: AREA MAP OF SHERMAN BAY

SHARPENING YOUR SKILLS

The following report situations require a formal report or a proposal. Using the report preparation form in the previous chapter to outline your approach, and employing the strategies for writing developed in Chapter Three, follow your instructor's directions to write one of these reports.

1. You work for Dolovich and Cowan, where several people have expressed an interest in upgrading their skills in professional communication, preferably through obtaining a certificate or diploma in the field. You have been asked to identify the options available to your colleagues, either through your local college or university or through an on-line certificate programme. Investigate at least three possibilities for your colleagues, and write up your findings into a formal report addressed to David Cowan. You will outline for these prospective students all they will need to know regarding the basics of the program (including admission requirements, duration, courses, any practicum or co-op experience or special courses, and advanced standing), as well as the general facilities offered by various institutions (student services, library facilities, recreational facilities, and any special assistance for mature students). You will also want to draw up a study of costs. Evaluate the programs according to their appropriateness for people who are working full time and who wish to achieve certification as professional communicators.

2. On occasion, a formal report will be preceded by a proposal for the written work that is to be submitted. This proposal is the advance work for your formal report and will give your instructor an opportunity to offer guidance for the report itself. The proposal should be descriptive, briefly outlining the report you will be presenting at the end of the course. Like the final project itself, the proposal should be presented as a formal report, approximately three to five pages long, typed or word-processed and double-spaced. It should include the following sections, with headings.

   a) *Report focus.* In this section, you must name the topic that you are researching and briefly describe its importance to the field. What makes it significant to your audience of fellow students, to members of the profession in general, or to me, as your professor? What do you hope to demonstrate or reveal in your report?

   b) *Purpose of or need for the report.* What is the primary purpose of your report? Why is such an investigation needed? Ideally, the report should contribute something to our understanding of the field. What do you hope to learn, discover, reveal, or contradict?

   c) *Main section headings or topics that will be covered in the report.* Given the topic you are researching, and the nature of the question you hope to answer, what are the major sub-topics you'll be discussing? On what elements will you focus? In other words, what headings do you expect to use in your discussion? You should briefly provide some reasons for your choice of focus, and show how this approach will best answer the question you have posed.

   d) *Outcome (results or conclusions) for your research.* Because you won't know for sure what your analysis will establish until it's completed, you will need to complete much of your preliminary research before writing the proposal. Your instructor may ask you to research and write a proposal even if you are not required to complete the final report. In either case, your proposal should suggest the probable outcome for the analysis you are projecting: what, exactly, do you anticipate this analysis will establish? What conclusions do you expect to be able to draw from your analysis? What, if any, contributions do you hope to make to the understanding of the profession as a whole? What recommendations will result from your report?

CRITICAL READING

In every chapter of this book, we have emphasized the importance of understanding the needs, interests, expectations, knowledge, and concerns of our audiences, and of considering the professional relationships we hope to build and maintain through our communication. We have learned that understanding the dynamic of rhetorical balance described by Wayne C. Booth can help us to establish credibility and create an appropriate relationship with an audience.

In this report, Burton Urquhart delves a little more deeply into the dynamics of rhetorical balance. Using the concept of "footing," an expanded version of rhetorical stance that includes five dimensions of the speaker-audience interaction, Urquhart analyzes ways of establishing the appropriate rhetorical balance in a significant communication task he faces. As he studies the challenges of writing a manual on public speaking for an audience of technical students, Urquhart reveals how complex it can be to establish that balance, and how an understanding of the dynamics of relation can help to improve our practical skills of message design.

RHETORICAL STRATEGIES FOR THE TECHNICAL WORKPLACE: FOOTING AND THE NOTION OF RHETORICAL BALANCE

Burton L. Urquhart

The project described in this report responds to the need for communication skills among engineering students, who face the challenge of finding employment in an industry that recognizes effective communication as a significant aspect of professional success. Although it seems obvious that they must be able to communicate effectively, new graduates of engineering programmes often have difficulty in doing so, as I discovered upon consultation with a number of professional engineers whom I interviewed in preparation for writing a handbook for student engineers on the subject of public speaking. The purpose of the current paper is to describe my preliminary conclusions regarding a suitable rhetorical stance for the handbook, based on my analysis of a survey of communicative situations faced by engineers.

When this article was written, Burton Urquhart was completing a master's thesis on the Communicative Demands of the Technical Workplace. His work involves an analysis of both the role of communication in the engineering profession, and the development of appropriate rhetorical balance for a student handbook on effective technical presentations.

Is another public speaking manual, specifically addressed to the needs of engineering students, really necessary? According to my subjects, the answer is a resounding yes. Students, they contend, typically have not been adequately prepared for the communication situations that professional engineers engage in each day. The engineers I interviewed confirmed that engineering students are leaving their years of training without the solid communication skills needed in the contemporary engineering profession.

There is little doubt that improving the communication skills of student engineers is an important goal; however, in order to achieve this goal, it is important to understand both the needs of the intended audience and the particular demands of writing a book of advice. Writing instructions in any field is complicated by differences between the knowledge and assumptions of the author and those of the audience, but my situation is made more complex by the fact that the audience and I are experts displaced in each other's fields: I am a communication specialist, not a technical specialist or engineer. As such, I am an expert in an area in which my intended audience lacks knowledge and experience.

Conversely, I will be instructing them on giving *technical* presentations, a specific application of public speaking skill in which I have no direct experience. The major challenge, then, will be establishing an appropriate stance, or relationship, with my technical audience. This paper is a preliminary analysis of the contexts of technical communication, using an elaborated model of rhetorical stance known as "footing." In this report I will present a description of the kind of footing required to accommodate both the instructional message and the needs of a technical student audience.

Before I propose a suitable footing for the manual, it will be helpful to understand how engineers engage and communicate with each other, and how they see their own communication needs. As a means of familiarizing myself with my intended audience, and as part of my research for a graduate thesis, I interviewed four mechanical engineers who had held positions in different areas and at various levels of the profession. I found that engineers require an environment in which cooperation and teamwork are smoothly coordinated, especially in larger engineering firms where collaborative projects are carried out with different departments.

Engineers must be flexible in adapting their communication styles and structures to many different situations and audiences. They need to be competent in different formats of communication, including fax or electronic mail memos to their clients; adept with different presentation styles, from informal structures all the way up to formal documents and reports; and flexible enough to employ any specific alternate formats preferred by their clients. It is also important for engineers to have the skills to understand and respond appropriately to differing communication demands and to adapt to new demands in a changing workplace.

Because the engineering industry is more competitive than ever, the persuasive element has become increasingly important. Engineering firms must be able to close as many deals as possible, and successful engineers need to be able to work effectively within complex organizational structures. Firms with separate management departments now expect engineers to present reports and other material in management- and sales-presentation styles to audiences with little technical expertise. The ability and willingness to adapt to an audience are of key importance in today's engineering profession, and these skills are nowhere more evident than in the development and presentation of an effective oral report.

The concept of footing, an expanded theory of stance and audience appeal, is described by George L. Dillon in *Rhetoric as Social Imagination: Explorations in the Interpersonal Function of Language* (1986). Based on his theory of the codes of engagement, Dillon argues that the relationship forged between writer and reader illuminates the communication interaction, the bulk of which is often not stated in the message itself. The unspoken "metacommunication" occurring between the reader and the writer is why Dillon classifies advice-giving, even in a written form, as interpersonal communication. Despite the fact that the writer is engaging more than one reader through the text, Dillon argues that the communication is interpersonal, or one-to-one, because each reader experiences the interaction individually.

As I analyze my communication situation and write my manual, I will have to consider the elements of footing. Though my findings will not be elaborated in the instructional manual itself, the analytical portion of the final report will deal explicitly with issues of audience and speaker positioning. I will also use other aspects of Dillon's "codes of engagement" to explore how my ethos, including the constraints of expertise, authority, and identification, will be established.

Dillon asserts that style, like footing, includes the writer's "individual self-definition" and is not shaped solely by the rhetorical situation. In advice books, which are in fact largely unsituated because of the lack of context for the relationship between writer and reader, there is a "construction" of that relationship, and therefore a footing between writer and reader. Writing under these circumstances is what Dillon calls "an act of social imagination," hence the title of his work (p. 15). Because the relationship, or footing, between these two constructions—writer and audience—can be managed in many different ways, it is necessary to develop a more detailed vocabulary of what constitutes footing, a concept similar to Wayne Booth's concept of the rhetorical stance (1963). Dillon introduces five comprehensive dimensions on which the footing can be analysed. These are impersonal and personal, distant and solidary [having solidarity with], superior and equal, confrontive and oblique, and formal and informal. The remainder of my analysis will consider these five elements as they relate to the demands of the advice manual I will be writing.

The first element of footing is how personal or impersonal the writer is with the reader (pp. 21–23). A common method of personalizing a text is to use personal pronouns, like "I" and "you." These help create a sense that the writer is speaking directly to the reader. In short, personal pronouns help to create involvement in the text, which can aid in eliminating the objective, non-interactive voice that often renders a text impersonal.

This involvement creates various degrees of personal connection between the writer and the reader. Although "I" and "you" can also be used in a general sense to mean anybody in society, a text can also be made to feel more immediate and personal to the reader when the relationship is dramatized; for example, a writer can pose questions to the reader to develop a dynamic example scenario, or present a familiar situation in a narrative format, with the reader and the writer as the characters.

However, devices like these must be used carefully. Too many questions begin to sound contrived; they may even increase the impersonal quality of the text and cause the reader to resist its advice. The reader may feel manipulated by too-obvious prompting and leading on the part of the writer, who has been positioned as controlling or "superior." This antagonism could be deepened if the writer is, or appears to be, either confrontive (threatening) or distant. By deftly using personal pronouns, questions, and dramatization, a writer can maintain a personal interaction with the reader, but these elements must be kept in fine balance to prevent the opposite effect of contrivance.

What can be learned from this aspect of footing that would apply to the writing of a manual such as the one I am working on? My research indicates that engineers are most familiar with an objective form of writing and reporting findings. Although they usually have a lot of personal and subjective interactions during the design process, when the reports are due they will write impersonally. The presentations based on their designs or findings are also largely objective, despite the immediate personal nature of public speaking as a medium. Engineers rely on their professionalism and accurate, objective research in order to maintain their reputations and, ultimately, their livelihood.

Since this impersonal, objective style is widely expected, using it is one way in which engineers project their competence to their clients and society in general. In order to establish my own professional credibility to an audience who is used to such a style indicating authority, I will also opt for a somewhat impersonal and objective footing in my manual. This will not only help to authenticate my credibility, it will also provide an example of the sort of final product that I am teaching them to produce.

Like the impersonal–personal continuum, the distant-solidary element of footing refers to the level of identification, or common ground, between writer and reader (pp. 23–27). Personal pronouns, particularly the first person plural, can contribute to the establishment of common ground by implying that the writer is in solidarity, or unified, with the reader. However, to establish this common ground, the pronoun must be used in the personal sense of "you and I" and not in a general sense such as "we engineers."

Once again, though, establishing solidarity is not as simple as plugging in "we" throughout the text; a metadiscourse, or undercurrent, of unity must already be present. Solidarity, as Dillon explains, is achieved by recognizing and responding to the values, concerns, and knowledge of the reader (p. 24). It is fully achieved when the writer actually approves of and endorses the reader's world view. Not only does the solidary writer share factual understanding with the reader, such a writer also accepts and shares the attitudes of the reader toward the subject and recognizes that affirming these attitudes may even contribute more than the message content to establishing an effective reader-writer relationship. These pre-existing attitudes about an issue, and the corresponding word choices, must be affirmed throughout the discourse in order to maintain common ground with the reader and to avoid the impression of distance in ideas and style.

Interestingly, the very acts of writing and reading themselves already assume some degree of similarity between the writer and the reader, since it is a normal human response when we have entered into communication with someone to look for similarities between ourselves and the communicator. Rhetorical theorist Kenneth Burke elaborates on this tendency in his concept of identification (Burke 1945). We communicate in order to overcome our differences and to ultimately, although never permanently, connect with other people. Burke asserts that such identification is the fundamental drive in all humans; everything that we do, especially linguistically, is an attempt at overcoming division in order to come together, or identify, with those around us.

At the same time, there is a recognition of the differences in knowledge and perspective between the writer and the reader, especially in the case of advice books on communication. The writer prepares the text on composition because an intended audience needs the information in order to increase the quality and effectiveness of their oral and written communication skills. The difference in knowledge between reader and writer imposes some distance between them, at least as regards their grasp of effective communication skills. Overcoming this distance is the task facing the writer of an advice book on communication. Dillon observes four devices that a writer can use to increase solidarity between herself and the reader.

First, using humour in the discourse is a common method of building a personal relationship with the reader. The author assumes that the reader will share a similar sense of humour and that humorous comments will garner a like response from the reader. The sense of shared attitudes and understanding is increased as the reader "gets" the joke or humorous statement without needing an explanation of what is funny. Humour engages the reader through its participatory nature. For example, the reader is flattered and made solidary with the writer, as Dillon explains, by being able to identify irony without being told the statement was ironic (p. 26). If it fails to create this identification, if the reader doesn't "get" the joke, humour can backfire and actually exclude the intended audience. For this reason, it must be handled carefully.

Second, the writer may use jargon as a way of explicitly including the reader in the interaction. However, like humour, jargon can exclude as well as include, and may actually prevent some readers from being able to follow and understand the message. For example, terms that are specific to a certain discipline are typically created to simplify and accurately specify discussion in a shared field of knowledge. If the reader knows what the term signifies, then his understanding of the text will be clear. However, jargon prevents those who are not familiar with the terminology from grasping the meaning of the discourse and has the effect of creating, or selecting, an audience made up only of those who are already in the "club" or discipline. Jargon unifies as long as writer and reader share similar expertise; however, it can create distance and even hostility when audience and speaker come from different disciplines.

Clichés can function in a manner similar to jargon. However, clichés have a broader audience appeal, at times even including every speaker of the language and member of that society. They can help to explain the writer's meaning in phrases or ideas that are familiar to most everyone. Instead of excluding, like jargon, clichés include many people in the discourse by illuminating the meaning in simple and common phrases. Nevertheless, these too can backfire, since a writer who uses too many clichés introduces the possibility of seeming insincere and uninterested in the precise meaning of what is being said, creating distance from the reader.

The fourth and last method of increasing solidarity is the use of comparisons in the form of similes and analogies. These devices allow the writer to connect a new solution or method to something that the reader already understands clearly, has a positive attitude about, or is interested in. Writers can use similes as an aid in the process of moving the reader to an altered point of view, or just a way to cross the boundary between the writer's imagination and way of thinking and the reader's conceptions of the issue being examined.

If appropriately employed, all four of these devices of solidarity—humour, jargon, clichés, and comparisons—can bring the reader and writer closer together, or to a more even and more personal level of understanding. My manual will have to create enough solidarity to engage the reader, so that I can then show the reader that improving presentation skills is possible with knowledge of message design and sensitivity to audience, combined with the experience gained by effective and consistent practice.

However, given the constraints of advice writing, only one of the four devices to increase solidarity is risk-free in a communication text for technical students. Jargon would not create solidarity because of the disciplinary boundaries I would be crossing. Using the technical jargon of engineering in this manual might create suspicion and resentment, because I could not be certain of using it accurately. I am assuming expertise in my field, holding my authority in the area of presentations. As a student of rhetoric, writing for technical students, I need to recognize and respect our differing knowledge bases or risk increasing distance by invalidly trespassing in their discipline.

Too much humour would also be inappropriate for this kind of book for this audience, not only because it is generally expected that textbooks will not be ironic or overly humorous, but because I would not be carefully demonstrating the style of writing or speaking that is desired in a technical presentation. My ethos—good will, good character, and good judgement—would be weakened, and a reader could not easily feel solidarity with an instructor who could not deliver the type and quality of work that is being taught. Humour will have to be used appropriately and not overdone.

Clichés also pose a danger in a discourse on delivering technical presentations. Such a presentation is supposed to keep the audience interested, and often even persuade them to take an action. However, clichés can drain the freshness and energy out of any communication by suggesting either that the speaker has not thought deeply about the subject or that nothing new is being said. A speaker or writer who uses too many clichés may seem too unconcerned with the audience to consider trying to make the diction and implied assumptions new, interesting, and memorable. Overusing clichés does not present a professional image or quality of work to clients who require an innovative solution to a technical problem.

It is important for a writer to create solidarity with the reader in order to persuade the reader, or even to establish a positive ethos. However, in the presentation manual I will be writing, creating solidarity will be a challenge. I am instructing a reader on how to prepare better presentations, which automatically introduces distance because I am sharing knowledge that the reader does not already possess. In other kinds of interpersonal advice books, the author can admit to still struggling with the problem that the discourse is trying to ameliorate, but the nature of what I am teaching precludes this strategy. If I am to be effective, I must demonstrate my command of the medium of public speaking. As a result, I must avoid devices that could potentially backfire, and instead choose strategies that will create identification without compromising my ethos. One such strategy may be to establish common ground based on the fact that, like my audience, I am a student who has learned some useful strategies for solving a problem we all share. Comparison is one possible method by which the boundaries can be crossed between the differing knowledge bases which are being forced closer together by the subject matter of the manual.

The third aspect of footing that Dillon presents is the level of superiority or equality in the relationship (pp. 27–30). Dillon contends that placement on this continuum is based on the level of "vehemence" found in the discourse (pp. 27–28), or the amount of conviction with which the writer asserts ideas. This vehemence, or conviction, is measured by two more features of the writer's stance, which are certainty–uncertainty and obligation–necessity. The first of these, certainty–uncertainty, describes the strength of conviction versus the amount of "hedging" that the author uses. The writer may hedge or qualify ideas with words such as "tends," "sometimes," "usually," and "perhaps," thus indicating to the reader a level of uncertainty or lack of confidence in the assertions. By contrast, the writer with a footing of certainty would use more authoritative, absolute terms, such as "always" and "definitely."

Obligation–necessity, the second aspect of vehemence, explores the manner in which the writer asks the reader to act. The writer can tell the reader either that she "must do something," out of necessity or that he "should do this," out of obligation to himself, friends and family, or society. Certainty and necessity, Dillon explains, are indicators of a superior footing, while uncertainty and obligation create a greater sense of equality between writer and reader.

The writer who wants to claim authority through superiority will communicate a vehement message to the reader. This writer may do so by using a couple of tactics to reduce the reader's involvement in the interaction and control the reader's opinion and will. In a vehement and assertive discourse, the writer will make sweeping value judgements, characterized by the writer's claim to "know" that a thing is so. Often this claim to "know" is without any evidence or support, and leaves little room to consider opposing or alternative viewpoints. As well as making generalized value judgements, the superior writer will also avoid participatory reasoning so that the reader does not have to be involved in drawing any conclusions. The superior writer keeps the power to make assertions and direct the meaning and outcome of the discourse; the reader is completely excluded from the interpersonal dynamic occurring in the writing and reading process.

Superiority can be qualified by employing equalizing devices in the discourse. Unfortunately, these devices, like other strategies of engagement, can easily miscarry and must be deftly executed. One possibility is for the writer to make references to his or her own performance in the area being discussed. For example, in my manual I could describe how poor my speaking skills were when I began my studies and how I struggled to improve them.

The danger of using such a device in an otherwise superior discourse becomes apparent even in such a personal example, as the writer may begin to sound falsely apologetic or pandering toward those who lag behind the writer in knowledge and experience. This potentially manipulative approach demonstrates how easily an equalizing device can become the agent of an even more superior footing. Dillon warns that "self-dramatization easily becomes self-glorification" (p. 30). Self-deprecating humour can also be used to create equality between the writer and the reader, but if this humour is taken too far or is differently interpreted by the reader, it will only serve to highlight the differences in knowledge between the writer and the reader. In the dynamic between myself and my readers, I will have a superior understanding of the subject matter, since the purpose of an advice-giving text is to impart knowledge. However, I will have to consciously avoid equalizing devices that could emphasize the superior footing at the expense of my ethos. Such choices would create a lack of credibility, most especially by my failure to accurately judge the situation and the audience.

The next aspect of footing is what Dillon calls the confrontive–oblique continuum (pp. 30–31). This continuum introduces another dimension in the construction of authority in footing. Whether the footing is confrontive or oblique depends on the manner and style in which solutions are presented to the reader by the writer. In composition texts, the more oblique writer may present information as a sharing of ideas and information that the reader will find useful, rather than challenging the reader's "face wants" with too much directive instruction. This oblique approach avoids openly challenging the reader.

By contrast, a confrontive writer constantly tells the reader what to think and feel, including why the reader behaves as he does. This tactic blatantly denies the reader's positive face wants, or the desire to be considered in a positive light. When carried to an extreme, the confrontive style can border on abusive, in that it does not allow for positive reinforcement of the reader. It also precludes even the semblance of a "two-sided" dialogue between reader and writer. The writer's thoughts, ideas, and solutions are presented as the only important or valid way of seeing and solving a problem.

There is, however, a difference between a confrontive and a superior footing. Dillon asserts that a writer can be both superior and impersonal without being confrontive (31). Even when adopting an impersonal stance, a writer does not have to give up any superiority in order to be non-confrontive. It is possible to write impersonally and superiorly while maintaining good ethos, by employing a more oblique approach. This indirect approach is the best option for my situation; an approach that emphasizes shared objectives will serve me best in a context that offers so many other challenges to ethos, such as crossing disciplinary boundaries.

I need to be impersonal to fulfill my audience's expectations for credibility, and superior in order to retain the authority to give advice on the topic, yet I must also recognize and respect the reader's positive face wants and desire to be thought well of. If I can achieve this rhetorical balance, my ethos appeals will help create a positive interpersonal communication experience. I will be able to demonstrate my good will, good character, and good judgement to my intended audience, which will raise their engagement with my message and advice.

The last element of footing identified by Dillon is the formal–informal continuum (pp. 31–32). An informal style is seen as more immediate, comforting, and appealing than a formal style. Informal style in diction and sentence structure can be advantageous in persuading people to follow a large course of change because it allows the writer to encourage and affirm the reader through the personal relationship that has been established. It is difficult to create the same personal bond with a reader through formal language and sentence construction.

However, formality has the advantage of connoting the writer's commitment to being explicit, clear, and trustworthy in thought and language. It can also convey the writer's willingness to take responsibility for what is being said. In my manual, I will have to maintain a moderate level of formality, because the students need to learn how to give effective formal and professional presentations. The constraints imposed by the textbook form also invite a relatively formal stance, but I could attempt to increase solidarity with the reader by introducing some informal qualities into my text.

The goal of the manual is to give the students solid skills in communication, so that when the report is due or it is time for a presentation on the job, they will already know how to deliver what is required, clearly and effectively. Dillon's model of the codes of engagement, or footing, offers specific insight into the important dynamic of speaker and audience, and provides an elaborated understanding of Wayne Booth's rhetorical balance. Using Dillon's systematized vocabulary of footing to study the challenges of a communication dynamic leads to a deeper understanding of the relationship between rhetors and their audiences.

Ultimately, by examining footing using Dillon's model, I am evaluating how ethos appeals are actually established in books that offer advice to a reader. From this evaluation, I can begin to assess how I should position myself with my audience in order to display good will, good character, and good judgement. Footing is a specific and concrete system for managing and understanding some of the more subtle aspects of the construction of good ethos appeals. It does so by dealing specifically with five different stances that combine to create the overall relationship between writer and reader.

The strength of Dillon's model lies not only in its comprehensiveness, but also in its flexibility and relevance to the demands of audience facing the writer of a handbook on communication. An analysis of the dynamics of footing will help the writer to use good judgement in creating the most effective audience–speaker dynamic. Good will is exercised if the writer concentrates on the audience while choosing the most appropriate footing on which to communicate the message clearly and effectively. Good character becomes apparent as the writer considers the reader's positive face wants and chooses not to sacrifice those wants for the sake of claiming superiority and authority.

Writing a handbook on communication in particular involves added complications, since communication is a very personal skill in which any reader has extensive face investment. My situation is made even more complex by the disciplinary boundaries being crossed. After completing this analysis, my projected footing is rather impersonal; typically distant, with some degree of solidarity demonstrated through slightly informal language and familiar comparisons; superior, with certainty, yet avoiding sweeping generalizations because the manual will discuss a specific topic; not confrontive, yet still superior and impersonal; and largely formal, with moments of informal footing, the only available footing for creating solidarity in composition books. As this report has shown, the five elements of footing closely interact with each other. That the elements of footing overlap each other illustrates the challenges inherent in completing a project like a technical presentation manual for students.

What can we learn about communication from a footing analysis like this one? First, that "metamessages" are communicated to the reader by the footing of a discourse. In other words, communication functions largely on a relational level, where what is unstated can be as powerful as what is explicitly stated, supporting Dillon's assertion that these implicit messages are often where the "real" communication takes place. Because the writing and reading process is interpersonal in this sense, a writer must consider footing so that the implicit nonverbal communication does not weaken, challenge or contradict the explicit messages of the discourse. Lastly, an analysis of footing confirms the importance of audience in the practice of rhetoric, and the role of ethos in creating an effective discourse.

WORKS CITED   Booth, Wayne. "The Rhetorical Stance." *College Composition and Communication* 14 (October 1963) 139-45.

Burke, Kenneth. *A Rhetoric of Motives*, rpt ed. Berkeley: University of California Press, 1969 [orig. 1945.]

Dillon, George. *Rhetoric as Social Imagination: Explorations in the Interpersonal Functions of Language*. Bloomington: Indiana University Press, 1986.

Things to Consider

1. What is the purpose of Urquhart's report? What inferences can you make about its intended readers? Are they technical professionals, non-technical professionals who have to communicate with a technical audience, or some other combination?

2. How would you describe the "footing" of this document? Is it formal or informal, personal or impersonal, direct or indirect, superior or equal?

3. How suitable is its chosen style—the language, the structure, the tone—to what it hopes to communicate?

4. How difficult was it to read this passage? Compare it with the information presented in Booth's article on "The Rhetorical Stance," on page 15. Which is more challenging? Can you suggest why that might be?

5. Did the additional details on the devices of footing help or hinder your understanding of rhetorical balance?

6. As readers with technical expertise, do you agree with Urquhart's conclusions about the most appropriate footing for his public speaking manual? Why or why not?

7. Re-read the article carefully, and write a summary of its findings that could be used if the article were presented in report form.

8. There are no headings in this article, but it would lend itself to their use. Following the SIDCRA structure provided for the informal report in Chapter Five, create appropriate headings for this report.

CHAPTER 7:

Oral Reports and Technical Presentations

LEARNING OBJECTIVES:

1. To master the four common types of oral presentations: impromptu; manuscript; memorized; and extemporaneous.
2. To learn how to prepare an oral report for extemporaneous delivery.
3. To master elements of vocal and visual presence so as to deliver your message effectively and confidently.
4. To learn how to incorporate visuals into your presentation.

More and more often, professionals in all fields are called upon to present materials orally, whether in workshops, seminars, design presentations, technical briefings, or staff meetings. Inexperienced speakers often find these public-speaking situations unnerving, and may even try to avoid them whenever possible. However, anyone who wants to advance in business or the professions will sooner or later have to face an audience.

An oral report or briefing, like a written report, should be carefully thought out, well organized, and clearly presented. It should be addressed appropriately and effectively to its intended audience, engaging their attention, motivating them to take the action recommended, and providing a means by which that action can be achieved. It should contribute effectively to the sense of human connection that is at the heart of all communication, and it should establish the speaker as a person of good judgement, good character, and good will.

Think of the last time you heard a talk or presentation that engaged and motivated you, that captured and held your attention, that challenged and delighted you. Chances are you can't name a single one. In fact, far too many speeches are lacklustre efforts in which the speaker is more concerned with surviving the experience than with actually communicating with the audience. Sadly, too many of us feel we have done a satisfactory job as speakers if we got through the speech without major embarrassment.

Because we so rarely hear a really outstanding talk, few of us understand what to aim for; as a result, we prepare inexpertly and inadequately, focussing on our own concerns for saving face rather than on connecting with the audience. Fear of public exposure, combined with a lack of really outstanding models for speaking, leaves us without skill in this indispensable form of communication.

However, as this chapter will demonstrate, no matter how uncomfortable you are at the thought of public speaking, you can learn to give an effective speech or presentation. No matter what kind of speech, technical talk, or presentation you are asked to do, it is possible to plan an effective talk, organize your materials carefully, and practice effectively. As an added bonus, you will find that the fear associated with public speaking will actually fade if you gain control over your subject matter and focus on communicating your message to your audience instead of on your own performance anxiety.

One big difference between oral presentations or reports and the written variety is the advantage of meeting your audience face-to-face. Though it may be intimidating to stand before a group of your peers or your supervisors, you should remember that it's also a lot easier to establish rapport with someone who is in the same room with you than it is to engage and motivate the unseen readers of a written report. If you can think of the opportunity for oral communication as an advantage rather than a burden, you will find it easier to prepare for the experience.

In order to take full advantage of the opportunity to speak directly to your audience, you will need to think about your speech in a different manner than you may have been accustomed to doing. A speech is not about you and your performance; it is about connecting and communicating with a group of other people. Public speaking is communication, and no talk need ever bore the audience if the speaker approaches the task with genuine enthusiasm and an intent to communicate with them.

Like all effective professional communication, oral communication depends on making an appropriate connection with your audience. It also means staying focussed on your communicative purpose, and taking care to present your message in an understandable and engaging manner. Finally, when you are preparing an oral presentation, you should also think about the constraints of the physical setting, and about the historical, interpersonal, or professional context in which your report will be given. As you prepare your presentation, and as you deliver it, you should be demonstrating your credibility and good will through your consideration of all these elements.

Preparing a Speech ✔

- Put the audience first.
- Focus on your purpose.
- Attend to your credibility.
- Identify the speaking context.

1. **Put the audience first.** This principle, as you know, is the key to all effective communication, but, if possible, it is even more important when you come face to face with your audience. It is also a principle that many speakers forget as they focus on their own anxieties and concerns. As you prepare, think clearly about how you will accommodate the needs and interests of those who are taking the time and trouble to come to hear your presentation: To whom are you speaking? Are you delivering material to your peers? Your subordinates? A group of visitors? The Board of Directors? What is the audience's interest in your project? How much information do they already have? What do they want or need to know? How much depth do they expect? As you consider these expectations and needs, adapt your report presentation as closely as possible to your audience's expectations. Recognize that some things you consider important may have to be left out if they are not as significant to your audience.

2. **Focus on your purpose.** Remember that you are giving this speech in order to communicate with an audience, that someone thought it was important for you to deliver your message in person rather than in written form. For this reason, you need to provide the connection that only a face-to-face meeting can achieve. As well, you need to think about why you are giving this presentation. What, exactly, do you hope to accomplish? Should you give a quick overview of your project, or should you present an in-depth analysis of your work? Are you expected to outline, support, or justify what you've been doing? Do you have to persuade your audience to accept a new point of view or course of action? Will you be subject to questions from your listeners? What are your own expectations? At the planning stage, you will need to shape your speech to reflect the face-to-face task it is meant to accomplish.

3. **Attend to your credibility.** Remember that you build credibility as a direct result of the quality of respect you demonstrate, both for your listeners and for the focus and structure of the speech itself. Your credibility depends on your demonstrating an understanding of the situation, of the audience's needs and expectations, and of your own role as a

professional. It will be portrayed in the level of preparation you bring to the presentation and in the quality of your delivery—confident, clear, and engaging. This chapter presents strategies for preparation and delivery that will assist you in establishing yourself as a credible, competent speaker who can connect effectively with the audience.

4. **Identify the speaking context.** One of the things that makes oral communication different from written work is the fact that it is presented face-to-face in a physical setting. A speech is not simply reading aloud to an audience who could easily have read the report for themselves. You are giving your information in person for a reason. Be sure to take full advantage of the opportunity to build a genuine connection with your audience.

   The size of your audience, the room in which you are speaking, and the limits of time are among the factors that will influence how you build that connection. Where will you be giving your presentation? How big is the room? What facilities are available? How far will you be from your audience? Will you be using a microphone? Overhead cameras? Computer slides? How much time will you have to make your connection with the audience? If you have prepared a forty-five-minute presentation only to find that you're expected to give a three-minute overview of your project as visitors are paraded past your desk, you'll have a difficult time—though not perhaps as difficult as if you're in the reverse situation! Make sure you know how much time you're expected to fill, and tailor your speech accordingly.

FOUR METHODS OF SPEECH DELIVERY

Four Methods of Speech Delivery

- Impromptu
- Manuscript
- Memorized
- Extemporaneous

Oral presentations vary not only in length and formality, but also in the style of delivery. There are four main types of speech delivery, and not all are suited to every occasion. Be sure to choose the one that is most appropriate for the requirements of your situation.

Impromptu

This kind of speech is given on the spur of the moment: the speaker is called upon to speak without warning and without any prepared notes. This type of presentation is most commonly used as an exercise in public speaking classes or groups, or it may occur when a party guest who wasn't expecting to be honoured is called upon to say a few words. For obvious reasons, the impromptu speech is usually short (under two minutes); a suitable topic for such a speech is one on which almost anyone could speak without having to prepare. The speaker isn't expected to provide new information; at best, she may simply give us a new way of looking at something we all know. This style of delivery is not suitable for any occasion where the speaker had reason to know that a presentation would be required, nor is it appropriate for any topic requiring specialized knowledge, since few people—including experienced speakers—can generate more than one or two minutes of coherent remarks, even on familiar topics, without advance preparation.

Manuscript

This presentation style involves reading word for word from a printed document that is the result of extensive research and thoughtful organization by the speaker. A manuscript speech has limited usefulness, and is most appropriate in situations where exact wording is important due to legal or political considerations (a judge, a lawyer, or an elected official issuing policy would probably use this form). This kind of speech is, under ordinary circumstances, not particularly effective, since it is difficult to maintain audience engagement when you're focussed on the manuscript in front of you instead of on your listeners. If you aren't focussed on the audience, you won't hold their attention, and if you lose their attention, you can't hope to move them.

As well, the act of giving a speech is an oral event, a dynamic interaction for both speaker and listeners. Unfortunately, written text just doesn't have the same sound, immediacy, or power of engagement that is evident in words spoken directly to an audience for whom the message is shaped even as it is being delivered. For this reason, reading from a manuscript is discourteous to the people who have come to hear you in person, since it tends to deaden the speaker-audience relationship. Even if you are very accomplished at reading aloud (most people are not), you may have trouble keeping your audience's attention during this kind of presentation.

Despite these warnings, you may be tempted to resort to a written manuscript in your talks because reading something you have written out in advance seems easier than talking more naturally with your audience. However, if you have ever had to listen to someone read aloud from a report or a paper that you could easily have read for yourself, you know how unpleasant that experience can be for the listeners. It's just plain boring to listen to, and a bored audience is not engaged, not motivated, and not persuaded. They may even be irritated or insulted by your lack of respect and engagement. No matter how much easier it might seem to prepare a manuscript in advance, boring your audience is not a style you should want to emulate. To keep your audience with you, you need to focus on communicating clearly with them and making a personal connection. Focussing your attention on the printed page at the expense of your audience connection will result in failure.

Memorized

Instead of being read to the audience, a speech can also be written out in advance and memorized. However, you may already have guessed that this process deadens audience connection just as surely as reading does—perhaps even more so. While you may be able to maintain the appearance of eye contact as you gaze over the crowd, your focus will be on the text in your head as you reach for exact phrasing, and not on the audience in front of you. If you've written something and then memorized it, it will sound like exactly that—a prefabricated message that does not respond to the audience's needs or contribute to the interaction.

Do not imagine you can memorize your speech and then fake the audience connection; experienced listeners will be able to hear in your voice the flattening of inflection and enthusiasm that mark a memorized presentation. As well, since you're reaching for exact wording rather than for ideas, a slip of memory will leave you gasping. You may not be able to recover your train of thought without repeating phrases you've already spoken. Finally, as you focus on remembering word-for-word what you were going to say next, you lose the contact that's vital to the communication interaction. Unless you are acting in a play, there is no reason to memorize your message verbatim; doing so will in fact cost you the audience connection that is necessary to effective, natural communication with your audience.

Extemporaneous

This form of delivery produces the most versatile and engaging oral presentation and is suitable for all but a very few situations. Although it is carefully and thoughtfully planned in advance, an extemporaneous presentation is not written out in full. Instead, the speaker works from a brief outline written on an index card, and expands the details from memory. An extemporaneous presentation is superior for most purposes because it allows you to respond to the immediate needs of your audience as the speech unfolds, and to adapt to unexpected circumstances; as you do so, you are able to establish and nurture a bond with the audience.

Don't make the mistake of thinking that an extemporaneous speech is unprepared or "ad-libbed." It isn't. It requires as much, or more, detailed planning and advance organization as a manuscript or memorized speech, but unlike those forms, the delivery of an extemporaneous presentation has a natural, spontaneous, engaging quality. It is by far the most adaptable and flexible of speech delivery styles, because it supports and enhances your interaction with the audience. Of the four types of speech delivery listed here, this is the one that you will find most useful. For that reason, it is the one we will explore in the remainder of this chapter.

PREPARING YOUR EXTEMPORANEOUS PRESENTATION

Like all good reports, extemporaneous presentations require meticulous planning, preparation, and practice. When they are done well, they appear comfortable and natural—so much so that they may even fool an inexperienced observer into thinking they are completely spontaneous. One of my former students recently gave a presentation in a senior class. A classmate astonished her by remarking, "I can't believe that you were able to wing it like that, and it came out so well!"

This young woman, a very accomplished speaker who has mastered the art of extemporaneous speaking, prepares meticulously, researches thoroughly, and practises her speeches repeatedly in order to achieve the command of her topic that enables her to speak naturally to her audience. All her effort, however, was invisible to her inexperienced classmate, who saw only the natural, comfortable delivery of an effective extemporaneous speech. That's what your audience should see also—a thoroughly prepared, credible speaker who knows the subject well enough to interact naturally with the audience. Remember: if your method of delivery calls attention to itself at the expense of your message or audience connection, your speech is seriously flawed.

The key to an effective extemporaneous speech is to create and maintain an effective relationship with the audience. Instead of reading from a manuscript, or delivering a memorized monologue, a speaker delivering an extemporaneous speech speaks naturally about the topic, based on the research and preparation that preceded it. The speech has a clear, explicit structure, and a clearly identified purpose. An extemporaneous speaker carries **one notecard** on which is written only enough, in outline form, to jog his memory of the points he wants to make.

An extemporaneous speech is *never* written out fully, even during your preparation stages; it is developed from the first as an outline only, and that outline is pared down as you practise the material until it is simply a cue to memory. Once you have arrived at your final bare outline, you should jot it lengthwise on a notecard no larger than 3" x 5."

Unless your speech is very lengthy (longer than twenty-five minutes), you should be able to fit sufficient information onto a single card. You may use both sides of the card, but don't be tempted to write down too much detail, and don't use more than one card unless the speech is longer than an hour.

Preparing your notecard well is one way of ensuring that your presentation will be successful, since it is the only text you will have with you at the front of the room. You should never read directly from the card unless you are citing quotations or statistics, and these should be used sparingly. Consult the card only as a reminder of your organization: it is a tool, not a crutch. The words on the card should be written as large as possible, in bold ink; you may wish to use highlighters or coloured ink to colour-code your main points. If you wish to include a brief quotation in your speech, or cite statistics, these may also be written on the card.

Why should you use a card rather than a page? A card is preferable, first, because its small size forces you to write down only main points, expanding the details from memory (that is, extemporizing) as you speak. Your outline should never contain a complete script of what you intend to say, and the small card will help to ensure you do not write too much. It is meant to prompt you and give you something to rely on should your memory fail you because of nervousness.

In addition, the small size of the card enables you to palm it quite easily, so that it is inconspicuous in your hand. By contrast, a full page is too large to be used unobtrusively and can actually serve as a distraction. It can rattle and shake if your hands tremble, communicating your nervousness to your audience and emphasizing it to yourself. Do yourself a favour: write your scratch outline for your extemporaneous speech on a card.

Before turning to the process of preparing the speech and making the card, we will spend a little time discussing topics for your presentation.

CHOOSING A TOPIC

Clearly, one of the first things you must do in preparing for a speech, whether it's in the classroom or the boardroom, is to select and properly focus your topic. The latitude you have for topic selection can vary considerably. In the classroom or on the job, you may be assigned a topic; more likely, you will be given at least some degree of choice. For instance, even on the job, you will sometimes be faced with opportunities to educate others about the work you do, the projects you are involved with, or the nature of the profession in which you work. You may be asked to give such a talk as part of a job search; you may be involved in outreach programs to local high schools; or you may participate in organizations such as Toastmasters.

Whether you are assigned a topic or are freely choosing for yourself, you should be careful to focus your topic in a way that is interesting and important to you and that can be made accessible and relevant to your audience. Also, if you can approach the topic in a way that unites your interests and those of your audience, your enthusiasm will show in your presentation, making it more dynamic and engaging. Speakers who are bored by their own speeches cannot help but bore their audiences with dull presentations. Remember: no topic is by nature boring; it becomes so only in the hands of an under-prepared, unenthusiastic speaker who has not found the link with the audience.

In order to discuss the process of topic invention, let's assume that you are free to choose your own topic and approach for your upcoming speech.

How do you choose something interesting for both yourself and your audience? First, you should strive to give your speech a sense of immediacy for your audience; if possible, it should address something that affects them directly. The American newspaper magnate William Randolph Hearst once said that a dogfight in your own neighbourhood is more interesting than a full-scale war half a world away. He knew the importance of giving the audience something relevant to their lives. You should keep this principle in mind, too.

> **A good speech topic**
> - can be made immediate and relevant to the audience;
> - engages your passionate interest;
> - offers a fresh approach or information.

Here's an example of immediacy. One of my students gave a speech designed to persuade his classmates not to use the school elevators when it was unnecessary. Because student tuition costs had taken a sharp rise just previous to his presentation, he used as his focal point the cost of maintenance contracts for operating the elevators at their current rate of use—more than a full year's salary for one extra professor. His research showed that reducing the use of the elevators would in turn reduce the cost of these contracts, thus saving the university money. By avoiding the use of the elevators, he argued, the students could contribute a cost-saving gesture that might influence the number of course offerings at the school, or help to prevent an additional scheduled tuition increase. This focus affected all the students in the class, and dealt with something of immediate interest and importance to them. They couldn't help but be interested.

Another student who was looking for a speech topic was walking past a small snackbar in one of the main hallways. She noticed with some annoyance the amount of debris that littered the surrounding area—candy wrappers, chip bags, pop cans, and pizza boxes. As she picked some of it from the floor to put into the trash, she found herself asking silently why nobody else was doing the same. She realized she'd found a subject for a persuasive speech, and she set about to investigate it: how much time did the maintenance staff spend cleaning up garbage that careless students left behind? What other chores were neglected because the staff were occupied by this task? In what way could her student audience benefit by taking the action she was asking for? The answers to these questions, which she obtained by interviewing the head of Maintenance Services and by doing some library research on the psychology of physical surroundings, provided the basis for a powerful speech persuading her classmates to put their own garbage into the trash cans, and to pick up just two pieces of litter from the floor each time they walked along the hall.

The second important quality of a good speech topic is that it reflects some interest and involvement on the part of the speaker. The famed public speaking expert Dale Carnegie reports hearing and evaluating approximately six thousand speeches per year at the height of his career. In one of his several books on public speaking, he emphasizes more than anything else *the urgent necessity of... having something clear and definite to say, something that has impressed one, something that won't stay unsaid. Aren't you unconsciously drawn to the speaker who, you feel, has a real message in his head and heart that he zealously desires to communicate to your head and heart? That is half the secret of speaking.*[1]

An effective speaker is passionate and enthusiastic about the topic, and it is that enthusiasm that connects with and convinces an audience. The most important thing you can do for yourself and your audience is to pick something that holds *your* interest as well as theirs, then find areas of common ground between your own interests and those of your audience, and build the structure of your speech on these. If you can't interest yourself in your topic, you're not going to succeed in holding the interest of an audience. You'll be bored, and that boredom will be evident to your listeners.

Speech topics are everywhere. As long as there are problems to be solved, new ideas to be communicated, and actions to be taken, there will be subjects for speeches. It's up to you to find one that you care about, and are committed to, and that will interest your audience.

There's one other rule to picking a topic that you should consider: you should bring something fresh and new to your speech, either in topic or in approach. Some topics, such as recycling and exercise, have been overworked. Try to pick something a little different from the same few tired topics. Chances are that, if a topic comes easily to your mind, it has likely crossed the minds of everyone else in your audience. Unless you can find some brand new information and focus on a specific connection to them, you will probably lose your audience in the first few minutes of your speech. Give them something unusual, original, and exciting. As an additional rule of thumb, if you heard about it on a television talk show, it's probably been overdone, and you should not choose it for your speech.

It is possible to give a familiar topic a fresh twist. One of my students decided to speak on becoming an organ donor. This topic, as important as it is, has received a lot of recent exposure, and it's hard to think of how to make it more immediate for the audience. But my student had a personal experience with organ donation, since her father had received a kidney transplant that saved his life. To open her speech, she held up an 11" x 17" colour enlargement of herself, her new husband, and her parents at her recent wedding ceremony. In a dramatic gesture, she tore her dad from the photo, telling her audience that, had he not had this transplant four yeas before, he would not have been alive to celebrate her wedding. This approach, based on her own direct experience, brought life to a topic that to most of the audience had been an abstraction.

---

[1] Dale Carnegie, *Public Speaking and Influencing Men in Business* (New York: Associated Press, 1955) 29.

Finding a Topic When You Can't Think of Anything

Many of us have a difficult time simply coming up with an initial idea for a speech topic. The suggestions below are intended to stimulate your thinking on topics of a general nature, though the same strategies can help you narrow down a technical topic or tailor it to a particular audience.

Strategies for Topic Invention

- Freewriting
- Brainstorming
- Challenging the status quo

Freewriting is a technique you can use to help generate ideas about speech topics. Simply sit down with pen and paper or in front of your computer and start to write—it doesn't matter what you write at the beginning. You can either begin with a specific topic area and try to jot down everything you know about it, or—if you're using the method to invent a topic area—you can begin with nearly any thought that pops into your head. The trick is to force yourself to keep on writing, no matter what, for at least fifteen minutes without stopping. Set a timer and don't pause for anything.

As you freewrite, think mainly about what you can say about the topic that will be of value to your audience. Even if you have to write something like "this is stupid! I can't think of anything to say!", you will find that pretty soon your thoughts will swing around to the task at hand and ideas will start to occur to you. Many people find this process to be a good generator of ideas when the topic is open, but it can also help you discover a fresh approach to an assigned topic. After you have generated some ideas for topics, you can go back over them to consider what possibilities they offer for developing your speech.

If you have to come up with a topic from scratch, you might want to try this **brainstorming** approach. It is similar to freewriting, but a bit more structured. Using your computer or a pen and paper, make a list of all the subjects you know about. Don't worry at first if they seem inappropriate or too broad for a speech; selecting and narrowing your topic to a specific purpose will come later. As you begin to brainstorm, you should write down all the ideas that come to you, no matter how silly they may seem. You can always cross the weak ones out later. Your list may look like this sample:

school	majors	teachers
sports	music	theatre
movies	books	dancing
art	hobbies	transportation
self-defense	clothes	education
advertising	stress!	exams
time out!	reading	

Once you've got a list, ask yourself if any of these subjects contain possibilities. You will need to find a "hook" that will catch your audience's attention and link the speech to them. Select a few that seem likely and set the rest aside. Once you've narrowed your initial list, take each of the possibilities you identified and brainstorm all the sub-topics that come to you one by one. For instance, let's say we have chosen "school" from our first list:

School—teachers	cost	courses
grades	requirements	special classes
exams	library	cafeteria
gym	clubs	meeting people
drama/music performances		student newspaper

Select from this second list any topics that offer interesting possibilities for you and your audience. If you need to, generate a third list, and then a fourth—each time becoming more specific. For instance, let's explore the topic of the student newspaper:

Newspaper—	
volunteer opportunities	controversies
movie reviews	articles about school
quality of writing	information about sports teams
weak reporting	interesting editorials
columnists	why do so many people hate it?
costs	journalism experience

This strategy works surprisingly well, and can actually generate more than one topic. After making just such a list, one of my students decided to give a speech encouraging her classmates to read the school newspaper regularly. Another chose to speak about why the class should boycott the newspaper entirely. And yet another encouraged people in the class to join the writing staff at the paper, citing the experience in journalism that could be gained.

This, of course, isn't the only topic that can work this way. One of my students gave an excellent speech encouraging her classmates to volunteer for the university's late-night escort program. Another offered a speech inviting the class to nominate a favourite instructor for a teaching award by pointing out how doing so might help to retain good teachers on staff and would pay something back to someone who had contributed in an important way to their education. Yet another student offered taught his classmates a do-it-yourself method for increasing reading speed and comprehension. By emphasizing the amount of reading they were faced with in the current term, as well as the overall reading load in their program, he was able to convince many of them to try the method he advocated, and he offered handouts describing the method in detail.

Third, it's also possible to make a good and interesting speech out of **challenging the status quo**. Most of us take certain social attitudes for granted: walk instead of taking the elevator, recycle, get more exercise, wear a certain brand of clothing, eat more vegetables. You can catch and hold an audience's attention sometimes just by taking an approach contrary to what the audience expects. The trick to this approach is to concentrate your efforts very specifically: not simply global resistance to commercial culture, for example, but a speech on why you should NOT wear a certain brand of shoes or jeans, on why you should NOT take the stairs instead of the elevator, or why you should NOT recycle. Of course, you will still need to support your argument and be sure your evidence is convincing. It's not enough simply to disagree with conventional wisdom; you must prove your case. However, many issues—even ones that everyone takes for granted—can be viewed from the opposite side. This approach can yield a really challenging and interesting speech if you think it through carefully.

The Best Topics Are Personal: Try This

As a final suggestion for finding a topic in situations where the choice is entirely open, you may find it useful to think through your experiences using the following technique. At least two areas of everyone's experience can profitably be "mined" for topics. First, consider recommending to your audience a product, a book, a movie, or a magazine that you have found entertaining or effective: I have heard excellent speeches inviting the audience to subscribe to the speaker's favourite magazine, recommending a favourite recipe, encouraging the audience to incorporate spinach into their diet. Any product that you have used that you think is good—perhaps a metal polish for removing rust from your bicycle, or a particular brand of jeans—can be made the basis for a persuasive or informative speech.

Second, you may wish to invite your audience to participate in an activity or join an association that you belong to. The activity may be socially significant, such as volunteering at the local hospital or donating to the food bank, or it may be personally relevant, such as participating in a new sport, joining a particular student club at your college, or trying a new hobby. I have heard some great speeches on rock climbing, juggling, participating in the 24-hour famine program, and taking a particular elective course. Once again, this method of topic invention is a good foundation for both persuasive and informative speeches.

Whichever of these methods you use for choosing a topic, or if you use another method, be sure to evaluate your choice in light of the course requirements, the probable interests of your audience, your own interests, and the demands of the assignment. Ask yourself what approach you could take to each subject, and consider where you might turn to research them further. If your presentation date is only a week away, you may not want to pick a subject that will require a trip to a distant library or a lengthy wait for inter-library loans.

RESEARCHING AND PREPARING THE SPEECH

Once you've identified your topic and your purpose, you will have to research your presentation. No matter how personal it may be, any speech is in some sense a research assignment, and to do well you must prepare in the same manner you use for researching reports or papers. Gather information that will help you to engage your audience and convince them to take the action you intend to ask for. Don't make the mistake of assuming you can give an effective speech, apart from a narrative of your own experiences, without conducting research: unless you are already a recognized expert on the topic, you will not be able to provide a convincing argument without research.

Speech Structure

Survey—Preview your purpose and main points.

Signpost—Link all your points to the purpose and to each other with clear transition statements as you move from point to point.

Summary—Restate your purpose and summarize the main points.

In fact, even recognized authorities rarely rely exclusively on their own expertise to persuade others; they bolster their arguments with evidence from other sources. Go to the library and check the catalogues. Ask the research librarian for help. Search for relevant (and reliable) materials on the Internet. Look for information that will offer some support for your ideas and assist you in supporting your argument. Don't expect to find ready-made arguments for your exact topic, however. Part of your job as a speaker is to link your research to the specific action you want your audience to take. Obviously you won't be able to find articles urging people to read the student newspaper of your college or university, or to take part in your college's student society—but you can find information about the role of newspapers in general and why they are valuable parts of our mass media, and you can find information about the value of campus involvement as a broader issue.

I want to remind you that, no matter how personal your speech may be, to do well you must prepare in the same manner you use for researching reports or papers. Once you've gathered the necessary information, you will need to list, in point form, several main ideas that strike you as significant. Consider your speech purpose and speaking situation and evaluate each of the ideas you've jotted down. Some will turn out to be useful for persuading your audience toward your goal; others will have to be discarded. Return to the "Three Ps" strategies for composition discussed in Chapter Three, and use those same methods to plan and prepare. However, do not write out your speech in full.

Instead, once you've completed your research and sifted the evidence, prepare a preliminary speaking outline. Your outline should have a clear and explicit structure, with a focussed introduction, a structured discussion, and a tight conclusion.

The structure of your speech should follow a survey-signpost-summary structure: first, survey or preview the purpose of your speech and the main points you will cover. Then, as you discuss each point in turn, provide verbal signposts to the audience to let them know where you are in the speech. Use the signposts to remind them of how each point links to the purpose and to the other points in your discussion. Finally, you should conclude with a brief summary, drawing all the points together in support of your purpose.

This overt structure may seem awkward to you at first because it's a more explicit method than you would ever use in a written report; however, in an oral presentation your audience will hear the material only once as you speak. Because they will need to pick up the structure and purpose as you go, and they will not be able to turn back to an earlier point for clarification, you cannot be subtle. State your purpose clearly at the outset, and use your transitional statements to remind the audience of that purpose as you move through the body of the speech. When you conclude, use the opportunity to review your points one more time, and to restate your purpose. An explicit structure is an aid to audience attentiveness and understanding, and will make your speech much more effective.

Practise out loud, and time your talk. As you work, you should aim for mastery of ideas rather than memorization of particular phrases or sentences. Work at an outline level only, and pare that down as much as possible as you practise, reducing sentences to phrases and phrases to single words wherever possible. Adjust anything that needs to be changed or reorganized, cut unnecessary details, and trim the outline as far as you can. Use words that have enough meaning for you to remind you of what you want to discuss. Keep practising until you can consistently finish within the allotted time.

When you have enough mastery of your information, but before you're through practising, you should make up the final version of your card. Your card should be so pared down that it would be useful only to the person who had done the research for the speech and who had practised delivering it. If it has enough information that someone who had not researched the material could present your speech, it probably contains too much detail.

MAKING THE NOTECARD

Use only ONE 3" x 5" notecard for your final speech outline. Remember, your card should contain only a brief outline that will serve to remind you of your main points. Each point you write down is intended to provide a cryptic signal to you to trigger your memory of the materials and help you organize your comments during the presentation. It isn't meant to record details or exact phrasing.

The sample card (Figure 7.1) is one I used for a half-hour presentation given at a dinner honouring D.K. Seaman, the benefactor of the Endowed Chair that I hold at the University of Saskatchewan. The audience was made up of dignitaries from the university and the engineering profession, including the Seaman family, the Dean of Engineering, and the President of the University. The process I used for preparing the outline and the card was the same one we are discussing here.

The phrase "How does someone... ?" which I used in my introduction reminded me that I wanted to begin my presentation by considering how a rhetorician ended up as a professor of engineering. Because this question has been put to me in many forms and is the one that was likely to be in the minds of my listeners, this introduction helped me capture their interest. The points listed under each section developed my presentation fully, and I was able to use personal anecdotes to frame my speech.

Don't be alarmed if you can't understand what is meant by all the points listed—if you could reconstruct the speech from the card alone, without having heard it delivered, the card would not have served its purpose. The

FIGURE 7.1   Sample Speech Outline on a Notecard

You Can't Get There from Here: On Being Comm Specialist in Engr

Introduction: How does someone . . . ?
    Anecdotes: Manny, Terry L.
    Understand going, understand came from

I. English — default
A. Beg. teaching career
    – comm, pub spk, tech wrt
B. Donnelly mystique — report wrt
    – scary
    – rewarding
C. discovered rht: Blicq

II. PhD
A. practical, prob-solv – like me
B. UW FIG
    – practical comm training
C. UL Comm Stream
    – 12 in 5 years
    – applied progs

III. Given background, what value?
A. Fill gap
    – D.K. Seaman
    – practical, creative
B. Curric challenges
    – 390 @ new courses
C. Long-term strategy
    – broaden theory
    – grow certif prog
    – student projects
D. Advise/consult
    – speeches, PR, fundraising
E. Continue research
    –CSSR, NCA, CCA, CATTW
    –publ

Conclusion: What value?
A. Manny, Terry not ask Q
B. Enrich, be enriched
    –experience   –vision
    –enthusiasm   –commitment

example is intended to show how individual and personal an effective card should be.

The sample shows the material on two sides of a 3" x 5" card, used as the foundation for a speech of approximately thirty minutes. Note that I use the card lengthwise to accommodate all my notes, and to keep myself from writing too much.

PREPARING A TECHNICAL BRIEFING

In your technical training courses, or on the job, you will not always have complete freedom in selecting your topic; more likely, you will be required to speak about a project you're working on, or some other aspect of your work. This kind of presentation, often referred to as a briefing, is similar to the kind of presentation outlined above, but may differ in the exact steps you follow to prepare your materials.

The exact shape and structure of your talk will vary depending on the dynamics of your communication situation. For example, as part of my professional role, I give many talks, seminars, and training workshops on communication. However, each is different from the previous ones, varying according to the demands of the particular setting and audience. Professional alumni of the college, new graduate student teaching assistants, undergraduates participating in a leadership programme, professional surveyors, veterinarians, engineers, or students of Oral and Written Communication all have different needs, interests, and levels of expertise which must be taken into consideration as I prepare my talk. Similarly, a professional engineer faces different constraints in presenting a talk to other engineers than in speaking to sales staff, managers, or to a group of students at a career fair.

Many professional talks are based on research and project reports which have already been prepared and circulated. The speaker in these situations must simplify and adapt the findings to fit the time available, the interests and expertise of the audience, and the purpose of the talk. A ten-minute presentation of research findings to a specialized audience cannot be expected to cover all the complexities of a lengthy research report. It can only hit the highlights, intriguing the audience enough to read the details in the report. If the audience for a design presentation is the clients or end users of the device, the speaker should keep technical details to a minimum and focus on how the project satisfies the original design constraints. Finally, in a talk to the marketing department, the research and development team must emphasize elements of the design upon which the sales staff can build a marketing strategy.

Let's consider the strategies used by Jennifer Varzari in preparing a briefing for the Board of Directors at Dolovich and Cowan on the training workshops her department hoped to develop for new employees. Because

Jennifer's presentation involves a specific aspect of her work, she does not need to brainstorm topic ideas. However, she does need to clearly define her audience, her purpose, and her speaking context. To do so, she uses a process similar to the topic invention strategies outlined at the beginning of this chapter.

Audience

Jennifer knows that her audience, the company's Board of Directors, have the power to recommend or veto her proposal. They are an important group in the company, with greater authority than she is able to command as manager of her department. Because they have already received and read her written proposal, she can assume that they are generally familiar with what she hopes to do. She can also assume that they are favourably disposed to the project, since they have indicated their interest by inviting her to attend their meeting to discuss the proposal and answer their questions.

Purpose

Jennifer must convince the Board of Directors that her proposal is worth implementing. She knows that they want to do what is best for the company, so she must show them that this project is to the firm's benefit. Because they have already read the proposal, her introductory remarks can be a quick overview of the project, emphasizing the company's need for improved training in this area. She can tailor her remarks to the probable interests of the board members, emphasizing the project's potential benefits and its reasonable implementation cost.

Speaking Context

Jennifer has been allotted a half hour at the beginning of the board's regular monthly meeting, which is held in an executive meeting room that seats twenty people. There is no lectern; Jennifer will be seated at a meeting table with the members of the group around her. It will be a relatively informal setting, and Jennifer is expected to present a brief introductory presentation followed by questions from the group.

Jennifer will want to bring support materials with her to the meeting—the survey results, copies of the materials to be used in the training, perhaps a cost-benefit analysis, and an outline of the implementation procedure. She should also prepare a card to help her frame her initial remarks. She can use as a notecard a shortened version of her proposal, but because her audience has already seen the full proposal, she must not simply read from it or repeat materials they have already read. Given the audience's interests, Jennifer should emphasize the company's need for report writing training, the failures of the current system, and the advantages of the new one. Figure 7.2 shows the outline Jennifer developed for her presentation.

Briefing to the Board

Intro: writing competence

I. Value of report wrt training
   A. Efficiency
   B. Professional image

II. Probs current system
   A. Inefficiency
      1) Lost time
      2) Repetition
   B. Empl dissatisfaction
      1) Survey

III. Advantages
   A. Employees
   B. Company
   C. Training staff

Conclusion
   A. Implementation

DELIVERY

Organization and preparation are, of course, very important in any oral presentation, but the effect on the audience will also be determined by the quality of your delivery. We all have been bored to near distraction by speakers who may have had valuable information to share, but who lost us with unimpressive or distracting delivery. Throughout this book, we have emphasized the importance of establishing an effective connection with the audience, in paying attention to their needs, expectations, knowledge, concerns, interests, and relationships, and of establishing professional and personal credibility. Your outline and preparation ensure that the quality of the message is addressed appropriately, but it is in the delivery that the other two keys of effective communication are established. In short, without a credible presence and an effective connection with your audience, your presentation will fail to communicate, no matter how well planned and researched it may be.

An effective oral presentation depends on engaging and professional delivery. Every element of your speech, including its delivery, should enhance and not detract from the overall impact. Sometimes this simply means not calling attention to weak spots, but just as often it means taking special care to create definite strengths in your presentation. In order to speak effectively, you need to take into account two major aspects of effective speech delivery—sight and sound. Your visual and vocal presence as you claim your place at the front of the room are the most important factors in successful delivery. Here are some of the elements of delivery that can make or break your speech.

Visual Presence: The Sight of Your Presentation

Many people don't realize how powerful a visual impression can be, and in an oral presentation it can be crucial. The speaker may be in front of an audience for anywhere from five minutes to two hours. The attention of audience members is concentrated on the speaker; listeners, without necessarily being completely aware that they are doing so, often take in every idiosyncrasy of the speaker's behaviour and every detail of appearance. You can test the validity of this point by asking yourself what small peculiarities you have noticed in your instructors—details of behaviour, expression, or dress. You'll be surprised how much you have noticed without necessarily intending to, and without being aware of doing so. For example, students in my classes, when asked, have even been able to tell me what colour socks one of their other professors had worn that same day.

While you can't control everything about your appearance as a speaker, there are some details you can take care of consciously that will help you to present a confident and capable visual presence to your audience.

1. **Dress appropriately for the occasion.** Wear clothing that is appropriate to the speaking situation. Don't wear clothing you will feel uncomfortable in, and avoid pulling at or adjusting your collar, sleeves, waistband, or any other part of your costume. Don't expect to be taken seriously if you show up in sweats. Even if your presentation is for a class, dress up a bit. Remember, too, that unless you are speaking about fashion, conservative dress is usually preferable to flamboyant outfits. You want to be memorable for the quality of what you say, not for the outrageousness of what you wear, and clothing that attracts attention away from your speech will undercut your effectiveness.

Create a confident visual presence

- Dress appropriately.
- Stay calm.
- Maintain eye contact.
- Employ appropriate facial expressions.
- Create energy with appropriate movement and gestures.
- Support your talk with visuals.

2. **Stay calm.** Approach the lectern calmly and pause briefly before beginning to speak, so as to give yourself a chance to catch your breath. Don't rush to the podium and immediately begin to speak. Give yourself time to relax and your audience a chance to get used to your presence. Likewise, don't rush away from the lectern just as your last words are leaving your mouth. Give the audience a few seconds to recognize

that your speech has ended. Allow time for questions if it's appropriate to do so.

3. **Maintain eye contact.** Your goal should be to meet the eyes of every member of the audience at all times—or at least to give this impression! Of course this is an impossible ideal, but if you keep it in mind, you will remember to make a visual connection with your audience. Eye contact is one of the chief means by which a speaker can create a bond with listeners; it helps to maintain their interest. Don't be afraid to meet your listeners' eyes. Although some people advise nervous speakers to do this, you should avoid fixing your eyes on a point on the back wall and delivering your presentation to it. The audience will not be fooled by this gambit, and may even begin to wonder what you are looking at so intently, and turn to stare too. You should also avoid looking too long at your card—if you have practised effectively you will not need to do this.

4. **Employ appropriate facial expressions.** People who are nervous sometimes betray their lack of confidence by giggling or grinning inappropriately, even when the speech is serious. Try to relax and keep your expression consistent with the material you are delivering. It is perfectly correct to smile when it is appropriate, but you should appear to be in command of your facial expressions.

5. **Create energy with appropriate movements and gestures.** You should appear comfortable and self-possessed. Stand straight, but not stiffly, keeping your body weight distributed evenly on both feet, using gestures to emphasize your points. As well, though you will likely feel vulnerable, don't lean on or hide behind the lectern.

   You should not be afraid to move about comfortably in front of your audience, but don't fling your arms about wildly, fidget, or shift uncomfortably from one foot to the other. Such extravagant movements are likely to detract from your presentation. (Your audience may begin to count your unconscious gestures—*Did you notice how many times she pushed her glasses up? Did you see him jiggling the change in his pockets?*) If you watch carefully, you will notice that a skilled speaker neither avoids nor overuses gestures and movement. Instead, such a person uses controlled movements and gestures to create energy in the delivery. An effective speaker knows how distracting unnecessary movements can be, and at the same time how horribly dull it can be to watch someone who does not move about at all.

6. **Use visual aids.** Of course, one of the most important of the visual factors of your presentation is your use of visual aids. When used effectively, these can make your presentation. Remember that they should be simple, readable, and well-timed. We will say more below about the appropriate use of a variety of visual aids.

Vocal Presence: The Sound of Your Presentation

Although in these days of television, sound may not seem as powerful as appearance, it is in fact the central element in any oral presentation, since your voice is the primary medium through which your information is transmitted. As with visual presence, your vocal presence must be confident and steady; it should not detract from your presentation. There are some common flaws to which first-time speakers are subject, but with awareness and practice you can eliminate them from your presentation style.

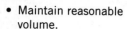

Create a confident and steady vocal presence

- Maintain reasonable volume.
- Speak clearly.
- Pace yourself.
- Avoid fillers or speech tags.
- Watch your pitch.
- Maintain a pleasant voice tone.
- Avoid grammatical errors, profanity, slang, or jargon.

1. **Maintain a reasonable volume.** Most inexperienced speakers speak too softly to be heard. Without yelling, be sure your voice is loud enough to be heard by everyone, especially those in the back row. If you can, practise your speech in the room where it will be delivered, having a friend sit at the very back to test whether you can be heard. If you absolutely cannot project your voice that far, try to arrange for a microphone. The members of your audience will not be attentive if they cannot hear you clearly.

2. **Speak clearly.** Enunciate your words carefully. Far too many speakers swallow the last half of their words, or run over them too quickly, making them extremely unpleasant to listen to and difficult to understand. Check your pronunciation too, particularly of words with which you are unfamiliar. Mispronunciations of important words will harm your credibility with your audience.

3. **Pace yourself.** Although you don't want to pause for too long between words, far more speakers are inclined to speak too quickly than too slowly. Don't rush through your material. Your audience will appreciate a brief pause here and there to allow them time to grasp your points.

4. **Avoid fillers or speech tags.** Don't say "um"! (Or "okay," "like," "you know," or "really.") These can be so distracting that an audience may actually begin to count them and thus lose the thread of your speech (Did you know she said "um" thirty-seven times in ten minutes?). Don't be afraid to simply pause if you need a few seconds to collect your thoughts; you need not make sounds all the time. If you find that you tend to use fillers when you make a presentation, the best way to eliminate them from your formal speaking is to practise eliminating them from your everyday speech. Consciously allow yourself to pause and take a breath instead of saying "um." If you can't always hear fillers in your own speech, ask your family and friends to gently alert you when they hear you saying the offending sound. Getting rid of fillers takes practice, but it is possible to eliminate them almost entirely from your speech.

5. **Watch your pitch.** A common weakness among inexperienced speakers is raising the voice at the end of statements as if they were asking questions. In fact, this voice tic *is* a form of questioning—a plea for the audience's constant support and reassurance. Unfortunately, when used repeatedly, it can make a speaker sound nervous and uncertain. State your points confidently, dropping your pitch at the end of each sentence.

6. **Maintain a pleasant tone.** Voice quality is another factor that can influence a speaker's effectiveness. Try to cultivate a voice that is pleasant to listen to: a voice that is piercing or nasal, for example, may irritate listeners and prevent your message from getting through. As well, try to project some animation into your voice. A deadpan delivery in a monotonous voice tone is just as annoying to the audience as a grating, nasal tone.

7. **Avoid any obvious grammatical errors, profanity, slang, or inappropriate technical jargon.** Audiences should be intrigued by your presentation, not put off by it. Slang and profanity are never appropriate, and professional jargon should be avoided unless the audience is made up of people from the same profession. Even then, it should be used judiciously. Remember that your most important task is to communicate your ideas to your audience. You cannot do this if your language is inappropriate.

VISUAL AIDS IN ORAL PRESENTATIONS

The impact you make on your audience can be enhanced by the effective use of visual aids; in fact, visual aids are one of the most effective means of demonstrating a point to your audience, since people tend to learn more easily and remember better when a demonstration accompanies the explanation. Even a large chart or overhead slide with your main points displayed for the listeners will help to fix your points firmly in their minds. Further, visual aids will not only help make your presentation clearer to the audience, but also will serve as an aid to your own memory.

Types of Visual Aids

- An object
- A scale model
- Photos, sketches, or drawings
- Charts and graphs
- Videos and films
- Blackboard or flip chart
- Demonstration

However, visual aids are most useful for communicating visual, rather than verbal, information. Apart from an outline of points, you should not present slides that offer dense text. The audience may become so preoccupied with reading from your slides that they will fail to make eye contact with you, or worse, you will be reduced to reading to the audience. Because a poorly designed or poorly used visual can actually compromise rather than enhance the effectiveness of your speech, you should:

- limit the number of visuals in your speech to a reasonable level;
- avoid using visuals for primarily verbal information;
- never use a font size smaller than 30-point for printed information on any slide or overhead.

When we think of visuals for a speech, most of us think immediately of slides, overheads, or computer-generated visuals. Actually, however, there are several different kinds of visual aids that you can use.

1. If you are discussing an object, bring it with you if it is large enough to be seen and portable enough to be carried around. Having the actual object with you will help to attract and hold your audience's attention during your speech. This is particularly effective if you are discussing a device or object you have designed. Having an actual prototype on hand not only clarifies the information you are presenting, but also boosts your credibility.

2. If you can't bring the object itself because of size or unmanageability, you may wish to provide a scale model, for example, a small-scale version of the CN Tower or a large-scale model of the DNA molecule. A scale model will assist your audience in visualizing what you are talking about and make your presentation easier to follow.

3. If the object is impossible to bring and no model is available, other visuals such as photographs, drawings, or sketches may be used effectively. These may be displayed as posters, slides, overheads, or computer projections, provided that they are large enough to be seen by the audience and clear enough to communicate your points.

4. Large, simple, and colourful charts or graphs can also enhance the communication of certain types of information. Like photos or drawings, these may be presented on posters, slides, overheads, or computer projections, and should be large and simple enough to communicate effectively with the audience.

5. Videos and films can also be useful, but you must be especially careful not to let them dominate your speech. Remember that they are there to support you, not to overwhelm you.

6. For some kinds of presentations, a blackboard or a flip chart is the best visual aid. If you want your presentation to be interactive, if you need to pace the presentation to the speed of the audience's understanding, or if you can't predict ahead of time exactly what you may need to write down, these means are ideal. Flexibility and pacing are two reasons why blackboards and flip charts remain popular for teaching.

7. If your speech describes how to do something or how something is done, demonstrate the process step by step, using real objects, models, or clear diagrams wherever possible.

Guidelines for the Use of Visual Aids

Decide on the type of visual aid you will employ (model, demonstration, chart, drawing, photograph, poster showing a list of main points, computer-generated slides, etc.) and prepare it in advance. If you are using an interactive visual such as a blackboard or write-able overhead, you may wish to write some of your material before beginning your speech and add additional relevant material as you go along. Whatever visual aids you choose, be sure that they are clear and understandable enough to be easily followed by your audience. Complex or overly detailed visuals, or visuals that are too small to read, will do nothing to clarify the information you are presenting and may just confuse your audience.

Guidelines for Using Visual Aids

- Prepare them in advance.
- Make sure they're large enough.
- Remember to show them as you speak.
- Use them sparingly.
- Speak to the audience and not to the visual.

Your visuals should be large enough to be seen by your audience. A 3" x 5" photo from your album may be interesting to you, but is unlikely to be of any value to your audience members who cannot see it from their seats. Passing the photo around while you speak is not a good solution, since doing so is disruptive and will make it difficult to keep your audience's attention. You will want to keep their eyes and interest fixed on you. Instead, enlarge the photo so it can be seen easily by everyone in the room.

Show your visual aids while you are speaking about them (believe it or not, some people forget to show their carefully prepared charts, drawings, or models at the appropriate moment because of nervousness or poor planning) and be sure to speak about them once you have displayed them. Nervous or inexperienced speakers sometimes display very intriguing-looking visuals and completely forget to refer to them during the presentation. Never let your visuals take the place of your spoken voice; never assume that they speak for themselves.

Visual aids should be used sparingly—don't overwhelm your audience with so much visual material that your presentation is lost. Remember that these are merely aids to your presentation. The speaker remains the focal point in an oral presentation and all visual aids should enhance that presentation. Too much visual material can detract from your presentation, and visual aids cannot by themselves serve as a substitute for an effective presentation.

Finally, while showing your visuals, remember to speak to the audience and not to the picture or chart you are discussing. Avoid turning your back to your audience as you speak; you want to maintain your relationship with the audience at all times through your speech. Be especially careful of this point if you are using a blackboard or flow chart as you speak, or if you are presenting visual material by computer projection. There is a temptation to look at the screen of your laptop instead of at your audience. When you practise delivering your presentation, be sure to practise using the visual aid in such a way that it will not cause you to lose your connection with the audience.

No matter what style of visual aid you choose to support your talk, you should remember that its role is exactly that: to support, and not to replace, the connection you make with your audience when you stand before them. Since communication is about relation as much as it is about content, your goal should be to engage the audience with you in exploring the matter you are discussing. For this reason, your best visual aid is YOU. The audience will pay attention to and better recall your talk if you make a human connection with them.

WORKING WITH COMPUTER-GENERATED VISUALS, OVERHEADS, AND SLIDES

In contemporary business and the professions, talks are frequently given using computer-generated slides and colourful overheads. It's easy to fall into the trap of thinking that a talk is automatically enhanced by flashy, impressive visuals with lots of technical enhancements. Unfortunately, this is rarely the case. A poorly prepared or ill-considered talk cannot be salvaged by flashy visuals, and many a good talk is ruined by speakers who allow these devices to take over. In such a case, the visuals can actually impose a barrier between speaker and audience.

Although it's very easy to put lots of verbal information on computer-generated slides, and to incorporate movement and colour, you should think twice about overwhelming your message with such gimmicks. Concentrate instead on creating an engaging, interesting speech that addresses the audience's needs and concerns, and on creating a human connection with them. Use the visuals to support this connection, and don't allow them to replace or undercut it.

Below are some tips for working with computer-generated visuals, overheads, and slides. We will begin with tips for preparing the visuals:

When Preparing the Slides

1. Make sure each slide is large and clear enough to be read easily from the back of the room. You should never use a type smaller than 30-point; 36-point is even better.

2. Avoid crowding too much text on a slide — a few lines is plenty. If you are tempted to reduce the font smaller than 30-point in order to squeeze in more text, you are putting too much on your slide.

3. Never, ever put every word of your talk on slides or computer-generated visuals: if you are presenting a complex argument and need to put some

of your message before the audience in writing, limit yourself to main points or highlights only.

4. Put only the necessary information or main points on the slide, and use your talk to elaborate the details.

5. Limit the number of slides you use—too many can be distracting. How many is too many? Unless your talk is a slide show, allow approximately two minutes per slide—which means that, in a ten-minute talk, you should prepare no more than five slides. There are, of course, exceptions to this rule, particularly when the information is highly visual, but in most cases, you will be able to use far fewer.

6. Avoid light print on dark background, except for headings; it is harder for the eye to process, especially from a distance.

7. If you need to use coloured fonts, be sure to allow sufficient contrast between font and background: avoid blue on green, red on orange or purple, yellow on white, or purple on red. These are impossible to read.

8. Avoid placing print on patterned or "busy" backgrounds.

9. Never make slides that will compete with you for attention: the slides will win, and they will compromise your audience relationship. Limit special effects such as animation; if overused, they may actually distract your audience and undercut your presence.

10. Never use a slide to present information that is primarily verbal; use visuals to communicate *visual* information.

11. Always proofread your slides for spelling, grammar, punctuation, and accuracy of facts and figures.

12. Prepare your talk so that it can be presented without the slides, in case of equipment failure.

Using slides in a talk can be tricky; they need to be incorporated smoothly into your presentation so that the two fit together as a seamless whole. For this reason, you need to incorporate your visuals into your practice sessions, so that you can operate the necessary equipment with ease when the time comes.

When you are delivering a talk with slides, you should try to check the system or equipment before your talk begins, to ensure that it is working properly, that it is focussed, and that your slides can be seen clearly from every point in the room.

Here are some tips for managing your visuals during the actual presentation.

During Your Presentation

1. If the equipment is not working, deliver the talk without the slides. Spend no more than three minutes of a twenty-minute talk trying to get the equipment to function; if your talk is shorter than twenty minutes, go ahead immediately without the slides. Delays to tinker with faulty slides or equipment can compromise your authority and credibility, they can cut into the time allotted to the speakers who follow you, and they can annoy the audience.

2. Never, ever read from the slide; if you find yourself doing this in a talk, you have very likely put too much information on your slide. Learn from your mistake and correct it for the next talk.

3. Maintain eye contact with the audience throughout the talk—look at them, not at the slides or computer screen.

4. Never let the slides dominate your talk; the audience should see and connect with *you* first, not the slide.

5. Be sure to discuss each slide as you display it, and to display each slide as you talk about the point it supports.

6. Never display slides that you aren't going to discuss—they may distract the audience from the talk.

7. Make sure that you do not block the audience's view of your slides while you are discussing them; try to place the screen where everyone can see it easily no matter where they are seated in the room.

Always keep in mind that your best visual aid is your own enthusiasm and energy. Be sure to move around freely, to smile, and to make eye contact with your audience. If you respond to their presence with vitality and engagement, they will be more attentive and involved. They will find your message more convincing and influential, and you more credible, if you make a human connection with them.

THE IMPORTANCE OF PRACTICE

Once you have organized your presentation and selected and prepared your visual aids, the next step is to practise your delivery. If you can, set up conditions as close as possible to those in which you will be speaking. Actually deliver your presentation several times, out loud, to master your timing, your command of your material, your delivery, and your use of visual aids. You will have enough to worry about as you step up to speak, without worrying about your command of the material. You will feel much less nervous if you are well prepared and will be able to concentrate on projecting a positive, confident image.

Have a friend or someone else you trust listen to your presentation and give you honest feedback. If you have access to a video camera, have your presentation taped and then watch it to see what you can improve. You can also audiotape your speech or practise in front of a mirror as a way of monitoring your delivery.

Practise delivering your speech out loud until speaking about this topic is as natural to you as breathing, but stop before you begin to memorize particular turns of phrase. Your audience will be able to spot a memorized presentation, and the quality of your speech will suffer for it.

In addition to improving your delivery, practice will also tell you whether your presentation fits the time you've been allowed. Think about how much time you have: a short time limit means you'll have to be selective about the details you include. On the other hand, you will need to make sure that you have enough material to fill the time allotment. As I write this, I have just awarded the first failing grade of the term to a five-minute speech assignment in my communication class. The speaker managed only two minutes and twenty-five seconds. She had chosen a poor topic and had done practically no research. Like many inexperienced speakers, she underestimated how nervous she would feel, and she assumed she would be able to ad lib material while she was standing in front of the class; she was mistaken.

Don't let this happen to you. An inexperienced speaker cannot possibly invent material on the spot that will be sufficient to present a coherent, organized, and convincing speech of five minutes. Without appropriate preparation and practice, most people can manage no more than two minutes—and some can't even last that long. Unless you practise your speech out loud, you will find it difficult to accurately gauge your time.

Practise your delivery, to be sure you have estimated correctly how long your presentation will take. You don't want to be in the uncomfortable situation of running out of material, or of being cut off because your speech is too long. If your speech is not long enough, you will need to develop your points with further research. If it comes out too long, you will need to cut some material out. Don't wait until the morning of your speech to find out that you didn't prepare properly—practise in enough time to make any adjustments that are needed.

When you are practising, **do not write out your presentation in full**. If you do this, you will have a tendency to read it during practice, or to memorize it, either of which will deaden your delivery and almost certainly bore your audience. Always practise from the single notecard that you intend to use in the final speaking situation; by the time of the presentation it will be familiar to you and will serve as an additional aid to memory, resulting in a much smoother presentation.

An oral presentation needs as much care and attention in its organization and preparation as a written report. Here, briefly, are the things to remember when getting ready for a presentation.

1. Choose a topic or aspect of a topic that you are interested in and familiar with; in a professional speech about some aspect of your work, select details and an approach that will appeal most effectively to your intended audience.

2. Focus your topic to suit your audience, according to their needs, their expectations, their concerns, and their prior knowledge.

3. Tailor your topic and your approach to the purpose and the setting of the presentation.

4. Prepare your topic so you can cover it adequately in the time you have been allotted for speaking.

5. Select and prepare at least one appropriate visual aid.

6. Practise your delivery, practise your use of visuals, and practise your timing!

7. Remember that every element in your presentation should support, and not detract from, your presentation. Avoid inappropriate gestures, mannerisms, or visual items that will distract your audience from the message.

8. Never let visual aids impose a barrier between you and your audience. Choose them wisely and use them sparingly.

SHARPENING YOUR SKILLS

1. Prepare a two-minute self-introduction in which you help your classmates and instructor to get to know you better. Identify yourself by name and major; describe your home town, family, and hobbies or interests; tell the class your goals for developing your communication skills; or outline your biggest concern about public speaking. Select any or all of these details, or something else of relevance to your audience to help give them a sense of your interests and character. Do not exceed the two-minute time limit.

2. Prepare a briefing on *one* of the report topics from Chapter Five or Chapter Six. Your presentation should be no more than ten minutes long and should employ at least one visual aid. Be sure your report has been adapted effectively to your audience.

3. Prepare a ten-minute presentation for your classmates, employing at least one visual aid, in which you provide them with tips on *one* of the following:

   a) Incorporating visual data into a report presentation (briefing or written format).

   b) Applying standard report writing structure—SIDCRA—to a specific report situation.

   c) Conducting library research for a report on any topic (identify a specific topic and show samples).

d) Distinguishing between a proposal and an ordinary report in purpose and organization.

e) Writing a self-evaluation report.

f) Organizing and presenting a briefing.

4. Choose a topic that you are interested in and knowledgeable about, and prepare a five-minute presentation for your classmates, employing at least one appropriate visual aid and providing some information that your audience is not likely to know already.

5. Prepare a five-minute persuasive speech in which you advocate a specific action that your audience can take. Some possibilities include reading a book or magazine you have read, participating in an activity or hobby you enjoy, joining an organization to which you belong, or trying a product, utensil, or study method that you have found useful. Be sure to focus on your purpose (moving your audience to the requested action) and explain the benefits to the audience of doing what you suggest.

CRITICAL READING

As we have discussed throughout this book, communicating effectively can be challenging. In our discussions, you have frequently been asked to look closely at the demands of sometimes tricky communication situations, and to come up with strategies for managing an effective audience appeal, a clear and coherent message, and personal credibility. In this short report to his communication instructor, Curtis Olson, a third-year engineering student, outlines the strategies he intends to use in presenting a short persuasive speech such as the one described in Assignment #5 (page 247). Olson has been asked to analyse and assess the communication challenge inherent in his upcoming speech assignment, and describe, in an informal report, the approach he intends to take to this task. As you read Olson's strategy report, consider how effectively he has used the principles of communication that we have been studying in this book, and the extent to which he has successfully anticipated and accommodated the probable attitudes and expectations of his audience of classmates.

STRATEGY REPORT FOR A FIVE-MINUTE PROPOSAL SPEECH

Curtis Olson

The purpose of this short report is to outline my strategy for preparing a persuasive speech, which will be presented to my engineering communication class.

During my five-minute speech addressed to my classmates and instructor, I will attempt to convince the audience that communication skills are an invaluable asset. By the end of the speech, I will urge the class to perform a specific action that demonstrates their valuing of communication skills.

GE 300 (Oral and Written Communication) is a unique class in the engineering program and develops skills which are not often emphasized. The strong focus on science and engineering classes in the program means that writing and communication skills are overlooked and undervalued, creating an exigence among some students. The challenge will be to maintain my audience's attention; the general lack of student interest in the course material demands a unique approach for presenting my message.

I will adopt the technique used by Jonathan Swift in "A Modest Proposal."[1] Using sarcasm and humour, I will begin my speech by taking the stance that many people in the audience have adopted: the communication class is irrelevant and a waste of time. Such claims will capture the attention of the audience, since they are used to hearing about the value of communication skills. I feel that in order to capture the attention of my classmates, I must acknowledge their emotional resistance to this subject and expose them to the flaws in that belief. As I continue to convince the class that communication skills are irrelevant, my own argument will become exaggerated and ridiculous. This approach will expose the dangers of undervaluing communication skills; as my position becomes more inflated, I will actually be encouraging the class to adopt the opposite beliefs: that communication skills are valuable and they should perform a specific activity to demonstrate this value. The recommendation that I actually want the audience to do will never be spoken, but rather implied.

An inherent risk in adopting this strategy is becoming the advertiser, as noted in Wayne Booth's "The Rhetorical Stance."[2] If I pay too much attention to my audience's reaction, I risk losing the message. However, the message which I am selling is an especially hard sell to this particular audience, and because of my approach, I run the danger of being taken seriously. Therefore, I believe that the risk of not paying enough attention to the audience could have potentially unsuccessful results.

A potential risk will be if I fail to make the argument ridiculous enough to have everyone disagree. To have the class not recognize the sarcasm in what I am saying will prove disastrous and leave some people confused. I will fail to communicate my message to the audience if they take my speech seriously and literally.

I realize that, due to my unconventional approach, the risks are large and the potential of failure is great, but this is the only method to possibly gain acceptance of my message. These risks are offset by the potential of this approach to gain attention, stick in the minds of the audience, and engage their sustained interest. From greater risks may come greater success.

As I attempt to correct the exigence, I will need to tailor my message especially for my audience. My recommendation is to take initial steps in the speech to befriend the audience and adopt a view stereotypical of engineering students: that our Oral and Written Communication course is a waste of time. Once the crowd has been won over, I will take the argument to the extreme and, in turn, convince the crowd to disagree with my point of view. In the end, the audience will adopt the message which I am implicitly promoting and explicitly denouncing.

NOTES
[1]Jonathan Swift's "A Modest Proposal," written in the early 1700s, is the most famous satirical work in English. In it, Swift proposes that the children of poor Irish families should be cooked and served as delicacies to rich English absentee landlords. Swift's essay caused outrage among the English, but was well-received in Ireland, where the former Dean of Dublin's St. Patrick's Cathedral is still regarded as a hero. Swift's essay has been widely reprinted, and may be found in many anthologies, including Jennifer MacLennan and John Moffatt, *Inside Language* (Scarborough, ON: Prentice Hall Allyn & Bacon Canada) 361–365.

[2]Wayne C. Booth, "The Rhetorical Stance." *College Composition and Communication* 14 (October 1963) 139–45. [Booth's essay can be found in this book, pp 15–23.]

Things to Consider

1. What is your reaction, as a student of communication, to Olson's approach? Given the audience he describes, is it well-chosen?

2. In your view, has Olson accurately assessed his risks?

3. This is a proposal speech. Has Olson clearly identified the action he hopes his audience will take? If so, what is it?

4. The audience for this short report is Olson's communication instructor; the audience for his speech will be primarily made up of his classmates. What inferences can you make from this analysis about Olson's perception of his intended speech audience?

5. How difficult was it to read this report? Is Olson's style appropriate to the purpose and audience?

6. Consider Olson's use of the principles of communication in analysing his task. Is there anything he has left out or taken for granted? What additional advice might you offer him for presenting such a speech to his engineering class?

The Application Package

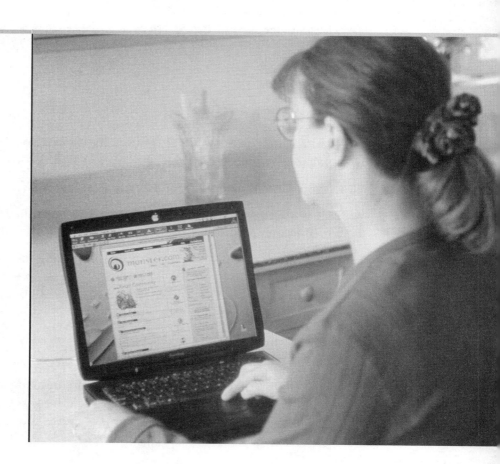

As your first step from college or university into the professional workplace, the job application package may well be the most important writing you will do in your professional programme. Whether you are competing for your first career position, an internship, or a plum placement, everything depends on communicating effective ethos, so your application must be as professional—and effective—as possible.

Like all other professional communication, the job application has a clear, specific purpose. In it (perhaps more than in any other kind of professional writing), you must acknowledge and respond to the reader's needs. The job application is a special kind of persuasive document; your task is to convince the employer of your suitability for the position for which you are applying. You must therefore focus not simply on what you have done or what *you* consider most important, but on what your prospective *employer* wants, expects, or needs to hear. You must also tailor your application to suit the job you're applying for.

Your submission must be appropriately adapted to its purpose in three respects. First, it must be adapted in content, so that it provides the information that the employer needs to make a decision to interview you. Second, it must be adapted to the needs and expectations of the employer. You can figure out what the employer is interested in by taking time to read the job advertisement carefully. You can then use that knowledge to shape your application so that you can draw from your experience to emphasize the skills that the employer will find relevant. Finally, your application should establish you as a credible and competent professional who understands the purpose of the application process as well as the demands of the job, and who has demonstrated care in preparing the résumé and letter. These factors will help establish the employer's confidence in your suitability for employment.

UNDERSTANDING THE EMPLOYER

There are two ways to gain a better understanding of the employer's needs. The first is to recognize the general qualities that all employers are looking for, and the second is to read the job posting carefully.

There are several general qualities that all employers consider desirable in job applicants—qualities that they want to see in the people they are considering for hire. These include stability, initiative, dependability, responsibility, loyalty, honesty, personability, energy, and enthusiasm. Stability is a quality of character marked by balance in your communication and behaviour. An employer will judge your personal and professional stability by the steadiness of your employment or school record, by how you deal with receptionists or support staff during the interview and probationary period, by your balanced approach to workplace demands, and by your reaction to stressors.

Initiative indicates a high level of motivation, the ability and willingness to see what needs doing and then do it. Initiative also means you can carry tasks to completion without constant supervision. Dependability is a measure of commitment to the job and to the quality of your work, no matter what the task. It is measured by punctuality, self-discipline, and reasonable professional goals.

Employers will also be looking for indications that you can take responsibility for your work, for both your achievements and your mistakes, and that you do not make excuses or blame others. They want employees who display loyalty to their profession, to their employer, and to their co-workers.

Needless to say, honesty is an essential quality, and misrepresentations or inconsistencies in the application process will destroy your credibility permanently. Employers also respond well to positive personality traits like energy, enthusiasm, and personability; they will judge these qualities by your courtesy, confidence, general friendliness, and tactfulness. In other words, employers are looking for evidence of convincing ethos throughout the application and interview process.

The second way you can learn about the employer's needs is to read the job posting carefully. Better yet, read several postings for jobs similar to the one you're applying for. Doing so will give you an indication of skills and qualities that are currently in demand, and allow you to emphasize the appropriate aspects of your own experience as you create your résumé. As you study the ads, you will notice patterns of skills and requirements that seem to be common to all. You can best pick out these patterns by surveying at least half a dozen ads for jobs in your field using a technique known as key terms analysis. As you read, look for skills that recur with frequency. The repetition of these terms can tell you where employers are placing emphasis. The patterns of skills identified in the job postings will give you an idea how to build your résumé and application materials so as to showcase those elements most attractive to the employer.

For example, ads for technical positions frequently emphasize qualities such as the following: "personal attributes that enable you to make a significant contribution... including the ability to listen to client needs and a commitment to strive to quality, team spirit, and creativity," "superior interpersonal/communication skills as well as superior skills in oral presentation, judgement/problem solving/decision making," "excellent customer service skills and leadership/coaching skills" or "the ability to persuade or influence others who may be angry, frustrated, rude, upset, difficult to work with, in order to reach mutually acceptable or workable solutions/agreements." All are communication skills, and these and more requirements like them can be found in ads for professional positions at all levels. The pattern of these concerns in job advertisements tell us that employers are particularly interested in applicants' communication skills. You should be sure to emphasize your own communication strengths on your résumé.

PREPARING THE RÉSUMÉ

A résumé is a kind of biographical summary that you prepare for an employer. It outlines the information he or she would likely want or need to know regarding your suitability for a job. You can think of a résumé as "you" on paper—or at least, the professional "you"—and you want to make the best possible impression. You should be aware also that the résumé you submit will not be read in isolation. It will be considered against as many as two hundred other applications for the same job. It is important therefore to think of the impact of your documents on the person who is screening the pile of applications; such a person is likely taking time away from other duties in order to complete this task, so your submission must communicate, in a reader-friendly way, the information that the recruiter needs. If the résumé fails to distinguish you significantly from the competition, or if it is confusing or unclear, it will be set aside.

Remember that the résumé and application are the first step in establishing a professional relationship with your prospective employer. In a résumé, even more than anywhere else, all the virtues of professional writing are important: understanding your reader and your purpose and applying the Seven Cs of professional writing—completeness, conciseness, clarity, coherence, correctness, courtesy, and of course credibility. Remember that the nonverbal aspects of the résumé—its organization and accuracy, its clarity, and its tone—all contribute to establishing an effective writer–reader relationship.

Visual impact (layout) also makes a big difference as to how you will be perceived. A pleasing balance between white space and print is important; a résumé should display its essential information without crowding or obscuring any part. As you have already discovered, effective layout actually assists in communicating your information—it's part of your organization. Your résumé should call attention to your individual strengths and should help to distinguish you from other applicants with similar experience. For this reason, I do not recommend using any standardized template, including the résumé "wizard" found in some word processing packages. A template may help you to organize your information initially, but its final effect is to produce a résumé identical in appearance and structure to that of everyone else who uses it. When several such résumés are read in sequence, as would be the case when the résumés are being screened, well-qualified individuals tend to get lost in the shuffle. As in all other communication situations, an effective message relies on the writer's good judgement in understanding the unique requirements of the situation and responding with an appropriately adapted message. It's very easy for a recruiter to reject a cookie-cutter résumé without giving it serious consideration, because, in the context of all those other cookie-cutter résumés, it appears to reflect little thought or effort on the writer's part. It is the person who has responded thoughtfully and creatively who stands out.

As you recall from Chapter Three, another element of an attractive and readable layout is the choice of font. Since readability is especially important in an application, you will most likely want to choose a font that is easy on the eye. In general, a serif font (such as Times) is easier to read than a sanserif font (such as Helvetica), although some of these can be quite clear if your résumé is short. Avoid script-style or gothic fonts or those that appear hand-printed. Choose an attractive 12-point font for your résumé, and use the same font for your letter of application so that the two will be visually unified.

Finally, there is one other non-verbal element that you should consider if you smoke or if you live with people who do. Do not smoke while you are preparing your résumé and application package, and do not keep your résumé paper in a room where smoking takes place. Paper readily absorbs odours, and cigarette smoke is no exception. While a smoker may not notice the scent in his or her work space, those in the non-smoking office environment who open the envelope will certainly notice it. Stale cigarette smoke is not only unattractive, it is also an allergen for many people, and it could cause your application to be rejected before it is even read. If you're not certain whether your own paper has picked up the scent of second-hand smoke, take your computer disk to school or to a printing service, and print your materials onto clean, fresh paper. It will be worth the slight cost to ensure that your submission makes a positive impression.

The résumé introduces you to the prospective employer and should make that employer want to speak with you. That is its only function, and you will be called for an interview only if it fulfils this purpose effectively. For this reason, it should make a very good impression. There are several rules of thumb for preparing a résumé; in addition to applying the nine axioms of communication and the Seven Cs of professional writing style, you should also keep the following elements in mind.

First, a good résumé combines organized content with positive visual appeal so that the arrangement of the parts actually becomes part of the organization. Second, an effective résumé presents its information according to two key principles: relevance to the job you seek and recency of the information; the experience that is most recent and most relevant is therefore placed first. The information on the résumé should be ordered so that these principles coincide and do not conflict. For example, my own background combines an interest in professional communication with industrial design, and I have fairly extensive experience in both areas. My most recent experience, however, is almost exclusively in communication. If I now decided to return to my emphasis on design, I would do well to subdivide my "employment" category into "relevant experience" (my design work) and "other experience" (professional communication). This strategy would allow me to place information about my design experience, which would be most relevant to my desired job, toward the beginning of the résumé.

Types of Résumés

There are three main types of résumé: the *functional* (or skills-oriented), the *chronological* (or data sheet), and the *analytical* (also known as the "crossover" or "targeted" résumé). The functional résumé is sometimes recommended for people who have little formal education or experience; because it emphasizes employable skills instead of positions held or training completed, it is useful for obscuring gaps in an employment or education history.

However, there are a couple of reasons why you might not want to choose a purely functional style. First, it often contains no dates, names, or locations that can be verified by the recruiter, so its claims appear unsubstantiated. What is worse is that, because the functional style is commonly used when the applicant's background is sketchy, employers may assume that all such résumés are being used to cover up undesirable information and may become suspicious about what is left unsaid. Thus, while this format may serve the *writer's* needs—to make a lack of experience or gaps in work history less obvious—it may fail to serve the reader well, since it doesn't usually provide the details an employer needs to make an informed assessment. As a result, because it does not effectively serve the needs of the person screening the applications, this kind of résumé may be set aside without serious consideration.

The more traditional chronological résumé has been widely used and is generally more readily accepted than the functional: it presents your training and experience chronologically, always beginning with the most recent and working backwards. The information is organized under fairly standard headings and focusses on facts and details, without elaboration or interpretation. Employers usually prefer this chronological organization over the functional format because it shows the applicant's actual experience and normally provides specifics that may be verified.

Because it accommodates the employer's expectations, the chronological format is generally a better choice than a strictly functional résumé. However, neither of these is perfect. The brief bare-facts outline provided by the chronological résumé alone may not be enough to convince a prospective employer to take a closer look at the applicant. The competition for good jobs is always intense, and many employers want more of a "sense" of the applicant before the interview. For this reason, the analytical or targeted résumé (which blends the chronological and functional approaches) is frequently a better choice. Because it combines the strengths of both forms and because it is adapted to the specific job applied for, the analytical résumé effectively accommodates both the employer's needs and the applicant's background. This is the résumé style you will learn in this chapter.

A Closer Look: The Parts of the Résumé

A résumé typically contains a survey of an applicant's education, work history, achievements, and skills. However, far from indiscriminately listing every job-related experience you have ever had, no matter what it is, an effective résumé is selective, both in terms of what it includes and in how that material is organized. A résumé, like other forms of professional writing, is a form of communication with a very specific purpose, and it is effective only if it gets you an interview. Since it is the *reader's* needs that will determine who is interviewed, the résumé should be designed and focused to meet those needs.

Although the résumé does contain standard sections, its contents are also to some extent flexible, and what you choose to include will depend partly on your experience and the position you are seeking.

The two principles that should govern what goes into your résumé are recency and relevance. In general, more recent experience precedes older information in a résumé; within categories, all details are therefore arranged in reverse chronological order (that is, starting with the most recent and working backwards in time). The résumé should also be arranged so that your most relevant experience is emphasized. As we consider samples of résumés, we will see how these two principles of "targeting" can improve the presentation of content, the adaptation to audience, and the establishment of credibility. For example, consider the résumé of Cameron Britten, (pages 267–271).

All résumés should include sections for personal information, education, employment experience, and skills. Experience and education are always arranged in reverse chronological order, preferably with dates displayed for the convenience of the reader. A few additional categories, discussed below, may be useful if they are relevant to your background and to the new position. The following is a list of categories usually included on a résumé.

Personal Information

This section includes your name, your address, and a phone number where you may be reached. You may also wish to include your e-mail address and a Web site address if you have them. Do not include an e-mail or Web site address that will soon be defunct, however; for instance, if your e-mail and Web site are college accounts that will disappear after you graduate, do not invite your prospective employer to contact you this way if you are close to graduation. Also, if you are using an e-mail address on a commercial server, you should avoid silly or suggestive names such as "sexybeast" or "megaforce." Your e-mail address, like everything else about your application, should project an air of professional competence, and—although they may be amusing—silly or off-colour names may portray you as sophomoric to a potential employer. Finally, if you choose to list a Web page on your résumé, it should not be one containing primarily personal information; it is better to include only material that is relevant to the professional relationship you are hoping to establish with your prospective employer. Web pages devoted to purely personal or recreational information may not be well received by an employer.

You will notice from the recommended examples that personal information should be displayed in an eye-catching position on the first page. You may even wish to use a desktop publishing program to design an appealing personal letterhead to use as the first page of your résumé. You can then use the same format for your letter of application. This type of letterhead can be very effective if well designed, but you should avoid cutesy graphics or too many combined effects such as boldfacing, italics, block capitals, and underlining. A slightly conservative design is more professional than one that displays a whole range of visual effects in a single heading.

Personal details such as age, height, weight, social insurance number, state of health, marital status, and citizenship do not belong on your résumé. Though in the past they were commonly included and are still occasionally recommended by people who are working with outdated information, they are no longer considered appropriate for inclusion on a résumé. If, despite these guidelines, you wish to include such information, place it near the end of your résumé—don't take up valuable space on the front page. You definitely should not include such sensitive information as religious affiliations, and racial or cultural origin, unless, of course, they are directly relevant to the job for which you're applying, or you have chosen to self-identify under employment equity guidelines. In this case, it is the principle of relevance that becomes most important. As a general rule, leave out any personal information that has no bearing on your ability to perform the job or that may invite prejudice; if a piece of information is not a professional strength, don't put it in your résumé.

Career Objective

The career objective statement, a phrase or sentence that states your career aspirations, is an optional part of a résumé. If your cover (application) letter is well-focussed and your experience is directly related to the job you're seeking, you may want to leave out the career objective, since in that case it would be redundant. Recent graduates of a college program that has specifically trained them for the position they seek may find a statement of objective unnecessary. However, a career objective statement can be useful if your experience is diverse; in this case, an accurate, well-worded career objective can focus your background. Here are some instances in which a career objective statement can prove useful.

- If you are aiming at a *specific* position or type of position and are not interested in any other kind of work, you may wish to focus on this aim in a career objective.
- If you have spent some time out of the workforce—to travel or to raise a family, for example—a career objective can explain the gap in your employment history.
- If you are changing careers, this statement can serve as a link between your aspirations and your experience.

If you decide to use a career objective statement, you must be clear and specific; avoid such statements as "I am seeking a challenging position that will make use of my skills and offer room for advancement." Can you think of any candidate in any job competition for whom this would *not* be true? Such a vague statement takes up room without adding anything valuable to your résumé. It is better left out.

Education

For a student or a recent graduate, information about your education usually comes immediately after the personal information, because it is your most recent, and likely most relevant, experience. Once you have been working in your field for a year or two, you may want to place your employment history before education on your résumé, because by then it, and not your education, will have become your most recent and most relevant experience. This is true in all cases except those in which a particular credential (degree, diploma, or certificate) is a minimum qualification for the job. In such a case, retain education as the first category to prevent being disqualified by a recruiter who makes an initial decision based on the first page.

All dated information is given in reverse chronological order, so that emphasis falls on the most recent and most relevant experience. Your educational history should include the following information, beginning with the most recent experience: dates attended; name and location of institution; diploma, certificate, or degree obtained or expected; and some brief detail about your particular program of study. Mention grades only if they are outstanding.

List everything back to (and perhaps including) high school, but no further. As well, if you have taken more than one diploma or degree following high school, or if you have been working for a period longer than two years following graduation from college or university, you should consider leaving out high school. Be selective: remember that the employer needs to know only what is relevant and useful in making the decision to interview you. Anything else is clutter.

Employment History

Beginning with your most recent position, in reverse chronological order list dates, place of employment, job title, and duties. Provide a quick outline of the job, emphasizing any skills you could bring to the new position. Again, always be forward-looking, focusing on the skills that will be needed in the job you want rather than on those demanded by your old job. As you plan, list all the duties you performed as part of your old job; then place them in order of relevance to the new job, deleting those at the bottom of the list. You may cluster similar or related jobs if you have had a series of them, or even delete some less important or irrelevant jobs. This strategy can reduce repetition and save valuable space on the résumé if you have a lot of information to include; it is demonstrated on the résumés of Cameron Britten and Kiro Hatzitolios, later in this chapter. In all cases, emphasize skills or knowledge that are relevant to the job you seek. For example, if I have worked for two years as a sales clerk, but for three different stores, I may still present this information in a single entry because all three positions were similar. I should provide dates only from the point I began my first position until I left my last position, then provide the job title (for example, Sales Personnel) and list the names of all three employers.

Skills

Employers want to know that they are hiring someone who does more than meet the minimum qualifications; they want assurance that the candidate they are considering is the best choice from among applicants with similar training or experience. For this reason, including an effectively targeted skills section on your résumé may give you the edge over other applicants of similar background. A segment that elaborates on your skills can be a useful device, especially for a recent graduate who may have little directly related employment background, because it provides an opportunity to emphasize relevant abilities, regardless of your past experience. Depending on your preference and the kind of experience you bring to the job, you may wish to title this segment as "Areas of Ability," "Areas of Expertise," "Special Capabilities," or "Employable Skills" instead of simply "Skills." You should take full advantage of this chance to promote your unique combination of skills, but make sure this segment of your résumé is focussed for the job you seek.

It can also be a good idea to cluster your skills under subheadings of some type, as demonstrated in some of the sample résumés that follow. You will notice a variety in the ways that the skills are clustered and presented in the samples; each person has selected a method that shows his or her skills to best advantage and that best suits the job sought. Like the writers of the sample résumés, you should choose the format that displays your skills and experience most effectively. Though you may place your skills section at the beginning of your résumé, most people prefer to put it at the end, where it supports the information given in the rest of the résumé. The prospective employer then comes on it after reading the details of your employment and education history.

Skills are generalized abilities that may be transferred with relative ease from one situation to another, and there are four main areas in which any employer is interested. These classifications will give you an idea of where to start when grouping your own skills. As you list your skills, be sure you can identify and describe a situation in which you demonstrated each one. An employer may ask you for such an example in an interview.

1. **Specialized or Technical.** Nearly all jobs require some type of specialized skills necessary to perform the job duties efficiently. These might be highly technical, such as operation of equipment (for example, a computerized milling machine), knowledge of specific procedures (such as WHMIS, CPR, drafting, quality control, or bookkeeping), mastery of certain kinds of software packages (Web design, desktop publishing, power point, or statistics packages), or any other specialized ability (public presentation, training skills). Do you have the skills required by the job you are applying for? You should state them clearly on your résumé.

2. **Practical.** General work-related skills can also set you apart from others with similar education or experience. These qualities emphasize how you handle tasks in the workplace. Although all the applicants for a given position may be trained in the field, not all are equally competent. Employers want to be assured not only that you can do what the job requires, but that you can do it efficiently and well. In this category you might include such qualities as punctuality, conscientiousness, organizational ability, problem-solving skills, ability to work to deadlines, and efficiency. Again, be sure you can think of a specific example from your past experience that illustrates each skill you name, in case you are asked for one in an interview.

3. **Communication and Interpersonal.** Skills in this area should be particularly highlighted, for the reasons we have been emphasizing throughout this book. Even when they are not listed on the ad itself, ability in writing, public speaking, listening, and critical reading are highly valued by employers, as are such interpersonal skills as tact, diplomacy, leadership, motivational skills, cooperation, and teaching ability. Essentially these skills are concerned with how well you handle your professional relationships with people. Employers are typically not interested in hiring those who cannot deal effectively with others. As with the previous categories, be prepared to provide evidence of your effective communication skills if called upon to do so.

4. **Drafting, Design, or Artistic Skills.** If they are relevant to your line of work, you may wish to include such abilities as drafting, design, technical illustration, document paste-up and layout, and photography. If you will need to demonstrate your skills in these areas, you should prepare a portfolio of your work to take with you to the interview. Your instructors in art and design courses can show you how to create an effective portfolio, and you can use your design ability to create an effective presentation of your work.

References

Prepare a list of two or three names of people who are willing to provide references for you. Opinion is mixed about whether you should include the list with your résumé or submit it during the interview. Some employers I have spoken with are suspicious of an applicant who does not automatically volunteer references, even when he or she promises to supply names "upon request." Others say they are not concerned about receiving the list of references until after the first interview is completed, because they normally do not contact references until that time, and then only if the candidate is to be offered a job. Still others put no stock in solicited references at all.

In light of such divided opinion, what should you do? All of the employers I spoke with agreed that providing references will never hurt your chances, whereas *not* providing them may do so. Thus, it may be safest to provide them if you can.

References should be obtained from former employers or others acquainted with your work, or from former teachers or professors. Letters or testimonials from friends, family, or fellow students are not appropriate for a job search. Personal references are of limited value to employers and should be avoided unless a character reference is specifically requested. You will find more information on references in the section on the letter of recommendation (page 295).

These are the common and essential sections of every résumé; however, there are some additional optional categories that may be useful to include in your résumé, because they allow you to tailor your application to your unique background and to the demands of the job you seek.

Additional Categories

Select any of these that seem appropriate, displaying your unique experience to best advantage. If none seem relevant to your experience or necessary for the job you want, you can certainly leave them out. If you have an accomplishment or experience that is relevant to your job search, but does not fit into any of the listed categories, you can invent your own appropriate heading. The résumé is a flexible instrument that you should structure according to the demands of the job you seek and your individual combination of experience.

Awards

In this section, list in reverse chronological order any awards received in school activities, giving dates, institutions, and titles of awards. If you have never won any such awards, leave out the whole category. If your awards are not school-related, you may want to place them in another category entitled "Achievements" or "Accomplishments."

Extracurricular Activities

This category is useful only for students or recent graduates; once you have been working for two or more years, this information will come to seem less relevant. List here any *significant* contributions to school-related activities; membership in academic or athletic clubs or teams, participation in student council, yearbook, or newspaper activities, and so on. If you did not participate significantly in such events or organizations, leave this category out. College experiences of this nature have greater longevity on a résumé than high school ones; as a general guideline, delete this section if it describes high school experience more than three years old or college experience more than six years old, unless your achievements are especially outstanding or relevant to the job you seek and have not been replaced or surpassed by anything else.

Additional Courses/Training

Use this category to list courses or certifications that are outside your main area of education but which could contribute skills to the job. Perhaps you are formally trained in mechanical engineering, but have continued your education with courses in management. Perhaps you are an electrical engineering technician, but have training in computer programming, or an engineering physicist with expertise in Web design. Perhaps you have taken a pilot's licence, first aid courses, mediation training, or a desk-top publishing certificate. If so, include this information on your résumé. You may not wish to include general interest or non-credit courses unless they are in some way relevant to your job search. Once again, list in reverse chronological order the dates, institutions, and titles of the courses.

Volunteer Experience

If you have held several volunteer positions and feel that they warrant consideration, you can create a separate category for them. If they are few (and relevant) you might want to include them as part of your experience instead. (Be sure that you do one or the other; don't include the same items in two different categories on the same résumé!) If you decide to create a separate category for volunteer work, list in reverse chronological order any positions where you would, under other circumstances, have been paid for your work. Finally, list volunteer positions only if they contribute meaningfully to your résumé.

Community Service

Include membership in service clubs, organization of community events, or service in civic positions, if these are relevant or significant. It is usually considered best to leave out organizations of a strictly religious nature, unless the job for which you are applying is with a religious organization. List dates, name, and location of organization, position title, and any relevant duties. If the organizations to which you belong are professional associations and are not unions (the Canadian Society of Chemical Engineering, the Canadian Council on Social Development, the Canadian Communication Association, and Canadian Healthcare Association are some examples), you can retitle this category "Professional Service" rather than "Community Service." If your experience is evenly mixed and significant in both professional and community service categories, you could include both on your résumé, or combine them. If the listings are primarily memberships in organizations of a professional or service nature, you may even want to call the category simply "Memberships."

Achievements or Accomplishments

Use this category to feature any relevant accomplishments not covered in any previous section. This section may include awards other than scolastic ones (for instance, Citizen of the Year), relevant certification of some form (pilot's license, WHMIS, or drafting certificate), publication of a book or article, presentation of a paper at a conference, or a special achievement in your work. Once again, in reverse chronological order, include date, name and location of institution, agency, or publisher, and nature of certificate, award, or publication. List such achievements only if they help qualify you for the job you seek, and title the section appropriately as "Publications," "Certificates," or "Achievements."

Web Pages or Desktop Publishing

Finally, if you have significant experience in developing Web sites or in publishing newsletters, pamphlets, or other documents using a desk-top publishing program, you might want to feature these in their own category on your résumé. For each entry, list the title, place, and date of publication, or URL. If they are few, these items can instead be included in one of the other categories such as "Activities" or "Achievements."

Layout and Visual Appeal

The appearance of the résumé—its visual appeal—is almost as important as its content. The résumé is seen by an employer before you are; it should therefore make a positive impression and establish your credibility. Just as you wouldn't dream of going to an interview with an unkempt appearance or a smudge on your face, you should never introduce yourself to a prospective employer via an unattractive, disorganized, or cookie-cutter résumé. Few employers will be interested enough to interview an applicant whose résumé is poorly constructed; to them a weak résumé suggests a lazy or unmotivated individual—not, of course, the kind of person they want to hire!

Remember that your command of nonverbal cues in your résumé, such as diction, sentence structure, grammar, spelling, punctuation, style, tone, and document design, can portray you positively to the employer as confident, enthusiastic, thorough, and reliable—or they can betray a lack of skill and command, and communicate the reverse. After all, if you cannot effectively manage the information on your own résumé, or if you have not taken care in preparing such an important document, the employer may doubt your motivation, your commitment to your tasks, or your understanding of the demands of the situation.

An effective layout builds visual appeal, and credibility, by creating a pleasing balance between white space and printed elements, with reasonable margins on all sides (1" to $1\frac{1}{4}$" at top, bottom, and both sides is standard). A consistent format also lends a professional appearance, as it does to all professional communication: major headings should begin at the left margin, lining up neatly beneath one another. They should all be presented in parallel format (all capitalized, for example, or all underlined). Subcategories should be indented so that they too line up consistently throughout the résumé, and they should also be consistently treated (all italicized, for example). Use consistent spacing between sections: for example, you might skip one line between subsections and two lines between major categories. This consistency is not only attractive, but, because the layout is also part of your organization, it actually helps the reader to make sense of the information you are presenting, and increases your credibility by revealing you as disciplined and capable.

Use capital letters, underlining, or boldfacing to set apart important details, but use these features consistently and sparingly. In general, you should not combine underlining with capital letters or with boldfaced or italicized type. Other combinations (boldfacing and block capitals, italics and boldfacing, italics and block capitals) are permissible, though these too should be used sparingly and consistently to help your reader make sense of the information on your résumé. A little of each goes a long way and overuse will destroy the effectiveness of these visual devices. For example, if too much of the résumé is printed in capital letters, it is not only harder to read, but important information no longer stands out.

In general, it is also not good practice to combine several different fonts in a single document, and you should especially avoid fonts that are difficult to read—gothic and script fonts, or those designed to look like hand-printing, are typically harder on the eye than a simple serif font; even some sanserif fonts are tiring if the document is lengthy. These should therefore be avoided in a résumé or application letter, which is at risk of being thrown out if it is difficult to read.

Avoid mixing font sizes as well; at most, you may want to use a slightly larger font for headings and a smaller one for information within categories. Finally, be sure to choose a font in a readable size for the body of your document—12–point type may take up more space than 10–point, but it is much easier to process visually. Never reduce the font to try to crowd more information onto the page. It is in your interest to design your résumé so that it's easy for the reader to find the information she needs to make a decision in your favour, and a cramped font will not leave your reader kindly disposed toward your application.

It should go without saying that visual appeal also means using a good quality printer, with sufficient toner or ink to produce sharp, clear type. Don't send out a résumé with faint print; the employer may simply discard the whole thing rather than struggle to read it. If possible, you should send an original résumé, tailored to the job, with every application you submit. Any photocopies you do provide should be perfectly clean.

There are nearly as many varieties of résumé layout as there are people to give advice, but some are better than others. How can you tell a good layout from one that is not so good? A good résumé format is easy to scan for important information and should be appealing to look at. Here are some characteristics of an effective résumé format. The examples that follow meet these criteria.

A good layout:

- maintains consistent margins and indentations throughout;
- uses white space effectively so that nothing is crowded or cramped, and everything is readable;
- is easily skimmed for main points. Use visual effects and indentation to make significant information easily available to the eye so that the reader quickly get the gist of the applicant's work history without having to read every detail (though necessary details are given too, of course);
- is professional-looking, but also flexible enough to be arranged to suit an individual applicant.

Never underestimate the importance of an attractive layout; your communication instructor may be willing to struggle to read your résumé to the end, but an employer does not have to do so. In fact, many employers can rapidly cut down a pile of résumés from two hundred to a short list of ten or twenty by immediately discarding unattractive résumés—without even reading them. Remember that, at least initially, the reader of your résumé is seeking reasons to *eliminate* your application, not reasons to keep it.

WHY RÉSUMÉS GET REJECTED

Recruiters reject application packages for many reasons, some of which might strike you as unfair. But in order to understand how the process works, and to give your own résumé a better chance to surviving the first cut, you need to recognize the context in which the submissions are read. Always remember that the reader has many résumés to consider and is looking for some key information. You should construct your materials to make that information readily accessible.

First, the employer is looking for evidence of competence, integrity, and the ability to work well with others. He or she will also be expecting to see some evidence of the person who wrote the résumé and will usually not be impressed by cookie-cutter productions. The recruiter will not read your résumé in isolation, the way you wrote it, but in the company of several dozen—even several hundred—other submissions. Thus, the reading of a résumé by its audience is always comparative. Finally, the recruiter is seeking a candidate who closely conforms to the expectations that have been laid out in the job posting.

Unless your résumé reveals a person who is both capable of performing the required job and willing to do what is required, a person who took care in preparing the résumé submission, it will in all likelihood be rejected. Here are some of the more common reasons employers give for rejecting résumés.

- Minimum required credentials not evident on first page (check the ad and make sure you display, in a prominent place, the credentials they require)

- Difficult to scan or find pertinent details, dates, or relevant categories
- No concrete details provided to support claims
- Listed skills are too general and not targeted to the job sought (check the ad, which will often tell you what skills the employer is interested in)
- Disorganized presentation (too many bullets, unclear divisions between categories, confusing margins, too many visual devices)
- Font size too small (remember that 12-point is easiest to read)
- Relevant information not emphasized
- Obvious lack of care: errors, inconsistencies, contradictions, grammar, or spelling mistakes
- No clear relationship to the job sought
- Too much irrelevant information
- Cookie-cutter structure
- Appears to have been prepared by a résumé service
- No cover letter accompanies the résumé
- Smart-alecky or too-cute message on the answering machine

SAMPLE RÉSUMÉS—INEFFECTIVE AND EFFECTIVE

The following résumés illustrate the principles discussed in this chapter (Figures 8.1 to 8.14). The first résumé of each pair is weak and in need of improvement; the second is an improved version. While the weak examples violate principles of relevance or recency and generally offer a poor layout, all of the samples of effective résumés follow the principles of effective layout. One way to check the effectiveness of a layout is to hold the résumé away from you. A good layout is appealing to look at even when it is held too far away for you to read the details, because the print and white space are balanced and the important information stands out from the rest of the material on the page. Study the before-and-after examples below, taking a critical look at what makes an effective layout and what looks unattractive.

The first set belongs to Cameron Britten. (See Figures 8.1 and 8.2.) Look carefully at both versions and at the commentary for each; can you see any additional weaknesses that Cam could improve?

Analysis of Figures 8.1 and 8.2

There are several problems with Cam Britten's first résumé. Like most people writing their first résumés, Cam has focussed on listing the things he has done rather than on the reader's needs or the purpose it should fulfill. Cam is justly proud of his impressive record of volunteer experience, but most of it is non-technical experience that has little value to the person trying to determine his suitability for a technical position.

Résumé

CAMERON L. BRITTEN

Current Address:	Permanent Address:
3987 Lois Lane	PO Box 234
Calgary, Alberta	Mudhen, Saskatchewan
T7U 4P1	S0R 1M0
(403) 424 2115	(306) 765 2121

FIGURE 8.1 (Continued)

CAMERON L. BRITTEN

3987 Lois Lane

Calgary, Alberta T7U 4P1

(403) 424 2115

EDUCATION:

(1999–present)

Western Plains University

Faculty of Engineering

Major: Mechanical Engineering

(1986–1999)

High School Diploma

Mudhen School

Mudhen, SK

EMPLOYMENT HISTORY

(2001 Summer)

Laborer/Farmer

Panelco Wall Systems

Construction of wood buildings

- •Cut and manufactured walls
- •Wall layout

(2000 Summer)

Pen Checker and Feed Truck Driver

Mitchell Dairy Farms

Basic farm operations:

- •Fed rations to cattle
- •Ensured herd health
- •Operated tandem silage truck
- •Maintained livestock facilities

(1992–1999)

Farm Laborer

Elmer Britten Farms

General farm labor (grain & cattle):

- •Including operatios of heavy machinery
- •Responsible for maintenance of livestock and livestock facilities

FIGURE 8.1 (Continued)

(1996–1997)

Grocer Store Employee
Mudhen Food Town

General operations of a Grocer Store

- Duties such as cashier
- Shelf stocking
- Deliver groceries
- Removal of garbage

ADDITIONAL SKILLS:

Communication Skills:

- Work effectively with the public

Computer Skills:

- Level 1 Auto-Cad
- Programming in Fortran 90
- Proficient in MS Word and MS Excel

Organizational Skills:

- Work productively in a group and alone
- Organize many events for WPESS
- Lead various groups and clubs
- Knowledge of Budget operations and organization

VOLUNTEER ACTIVITIES:

- Member of the Western Plains Engineering Students Society (W.P.E.S.S.)
- Social-Vice President for W.P.E.S.S.
- Internal Fundraiser for the W.P.E..S.S.
- Coordinator for Engineering Light and Magic
- Member on the Social Committee for W.P.E.S.S.
- Member of the council for organizing the annual "En-Nurgy" dance held by
Faculties of Engineering and Nursing.
- Vice president of the Student Representation Council for Mudhen School
- Graduation council for organizing the Gr 12 Graduation
- Main organizer in Convoy 97 a Ski Trip of the graduating students of Mudhen High

ACTIVITIES AND INTERESTS:

- Sports:
active in Campus Rec Hockey and previousl a member on a Senior hockey team,
volleyball, badminton, sky-diving and water sports
- enjoy working with large animals such as horses and cattle

REFERENCES: Provided on request

CAMERON L. BRITTEN

3987 Lois Lane, Calgary, AB T7U 4P1

(403) 424 2115 <cabri@altamail.ca>

GOAL   An internship position in mechanical engineering.

EDUCATION

BSc Program, Mechanical Engineering   (1999–present)

Western Plains University, Calgary, AB

•Currently completing third year of a four-year program

RELEVANT EXPERIENCE

Design Team Leader   2001

MEng 451—Engineering Design

•Designed and fabricated an automated tray-mover for medical research

•Prepared comprehensive design report and public presentation

•Led a team of five

Construction Worker   1999 (Summer)

Panelco Wall Systems, Calgary AB

•Cutting and manufacturing wall components and systems

•Wall layout

Design Project   2000

MEng 251—Introduction to Engineering Design

•Part of a five-person design team

•Developed and built track switcher for model train system

•Contributed to writing the design report and planning the presentation

RELEVANT SKILLS

Engineering and Fabrication Skills

•drafting, computer and mechanical

•welding and metal shop

•wood construction and fabrication

•experience operating heavy equipment

FIGURE 8.2 (Continued)

Computer Skills:
- Level 1 Auto-Cad
- Programming in Fortran 90
- Proficient in MS Word and MS Excel

Organizational and Communication Skills:
- Work productively in a group and alone
- Extensive leadership experience
- Experienced and effective public speaker
- Knowledge of budget operations and organization
- Work effectively with the public

SUMMER EMPLOYMENT

Feed Truck Driver and Pen Maintenance   1998 (Summer)
Mitchell Dairy Farms, Mudhen SK

Farm Laborer   1992-1999
Elmer Britten Farms, Mudhen SK

Grocery Clerk and Cashier   1996-97
Food Town, Mudhen SK

VOLUNTEER ACTIVITIES (selected)
- Member of the Western Plains Engineering Students Society (W.P.E.S.S.)
- Social-Vice President for W.P.E.S.S.
- Internal Fundraiser for the W.P.E..S.S.
- Pathfinder Coordinator for Engineering Light and Magic Science Show
- Member of the Social Committee for W.P.E.S.S.

REFERENCES

Dr. G.M. Hardbody   Mechanical Engineering Department
(403) 555-4750   Western Plains University
                         P.O. Box 666
                         Calgary, AB T7N 0W0

Jim Horowicz   Panelco Wall Systems
(204) 2887652   Calgary, AB   T1W 1F8

You will notice that Cam has included a cover page on his résumé. This serves no useful purpose since the information is repeated on the first page of the actual résumé, and it increases the length unnecessarily. Cam should discard the superfluous first page and proofread more carefully to catch mistakes in grammar and spelling. He needs to cut down on the information that is of little benefit to the potential employer, and add some items from his engineering program that would catch the eye of someone hoping to find a research assistant.

Cam's new résumé (Figure 8.2) focusses on his goal of an internship position, and features information that would be of value and interest to a potential employer. The format is much more readable, and Cam is able to establish his credibility through effective selection and organization of information relevant to the position he seeks. He has downplayed the social emphasis that distinguished his first attempt, and has instead included some details about the projects he has worked on. This new résumé has also been carefully proofread and will help to establish his credibility with his potential employer.

Notice how the categories Cam has selected, along with the new ordering of these categories, focus on the qualities and skills the employer is likely seeking. Remember that the first time through the employer will quickly scan the résumé to determine whether it should be retained for further consideration. Because the new design directs the employer's attention to the skills and experience that have prepared Cam for a technical internship, he has a much better chance of making it past the first cut.

Notice also how Cam has refined and targeted his presentation of skills to the probable interests and needs of the employer. By carefully considering the purpose of the résumé, the needs and interests of the potential employer, and his credibility as a candidate, Cam has been able to redesign his résumé into a more professional and effective tool, and he has brought himself much closer to the internship he's hoping for.

Analysis of Figures 8.3 and 8.4

The first draft of Shawn Horton's résumé (Figure 8.3) has some flaws that undercut his credibility. Like Cam before him, Shawn has failed to address the probable needs of his potential employer, and he has not effectively targeted his résumé to the job he is seeking. Shawn created his résumé using a template that accompanied his word-processing package; unfortunately for him, the effect of this template has been to make him appear very much like every other candidate in the competition.

Shawn has also fallen into the trap of including in the résumé every piece of information about his background, even if it's not relevant to the employer. For example, Shawn began his studies in Arts and Science and entered engineering after his second year, a common path at many universities. This information should not be featured separately on the résumé, since what counts from the employer's point of view is the programme from which he received

FIGURE 8.3 How effective is this first draft of Shawn Horton's résumé?

Shawn Horton
1234 – 5ᵗʰ St. W
Fredericton, NB
E8J 0S3
(506) 345-6789

EDUCATION

2002-04	Faculty of Engineering, *Central Maritime University* Mechanical Engineering
2001-02	Faculty of Arts and Sciences, *Central Maritime University*
2001	Graduated with distinction, and as class valedictorian from Willis Bowler High School, Hindenberg, NB
2003	St. John's CPR, and Safety Oriented First Aid. TDG, WHMIS H2S Alive - H2S gas safety training.

EMPLOYMENT HISTORY

May 1 to Sept 10 2003	*North American Resources, Saint John, NB* *Summer Student.* Duties included the operation of various water injection plants and oil batteries. A large amount of responsibility and trust was placed on my behalf, as I spent most of the time working independently (having to make my own decisions), with expensive and dangerous equipment. I was also entrusted with a company vehicle for the summer.
May 2 to Sept 8 2002	*Bob's Building, Saint John, NB* *Assistant Carpenter.* Free use of all tools. Often worked independently.
2002	*A Buck or Two, Edmunston, NB* Ran till, stocked shelves, aided customers.
2002	*Oscar's General Merchants, Edmunston, NB* Ran till, stocked shelves, aided customers.

FIGURE 8.3 (Continued)

Shawn Horton
1234 – 5th St. W
Fredericton, NB
E8J 0S3
(506) 345-6789

SKILLS

Team skills	Enjoy working with parteners. Am able to work efficiently within a group, and able to motivate if necessary.
Independent	Able to make independent decisions based on my evaluation of the situation.
Intelligent	70% avg. last term in Engineering (full course load).
Leadership	Ability to take charge when it becomes necessary. Have organized campus intramural teams, high school SRC vice-president, coached junior soccer, junior leader in 4-H, Social Coordinator for Clegg Hall Residence, engineering "Peer Mentor Program."
Computer skills	Experience with a large amount of applications (both DOS and Windows,) programming (FORTRAN, C/C++), hardware (from drivers to installation).
Problem solving skills	Find solving problems enjoyable and rewarding.

EXTRA CURRICULAR

Sports	Soccer, hockey, fastball, volley ball, cross-country, mtn. biking, skiing.
Interests	Electronics, programming, mechanics, music, reading.
Organizations	UCESS, 4-H Alumni, National Geographic Society
Volunteer	Meals-on-Wheels, coached junior soccer

REFERENCES
available on request

FIGURE 8.4 Shawn has redrafted his résumé to better present himself and his skills to an employer. What improvements can you see?

SHAWN HORTON

1234 – 5th St. W, Fredericton, NB E8J 0S3

(506) 345-6789   **<shorton@atlantis.nb.ca>**

EDUCATION

2004   **BSc, Mechanical Engineering**
                Faculty of Engineering, Central Maritime University
                • 70% avg. last term in Engineering (full course load)

2001   **High School Diploma** (with distinction)
                Willis Bowler High School, Hindenberg, NB.
                    • class valedictorian

CERTIFICATES

2003   St. John's CPR, and Safety Oriented First Aid.
                TDG, WHMIS
                H2S Alive - H2S gas safety training.

RELEVANT EXPERIENCE

Summer   **Engineering Summer Student**
2003   North American Resources, Saint John, NB
                    • operated water injection plants and oil batteries
                    • worked independently with expensive and dangerous equipment
                    • entrusted with a company vehicle

Summer   **Assistant Carpenter**
2002   Bob's Building, Saint John, NB
                    • skilled on various construction and woodworking tools
                    • frequently worked independently.

FIGURE 8.4 (Continued)

Shawn Horton (506) 345-6789

DESIGN PROJECT EXPERIENCE

2004   **Free-Motion Wheelchair Stabilizing Mechanism**
- judged top design in internal competition
- involved analysis, testing, and fabrication of prototype
- formally presented to a public audience

SKILLS
- teamwork
- troubleshooting and problem-solving
- leadership
- computer programming (FORTRAN, C++)
- public speaking
- clear and effective writing

ACTIVITIES
- Mentor for Faculty of Engineering "Peer Mentor Program"
- Social Coordinator for Clegg Hall Residence
- organized campus intramural teams §Meals-on-Wheels volunteer
- coached junior soccer
- Student Council vice-president, high school
- junior leader in 4-H
- enjoy sports, including soccer, hockey, fastball, mountain-biking, skiing

REFERENCES

Professor C.S. Major (506) 555 6789
Department of Mechanical Engineering
Central Maritime University
PO Box 5005, Truro, NS B2S 4M3

Professor D.A. Broker (506) 555 0987
English Language and Linguistics
Central Maritime University
PO Box 5005, Truro, NS B2S 4M3

the degree in the end. As well, the way he has listed the details takes up unnecessary room without communicating any significant information and in fact could potentially create the impression of a spotty school background.

Like Cam, Shawn needs to run his spell-checker and proofread more carefully. As well, he needs to consider the effect on an employer of some of what he has said on his résumé: under skills, for example, he has listed "intelligent," a claim that might strike the reader as immodest or even arrogant. Remember that the employer is seeking confidence, not arrogance, and is not likely to be impressed by someone who labels himself "intelligent," "gifted," or "impressive." And consider this: What if the employer graduated from engineering with an average of 95%? Is such a person likely to find 70% impressive? What about an employer who graduated with 50%? Would this person feel that grades are a reliable indication of intelligence? Since it can potentially be taken negatively, Shawn would do well to eliminate this problematic statement entirely.

The improved résumé (Figure 8.4) is easier to scan, and is targeted to catch the reader's attention; like Cam, Shawn has chosen to feature the design experience gained in his classes. He might have chosen to leave out his grade average entirely, but he feels it's important to mention; however, instead of featuring it under the heading of "intelligent," he has presented it in a less problematic way, along with his degree as part of his education. Note that Shawn has moved the industrial certificates into their own category, since they aren't properly classified as "education," and he has focussed his skills to better serve the employer's needs. To assess the effectiveness of Shawn's targeting, turn to his letter of application (Figure 8.10) on page 289.

Analysis of Figures 8.5 and 8.6

Ghirmay Zakaluzny can effectively present his information on one page, but he hasn't done so here. (See Figure 8.5.) It's hard to scan, and is cluttered even in this short space with information that is hardly relevant to his job search. In particular, the space that is used for the personal information is wasted—the details of age, marital status, and health are not considered acceptable, and his personal contact information need not take up five lines. Finally, there is no need to label the document "résumé," since it's pretty clear what it is. In all, he has taken up fourteen lines in the most prominent position on the resume to communicate information that could be communicated in three lines, or deleted entirely.

The skills on this résumé need to be reconsidered and ordered in a more reader-friendly way, and the whole thing needs a visual overhaul to make it more appealing to the eye.

Ghirmay's new résumé, though no longer than the original draft, is much more effectively presented than his previous attempt. (See Figure 8.6.) It is much clearer and visually more attractive. Ghirmay has eliminated the

FIGURE 8.5 What changes would you recommend to make Ghirmay's résumé clearer and more reader friendly?

Résumé

GHIRMAY ZAKALUZNY
5321 - 56 Ave, #234
Vancouver, BC
V8U 1Q3
Tel. (604) 234-0981

Personal:
Age - 24 years old
Height - 5'8"
Weight - 145 lbs
Health - excellent
Maritial status - single
S.I.N. - 612-345-678

Education:
Second year of 4-year Electrical Program
West Coast College
Burnaby, BC

Grade 11, Surrey Composite
Surrey, BC

Occupation:
Second-year apprentice electrician
Experience: - 3600 hours in electrical trade
- industrial installations
- commercial installations
- residential and rural construction
- overhead and underground service installation
- good background in motor control systems, heating and cooling systems, fire alarm systems
- some general maintenance and trouble-shooting experience
- knowledge of program development for Allen Bradley programmable logic controllers, series 5

Related experience
- good mechanical and electrical draftsman
- can read and work from schematic wiring and architectural diagrams
- can convert schematic drawings to P.L.C. logic

Workshops:
- C.P.R. certification
- Power Actuated Tools Certification

Interests:
Photography, swimming, fishing, hiking, camping, skiing

References:
References will be provided upon request

FIGURE 8.6   What improvements has Ghirmay made in both format and content that will make his résumé standout to an employer.

GHIRMAY ZAKALUZNY
5321 – 56 Ave, #234, Vancouver, BC   V8U 1Q3

(604) 234-0981   e-mail: <gzaka@empatico.bc.ca>

EDUCATION

2000-2002   **Electrical Engineering Technician**

West Coast College, Burnaby, BC

EXPERIENCE

2002-Present   **Apprentice Electrician**

City Electrical, Vancouver, BC
- 3600 hours logged
- industrial and commercial installations, overhead and underground
- residential and rural construction

RELATED SKILLS

- good background in motor control systems, heating and cooling systems, fire alarm systems
- some general maintenance and trouble-shooting experience
- good mechanical and electrical draftsman
- can read and work from schematic wiring and architectural diagrams
- can convert schematic drawings to P.L.C. logic
- knowledge of program development for Allen Bradley programmable logic controllers, series 5
- Power Actuated Tools Certification
- C.P.R. certification

INTERESTS

Photography, swimming, fishing, hiking, camping, skiing

REFERENCES

Provided upon request

unnecessary personal details and the header "résumé," and has selected details on the basis of relevance. The new design is more eye-catching than the boring presentation he had previously used, and is more likely to attract an employer's attention and confidence. This résumé communicates more than the information it contains: its nonverbal cues communicate a confident, focussed individual who will likely rate an interview.

Analysis of Figures 8.7 and 8.8

Figure 8.7 is an example of a very disappointing résumé from someone who is actually very capable. Not only is the format confusing and hard to read, it fails to communicate the writer's contact information in a prominent place. As well, it contains numerous irrelevant details and is not effectively targeted to a technical job.

This writer should display his contact information prominently on the first page, so that the employer is familiar with his name right from the start. He should de-emphasize or delete the irrelevant information, leaving only that which demonstrates how he has prepared himself specifically for a career in chemical engineering. He should also focus the remaining information to accommodate the employer's needs, and he should polish his format to communicate important details more clearly and effectively. Turn to the improved version (Figure 8.8) to see how effectively he has accomplished these goals.

Kiro has improved his résumé dramatically. It now communicates professional credibility in its content, organization, and design, and is more likely to attract the attention of an employer who hopes to fill an entry-level engineering position. Kiro's strengths are much more visible in this well-designed résumé that will distinguish him from other applicants who have relied on a cookie-cutter approach. Notice in particular how Kiro has selected categories that display his background to greatest advantage, and how he has arranged them in order of relevance to the job he seeks. Finally, the layout allows the reader to scan quickly to achieve an overview of the candidate's background, but also allows for additional information to support and elaborate the information provided in the headers. The new résumé is not only more informative and selective, it is also more readable and more credible than Kiro's initial attempt.

WRITING THE APPLICATION (COVER) LETTER

A résumé is always accompanied by a letter of application, also called a cover letter. If you are submitting your résumé as an electronic attachment, your e-mail message may function as your cover letter and should follow the same guidelines as those used for a paper submission. Like all professional

FIGURE 8.7   What flaw can you spot immediately in this weak résumé ?

Abilities and Skills

-familiar with use of an Excel spreadsheet

-honesty and integrity are regarded highly

-have excellent social skills

-have typing training

-have experience dealing with dangerous goods

-familiar with the operation of a forklift

-possess an air brake endorsement on a class 5 license

Education

June 2000	Grade 12 – with honours
	Rosston Central High School
Sept 00–Dec 00	All Gospel Bible Institute
	Bible Knowledge Program
	Glory, SK
Sept 01–Present	Tyrell University
	College of Engineering
	Pursuing a Bachelor's Degree in
	Chemical Engineering

Personal Interests

-enjoy basketball and other such sports

-enjoy playing pool

-enjoy walking, biking, swimming, and other physical activities

-enjoy listening to music

Reference

Kurt Dedalopolous

President of Dedalopolous and Associates - Biggar

(306) 882-3575   (Home)

FIGURE 8.7   (Continued)

Work Experience

September 2003–August 2004   Chem-Tech of Calgary

Position: Intern

Duties: - Apply advanced computing skill to design and imple-
ment new software technology in the Delphi (Object
Pascal) programming environment.
- Identify and solve engineering and software problems.
- Pinpoint sources of error by debugging the software
code and examining simulation results.
- Work as part of a high performance team committed to
producing high quality engineering software.

May 2003–August 31 2003   Town Transportation Department
Lougheed Transit Services

Position: Utility Person - Maintenance

Duties: - Responsible to clean the interiors of city buses
- Assist in the changing and retorqueing of tires
- Perform general custodial duties in the garage and
maintenance shop
- Safely operate the buses about the city and in the
garage
- Assist in ensuring that bus shelters are clean and free
of safety hazards

Supervisor: Devorek Andrusiak   (306) 864-4245
May 2001–September 2001
April 2002–September 2002   Lougheed Transportation Company
Lougheed Depot

Position: Express Services Attendant I

Duties: -Responsible to assist in the loading of buses.
- Responsible to assist in the sorting and putting away
of freight.
- Responsible to serve customers at the front counter.

FIGURE 8.7 continued

February 2000–August 2000 December 2000–May 2001	Biggar Petro Canada Service Station and RTC bus depot
Position:	Manager of Depot / Gas Attendant
Duties:	- Responsible to make sure that people and business are informed of newly arrived parcels. - Responsible to send, receive, and sort freight. - Responsible to give information upon request concerning arrival and departure times of bus routes, passenger ticket prices, and shipping costs. - Performed Gas Attendant duties and some custodial duties.
September 1998–February 2000	Biggar Burger Queen
Position:	Cook
Duties:	- Preparation of hot food and ice cream products. - Operation of cash register. - General Cleaning.
July 1997–June 1998	Biggar Co-operative Grocery Store
Position:	Grocery Packer
Duties:	- Bagging groceries and taking them to customer's vehicles on request. - Stocking shelves and working with fresh produce and dairy shelves.

For further information please contact me at:
  Phone: (306) 373-3313
  Email: <klh117@stdnts.tyrellu.ca>

FIGURE 8.8   How much better is Kiro's improved résumé?

Kiro Hatzitolios

#2 – 2710 Main St., Regina SK, S4R 0M3

Email: <klh117@stdnts.uregina.ca> Phone: (306) 343-0213

EDUCATION

Tyrell University, Lougheed, SK   2001–Present
BE, Chemical Engineering

Academic achievements: • member, Golden Key National Honour
                             Society
                           • improved my academic performance each year
                           • 84% avg. last year
Relevant Skills   • work productively in a fast-paced environment
                           • strong problem-solving ability
                           • advanced programming skills
Graduation: • expected May 2005

RELATED EXPERIENCE

Chem Tech Ltd., Calgary, AB   2003–2004
Intern Engineer in Batch Design Kit (BDK) Development

Responsibilities: • design and implement new software in Delphi
                             (Object Pascal) programming environment
                           • identify and solve engineering and software
                             problems
                           • pinpoint sources of error and debug software
                           • part of a high-performance team producing
                             engineering software

Tyrell University, Lougheed, SK   2001 (Summer)
Undergraduate Research Assistant, Chemical Engineering

Responsibilities: • develop software package to model a Packed
                             Bed Biofilm Reactor
                           • create a proposal outlining the goals of the
                             project
                           • correspond frequently with Saskatchewan
                             Research Council
                           • prepare final report and formal presentation
                             describing the software and proving the
                             integrity of the mathematical model

FIGURE 8.8 (Continued)

Kiro Hatzitolios (306) 343-0213 /2

PRESENTATIONS AND TECHNICAL PAPERS

A Computer Model for Characterizing the Packed Bed Biofilm Reactor

Canadian Society for Chemical Engineering Conference 1999

Saskatoon, SK (October).

Third Place Winner

Communicating Value: Rhetoric and the Undergraduate Degree.

Communicating across Boundaries Conference 1999

(Panel participant) Tyrell University, Lougheed, SK (September)

ENGINEERING PROJECTS

Knock Out of H$_2$S in Vent Gas   Ongoing

GEng 465, Tyrell University

Responsibilities:
- design a method of removing H2S from the vent gas of a local canola crushing plant
- work with all involved to provide an effective, economic, and practical solution
- present final design to colleagues, college professors, and industry professionals

Preliminary Economic Feasibility Analysis   2002

GEng 365, Tyrell University

Responsibilities:
- evaluate economic feasibility of trucking coal from a mine to seaport
- merge reports of members of design group to create final report
- carefully show all calculations and assumptions

REFERENCES

A list of those who have agreed to supply references is attached to this application.

correspondence, a job application letter must follow the Seven Cs: it must be complete, concise, clear, coherent, correct, and courteous, and it must also establish your professional credibility.

Your paper submission should also follow the format of a standard letter—preferably the full block style. If you have developed a letterhead-style header for your résumé, you should employ the same header for your letter of application so that the two are unified visually. As well, use the same font for the text of your letter as you used in the body of your résumé, and in the same size.

Two Types of Application Letters

The *solicited* letter is written in answer to an advertisement for an available position, while the *unsolicited* letter is written to a company or organization in the hope that a suitable position is currently, or is about to become, available. The advantage of the second, if your timing is right, is obvious: there will be less competition than for an advertised position. However, the risk is that there may not be a job available.

Whether written in response to a specific posting or written "on speculation," both letters perform essentially the same task, and the process of writing them may be broken down into steps. How many paragraphs each step takes will depend on the background of the individual and the nature of the job being applied for, but as a general guideline, you may assume that each step is treated in a separate paragraph.

Remember that the application letter, like all professional writing, must be carefully directed to your reader's requirements and expectations. Its sole purpose is to engage the reader's attention and direct it toward your résumé. Thus, what you might want to say is not as important as what the employer needs to hear, and what you choose to put into the letter should be conditioned by the information the reader needs to make the decision to interview you. As you write the letter, you should draw on the information you gathered from your survey of advertisements and position descriptions; use that knowledge to shape the way you present your information. Always ask yourself what questions an employer will want your application to answer. Targeting the letter effectively to the reader and the job will help to demonstrate your own good judgement and professional credibility.

The following steps can serve as a guideline; you could plan to allow one paragraph for each step, though each step may be shorter or longer than a paragraph.

Step 1. The first thing an employer will want to know upon receiving a letter is "What is this about?" As in all professional communication, you need to put your main message first; begin by identifying your reason for writing. If you are answering an advertisement, state the title of the position you are applying for, quoting the competition number if there is one, and the source and date of the advertisement. For an unsolicited letter, state clearly the type of work you desire and enquire whether such a position is currently open or soon to become available. You may wish to use a "re" or subject line for both types of letters, identifying by title or type of position sought.

Step 2. The next thing the employer will want to know is, "What qualifications does this applicant have for the job?" Whether your letter is solicited or unsolicited, you should next provide a very brief outline of the highlights of your background and your reasons for applying for this position, adapting the information to the job requirements. This section need not be elaborate, since your résumé will take care of the details, but it should provide some legitimate reason why you would be a suitable candidate for the position, based on how well your skills and education fit the position as described in the advertisement. Remember that the purpose of the letter is to direct the employer's attention to your résumé, which accompanies the letter, and to motivate the employer to read it. In this paragraph, you may wish explicitly to refer the employer to the résumé.

Step 3. An employer's third concern will be what makes you the best choice from the list of similarly qualified candidates for the job. You will want to be specific about the close match between your skills and those sought by the employer. You should highlight some particularly relevant skills and details from your résumé, qualifying these with brief examples appropriate to the job you seek. This is where you really emphasize your appropriateness for the position; show the employer what you can do for him or her and emphasize your strengths.

Step 4. If you have not yet referred the employer to your résumé, do so at this stage; mention also that you have attached letters of recommendation or other documentation (if you have done so) or invite the employer to contact the references you have listed.

Step 5. Close with a statement of confidence in your abilities; thank the employer for considering you and ask for an interview. You might even say that you will call the employer on a specified date to set up an interview. This is a good move if you can carry it off. However, if the idea of cold-calling an employer to request an interview makes you uncomfortable, and you really can't imagine yourself making such a call, don't say that you will do so. Instead, request an interview at the employer's convenience. Provide a telephone number where you may be reached or where a message may be left, and then make sure that the message on your answering machine conveys a suitably professional image. Figures 8.9 and 8.10 are effective examples of the solicited and unsolicited application letter.

FIGURE 8.9 How effectively does this solicited letter of application target the employer's needs? What risks does it take? Would it work for all applicants? Why or why not?

#2 – 2710 Main St.
Regina SK, S4R 0M3

3 March 2003

Wayne Chiu, Manager
Chem-Tech Ltd.
3267 Periphery St
Calgary, AB T2X 6L7

Dear Mr. Chiu:

Re: Chemical Engineering Internship Opportunity

If you are looking for a committed chemical engineering student with strong technical and lab skills, you'll want to put me on your interview list. In addition to a solid academic record (85% overall this year, 87.5% in my engineering courses), I bring an unusual level of determination and self-discipline. Let me give you an example of what I mean: when I began my engineering studies, my lack of high school calculus put me at a disadvantage. Consequently, I left first year with a 62.5% average in math. However, once I saw what was needed, I rolled up my sleeves and got to work, and within a year I had pulled up my grade to 85%—an increase of more than 20% when others' grades were falling. This is the same level of discipline and commitment I can bring to an internship position with Chem-Tech.

In addition to determination, I can bring solid technical skills developed from course work in my engineering program. I have learned to think through problems logically and to propose solutions based on sound application of mathematical and chemical principles, and have performed at the top of my class in fluid flow, thermodynamics, and mass balances. I am experienced in object-oriented programming, and have used both Delphi and Visual Basic.

My communication skills have been sharpened both through a course in oral and written communication and through my work experience to date. I have functioned as part of an effective unit, and I'm comfortable working with a variety of people.

As my resume illustrates, I am ready and willing to put my skills and enthusiasm to work for you. I would very much appreciate the opportunity to speak with you in an interview, and may be reached at (306) 343-0213 or by e-mail at <klh111@stdnt.uregina.ca>.

Sincerely,

Kiro Hatzitolios

Kiro Hatzitolios

FIGURE 8.10 What features of Shawn's letters enable him to target the needs and expectations of the employer? Compare his letter with the one written by Kiros Hatzitolios; what differences can you see?

1234 - 5th St W
Fredericton, NB E8J 0S3

June 27, 2004

Eastern HR Consulting Ltd.
FAX: (506) 215-2151
Email: careers@easternhr.com

Re: **Quality Assurance Specialist**

Based on the qualifications outlined in your posting for this position, I believe I may be the person you're seeking. Please add my application to the list.

I graduated in June with a degree in Mechanical Engineering from the University of New Brunswick, where I've been fortunate to receive a first-class technical education that has provided a strong knowledge base and advanced problem solving skills. My technical course work has included materials and manufacturing, mechanics of materials, and two courses in engineering and industrial design. Lab work involved analysis and fabrication to client specifications, including trouble-shooting and diagnostics.

In addition to this technical education, I believe communication and report writing are also highly important. I've had the opportunity to extend my communication and leadership skills through advanced study and through participation in the UNB Toastmasters Organization, where I sit on the executive. Engineering students are required to take an oral and written communication course as part of their program of study; however, I was able to take advantage of two advanced courses in communication theory and practice, from which I gained additional experience in writing and public speaking, as well as in understanding the dynamics of communication interaction generally.

If you are seeking a capable communicator who brings solid technical background in product fabrication, I would be pleased to hear from you when you begin interviewing for this position. I may be reached at (506) 345 6789 or by e-mail at <shorton@atlantis.nb.ca> Thank you for considering my application.

Sincerely,

Shawn Horton

THE APPLICATION FORM

Completing an application form is relatively straightforward once you have finished preparing your résumé. Though it may seem initially like an unnecessary overlap in information, there are good reasons for completing an application form in addition to your résumé. In some firms, the two are kept on file in different locations: the résumé goes to the department in which you would be working if you were hired, while the application form is kept on file in the personnel department. Thus, some companies will require that you both submit a résumé and complete an application form. In fact, some prefer to receive only a form, because it standardizes information and makes comparison between applicants easier. Because you may be asked to present your credentials using a standardized application form, you should learn strategies for presenting your information to advantage.

Forms vary considerably in their thoroughness. Some are little more than data sheets that allow no room for elaboration. These are normally used in conjunction with the résumé and cover letter. Others are more detailed, with sections that allow you to present additional information. It is the latter kind with which we are concerned here.

Some employers use a form produced by their own company, whereas others use a standardized form such as the one created by the Canadian Association of Career Educators and Employers. A copy of this detailed application form may be required by the firm you are applying to; if so, you can find it on line at <http://www.cacee.com/english/students/cacee_application.html>.

The sample application form (Figure 8.11) is similar to the CACEE form in the level of detail it requires. Though completing this kind of form can be demanding, it generally offers an advantage over briefer versions because it allows room to communicate any special skills and abilities that might be overlooked on a shorter form. It can thus help you to make a stronger impression on the employer. However, this advantage exists only if you take the opportunity to showcase your unique abilities and to make yourself stand out from the competition. Be sure to take the time to complete all the sections thoughtfully and carefully.

Your instructor will likely provide you with a full-sized version of the application form shown in Figure 8.11, or you may be asked to complete the CACEE form as part of your résumé submission. In filling out either form, complete all sections as fully as possible, taking special care to fill in areas that ask for elaboration on the form's standard questions. On the form below, you will find such areas under the headings "Career Goals" and "Areas of Expertise." It is important to take advantage of these sections, not only because they allow you to distinguish yourself from other applicants, but also because employers frequently complain that applicants fail to fill out forms completely.

FIGURE 8.11 A comprehensive application allows you to showcase your abilities.

APPLICATION FOR EMPLOYMENT

Name:_____
  (last)   (first)   (middle)

Address:_____

_____

EMPLOYMENT HISTORY: List in reverse chronological order, beginning with most recent.

<u>From</u>   <u>To</u>   Position <u>Held</u>   Name <u>of</u> <u>Firm</u>

_____

  Duties:_____

_____

_____

_____

<u>From</u>   <u>To</u>   Position <u>Held</u>   Name <u>of</u> <u>Firm</u>

_____

  Duties:_____

_____

_____

_____

<u>From</u>   <u>To</u>   Position <u>Held</u>   Name <u>of</u> <u>Firm</u>

_____

  Duties:_____

_____

_____

_____

<u>From</u>   <u>To</u>   Position <u>Held</u>   Name <u>of</u> <u>Firm</u>

_____

  Duties:_____

_____

_____

_____

FIGURE 8.11 (Continued)

From   To   <u>Position</u> Held   <u>Name of Firm</u>

_____

     Duties:_____

_____

_____

_____

Do you type? _____ wpm speed _____

Do you drive? _____ Licence _____

Do you speak French? _____ fluently _____ well _____ some _____

Do you write French? _____ fluently _____ well _____ some _____

Are you computer literate? _____ If yes, explain _____

_____

CAREER GOALS: Briefly describe the nature of the work you are interested in and the

origin of this interest:_____

_____

_____

_____

_____

_____

_____

_____

_____

_____

_____

_____

_____

_____

FIGURE 8.11 (Continued)

EDUCATIONAL HISTORY: List in chronological order, beginning with most recent and working back

From _____ To _____ Institution _____

Program Title. _____ Date _____ Diploma Rec'd _____

  Details: _____

_____

_____

_____

From _____ To _____ Institution _____

Program Title. _____ Date _____ Diploma Rec'd _____

  Details: _____

_____

_____

_____

From _____ To _____ Institution _____

Program Title. _____ Date _____ Diploma Rec'd _____

  Details: _____

_____

_____

_____

RELATED ACTIVITIES: Describe any school- or community-related activities in which you have taken part, including offices held:_____

_____

_____

_____

_____

_____

_____

_____

FIGURE 8.11 (Continued)

AREAS OF EXPERTISE: Elaborate on the above factual material by outlining briefly any skills you have developed from your experiences, any strengths you can bring to your position, or any information not already covered above.

_____

_____

_____

_____

_____

_____

_____

_____

_____

_____

_____

_____

_____

_____

_____

_____

_____

_____

_____

To my knowledge, all of the above information is true and accurate.

Date _____ Signature _____

It is especially important to complete the flexible categories of "Career Goals" and "Areas of Expertise." These are areas that you can use to your advantage, elaborating on strengths that may not have shown up clearly in the rest of the application form. Whether you are working with the form below, or using the on-line CACEE version, read the form carefully before completing it fully. When completing the narrative sections of the form, do not simply cut and paste from your résumé and cover letter. Instead, re-write the information in other words, so that an employer reading the submission will see that you have taken the time to complete the form thoughtfully. If you are completing the form by hand, write in ink. Check your spelling and grammar, and the accuracy of all your details. Finally, be sure to sign and date the form where indicated.

THE LETTER OF RECOMMENDATION

At times, employers will ask for references from people who have known you professionally, either as an employee or as a student. Sometimes you can simply append a list of names and contact information to your résumé, but at other times you may need or want to provide a written record of the person's impressions of you.

For a number of reasons it is a good idea to request a letter of recommendation whenever you leave a job or an educational institution. If you apply for a job in an area far away from your previous location, a prospective employer may not bother to contact a referee who is a long distance away, and may instead hire someone whose references are easier to check. A letter might be sufficient to allay this concern. As well, people who have been familiar with your work move on, retire, or get promoted, or just plain forget you. A letter written when your performance is current and fresh in the person's mind is more desirable than a vague recollection written long afterward, which may not be as enthusiastic. Finally, the act of writing a letter for you creates a record of the referee's impressions of your experience and capabilities. Then, should you need to contact that same person later for a reference, you can provide her with a copy of the earlier letter, which will refresh her memory should it have been a long time since she worked with you.

You should know what a letter of recommendation ideally includes because you will probably be asking people to write them for you; eventually you may even be writing them for others.

A letter of recommendation is usually written by someone who knows you in an employment or educational context; people who have supervised you either at school or on the job will be considered by potential employers as reliable sources of information about your work habits and attitude. Most employers no longer accept personal references, and you should not include one of these unless it is specifically requested. In such a case, the personal letter of reference should be written by someone who can evaluate you objectively and who will be viewed by the employer as a credible source (from the employer's point of view, credibility may be at least partly a function of the person's status in the community). A physician, a member of the clergy, or a professional in your chosen field is usually a good choice for a personal recommendation, while a relative or friend would be considered biased in your favour and therefore unreliable as a reference. Remember, prospective employers want information regarding your ability to do a job, and they will want this information from as objective a source as possible.

When you are arranging for letters of recommendation, it is courteous to allow your referees plenty of time to prepare them. Since your former employers and professors are likely to be very busy, and a thoughtful and careful letter takes time to compose, last-minute requests for letters may lead to disappointment. At the very least, such requests are not appreciated by your referees. Always allow sufficient time for your referees to write as carefully and thoughtfully as you, and they, would like.

In general, an employment reference comes from a former employer or a professor who knows you well. No matter who is writing the letter of recommendation, its contents are approximately the same. The writer of a letter of recommendation, just like any other letter writer, should adhere to a standard letter format, and should take care that the content and focus of the letter demonstrate the writer's own credibility. He or she must also consider the needs of the reader and provide answers to the questions the employer is most concerned about.

1. *How long, and in what context, have you known the person you are writing about?* The writer must indicate the relationship (supervisor? employer? instructor? academic advisor?) to the job applicant, and the length of time that he or she served in that capacity.

2. *What is your estimation of this person as an employee or a professional or, if it's an educational reference, as a student?* The writer should provide some specific examples—mention grades achieved, work completed, duties performed in the position, record of advancement, quality of work, or outstanding achievements.

3. *What is your estimation of the subject's personality?* Employers want to know what kind of person they are considering for a position. Will she get along with others? Is he flexible? Cooperative? Reliable? Personable? Stable? Motivated?

4. *Close with a strong statement of recommendation for the person* and invite the employer to contact you for further information. A reference writer who can't strongly recommend the applicant should not be writing the letter in the first place.

The format for a recommendation letter is identical to that of other professional letters. Naming the person about whom you are writing in a "re" or subject line will help the reader more easily identify who is being described in the letter. Figure 8.12 demonstrates an ineffective letter of recommendation; Figures 8.13 and 8.14 show effective letters.

POINTS TO REMEMBER

If you ask someone for a recommendation, keep these points in mind.

1. Be sure that the person approached can and will give you a positive recommendation; a lukewarm or unenthusiastic letter is as bad as a negative evaluation. Approach only those who know your work and who are in a position to offer a positive assessment of it.

2. Many people are uncertain about what to include in a letter of recommendation. If the person is unsure of what to say, don't be afraid to make suggestions! Ask the person to comment upon the three major areas above and emphasize any qualities that are important to the job you will be seeking.

3. Provide your writer with the correct name and address of the person to whom the letter is addressed, if it is to be sent directly; if not, ask for a general letter addressed "To Whom It May Concern."

4. Provide any other information or forms that are necessary, including a copy of your transcript or updated résumé.

5. Offer to supply pre-addressed envelopes and postage if the letter has to be mailed directly to the employer.

6. Recognize that the person is doing you a favour; be sure not to take this support for granted; always thank those who have written you letters.

Similarly, if you are asked to write a letter of recommendation for someone else, try to adhere to the following principles:

1. Avoid writing letters of recommendation for someone you cannot recommend positively. If you don't feel you can write enthusiastically in the person's behalf, politely suggest that he or she contact someone else.

2. Follow the guidelines above to help you decide what to say in your letter of recommendation. As well, ask the applicant to provide you with a copy of the job posting, an up-to-date résumé, and any additional information about what is expected.

3. Be sure the applicant has provided you with the correct name and address of the person to whom the letter is addressed, if it is to be sent directly; if not, address your letter "To Whom It May Concern."

TAKING YOUR JOB SEARCH ON LINE

These days, it seems that more and more employers are using the Internet as a means of recruiting professionals to their firms. Preparing an electronic application package requires as much care as creating hard copy, but there are a few differences that you will likely want to consider.

Posting Your Résumé to a Web Site

If you have a Web site of your own, you may wish to display your résumé there. If you do decide to post an electronic version of your résumé, you

FIGURE 8.12 What qualities make this letter unsuitable for Shawn to include with his applications?

Bob's Building 2450 Schindelhauer Hill Road •
Saint John, NB• E6Y 4R7• (506) 234 5098

To: Whom It May Concern

From: Robert Clowney, Jr., Foreman and Manager

The purpose of this correspondence is to verify the employment of Mr. Shawn Horton with our firm.

Mr. Horton, was in our employ during summer, 2002. During this time he performed the duties assigned to him. His personnel file contains no records of complaints received about his work.

Mr. Horton seeks another position. Because of the usual vagaries of construction considerations, not the least of which, the fiscal restraints have mitigated against a permanent position at our facility. I share Mr. Horton's anxiety for his security.

Sincerely,

Robert Clowney

Robert Clowney,
Jr Foreman and Manager

City Electrical
4531 Victoria Avenue
Vancouver, BC V7J 4T6

April 12, 2004

To Whom It May Concern:
Re: Ghirmay Zakaluzny

As Manager of City Electrical, I have had the pleasure of supervising Ghirmay during his apprenticeship from the Electrical Engineering Technician Program at West Coast College.

Since joining our firm two years ago, Ghirmay has been consistently dedicated and eager to learn. He has willingly taken on all jobs assigned to him and has been quick to learn new techniques. Despite the rapid expansion in his skills, he is both efficient and reliable, and has performed well above the recommended standard for apprentice electricians. He has even offered some valuable suggestions for improving our job tracking and scheduling process.

I have never had even a moment's concern over Ghirmay's work or his professionalism. He is cooperative, dedicated, and personable, and he interacts effectively with customers and staff alike. His good humour is apparent in everything he does, and his skill in handling customer enquiries and complaints has earned him extra respect from his co-workers.

I can recommend Ghirmay, without hesitation, as an outstanding employee and co-worker, and would be pleased to elaborate on these comments by phone. I may be reached at (604) 897 0612.

Sincerely,

Mike Dharmaratnan

Western Plains University
PO Box 666, Calgary, AB T7N 0W0 (905) 987 5643

Dr. P. Ramamurthy
Department of Mechanical Engineering
McMaster University
Hamilton, ON L8N 3T2

April 2, 2004

Dear Dr. Ramamurthy

Re: Cameron Britten

In response to your e-mail of March 25, I am happy to recommend Cameron for a research assistantship.

I had the pleasure of teaching Cam in several courses in mechanical engineering, including MEng 351—Engineering Design II. I was impressed from the start with Cam's technical competence and trouble-shooting skills; he was able to grasp the problem quickly and isolate the issues that would be most challenging. He assumed leadership of his design group, and they produced one of the better solutions to their particular problem, building a full-scale prototype of their design for an automated tray-mover for a medical research lab. I believe that Cam and his team have been working with the client to create a modified version of the initial device that will be used in the lab. This group's design is among the best student work I've encountered in ten years in the program.

In addition to his technical skill, Cam also communicates effectively with others, particularly in a team environment. His presentations are also engaging and technically sound. He is not as strong a writer as some, but takes direction well and willingly edits. I am confident that, with time, his writing skills will develop the polish necessary for advancement in professional engineering.

If you're looking for a technically outstanding research assistant, a willing and disciplined worker, and a personable individual, I recommend Cameron Britten most highly. Please call me if you'd like to discuss his skills further.

Sincerely,

G.M. Hardbody
Department of Mechanical Engineering

should make sure that the page is readable and attractive, just as you would do for a print version, using visual devices to help organize the information.

One major difference, however, between the print version and the electronic version of the résumé is the potential for violations of your privacy. An employer to whom you have sent a hard copy does not have a legal right to distribute or divulge the information on your résumé to anyone else without your explicit permission. By contrast, a résumé on a Web site is vulnerable to anyone who has access to the Internet. Once you have "published" online, your work history and educational experience, along with every other piece of information contained in your résumé, is no longer private.

This can be a good thing if it brings you to the attention of an employer; however, you should realize that few serious employers will seek prospective candidates in this fashion, especially for entry-level positions. As long as potential candidates outnumber available jobs, there is really little incentive for a busy employer to cruise the Internet seeking qualified entry-level candidates; generally it is more effective for them to post the job and wait for the candidates to come to them. In any case, placing private information on public view can have some negative and even potentially disastrous consequences if someone decides to misuse the facts you've provided, so you should exercise some caution about what you place on line.

In my view, the potential risks involved with supplying so much information to possibly unscrupulous readers far outweighs the possible benefits. If you decide that, despite these concerns, you would like to post your résumé on your Web site, I strongly suggest that you remove your home and work addresses and supply only your e-mail address (a good idea is to provide a hotlink to your e-mail address). As well, you may want to provide incomplete information about your credentials, to prevent others from securing copies of your degrees or transcripts or claiming these for their own, as recently happened in a highly-publicized case at a Canadian university. Criminals have also been known to make use of personal information to establish false identities, potentially destroying the victim's credit rating, job prospects, or clean record. For your own safety and security, you should not provide any information that might make you vulnerable to unwanted harassment or credential theft.[1]

---

[1] The problem of identity and credential theft is widespread, according to several sources. A recent news report declared it the third most reported consumer crime in Canada; the same report noted that it is the number one consumer crime in the U.S. CBC News, The National. Broadcast. Toronto: Canadian Broadcasting Corporation, 30 September 2001. See also Michael Gilbert, "Identity Thieves." <www.abanet.org/journal/oct98/10FIDENT.html>. October 1998.

Searching the Job Market On Line

Although posting your résumé to a Web site and waiting for prospective employers to come to you is risky and most likely futile, the same cannot be said for using the Internet as a tool for your job search. Businesses large and small now operate Web sites, and many of these maintain a listing of position openings, as do both provincial and federal governments. Most newspapers also operate Web sites, and many make their classified advertisements, including job ads, available on line. Another possibility for an on-line job search is professional and trade journals and magazines, many of which regularly feature job ads for career-specific positions. There are also databases to which you can be invited to subscribe on line, which feature current job listings in a given field (not all of these are completely reliable; you should probably make it a policy not to pay for access to job listings).

One very nice advantage of using the Internet to look for job opportunities is that you can locate openings not just locally but also in other cities or provinces, or even in the U.S. or other foreign countries. Once you've found a position that interests you, you may even be able to submit your application on line by sending your résumé and application letter as document attachments to an e-mail message.

If you are submitting an application electronically, you should be aware that many of the firms that accept electronic applications use search engines, rather than human readers, to screen résumés. You will want to be sure to include in your submission the key words employed in the posting for the job, or those that are currently "hot" in the field in which you are applying; if the search engine does not pick these out, the résumé will be discarded.

As well, since formatting may not survive the translation from one word-processing program to another or from html format to text, you should strip as many of the formatting codes as possible from your document. Try to simplify the layout and design so as not to rely on textual features like bold-facing, italics, or a variety of fonts, since these won't translate effectively and may render your carefully worded application into gibberish. Simple effects like underlining and indentation will usually survive the translation into plain text format, and since this format can be opened successfully in any program, it's a good idea to send your document as a text attachment.

Finally, try to keep your résumé as brief and simple as possible; the longer the document, the more likely it is that there will be a problem with transmitting it, and the less likely that it will receive appropriate consideration at the other end. As well, always be sure to submit a cover letter along with the résumé, even for an electronic submission, since to neglect this important part of the application is to risk rejection. Some people prepare a letter that is attached along with the resume as a plain text file; others prefer to use the e-mail message as the cover letter. While there are currently no fixed standards for electronic submission, for reasons of consistency you may wish to write a letter and send it as an attachment. This way, it's easier for the recipient to simply save and print without having to translate from an e-mail program to a word processor.

If you are looking for a position in another province or country, you should probably remember that some employers may still be reluctant to consider applicants who have to travel a long distance to the interview and may for this reason prefer local candidates. Nevertheless, the access to job listings provided by the Internet can be a major convenience for the applicant. While the Internet should probably not be the only resource you use for searching out employment opportunities, it is a useful source of information that can aid you in your job search. Even if you use it only as a research tool for gathering information about potential careers, the Internet can be a valuable addition to your job-seeking strategies.

If your application has been successful, the employer will be interested in talking with you about the position and will invite you for an interview. This second stage of the job search process is every bit as important as the résumé preparation stage, and will be dealt with in the next chapter.

POINTS TO REMEMBER

1. The job application, and the interview, are opportunities for you to display your suitability for the job.

2. Be sure to identify your reader's needs, values, and expectations in order to "target" your application effectively; study job postings in your field to help out with this process.

3. Prepare your résumé and application letter carefully to display your strengths, not reveal your weaknesses.

4. Use the principles of relevance and recency to organize your résumé, and arrange information within categories in reverse chronological order.

5. Always proofread, and remember to apply the Seven Cs of professional writing.

6. Word-process your application materials and use a printer cartridge with sufficient ink or toner to produce a dark, sharp image. If you must use a typewriter, make sure it has a new ribbon.

7. If you are submitting an application electronically, simplify your document design, use a plain text format, and be sure to include both a résumé and a letter as part of your package.

SHARPENING YOUR SKILLS

1. Collect as many job advertisements as you can find from newspapers, flyers, Web sites, or the placement office of your college. You can look only at jobs related to your field or scan a broad cross-section in all fields. Study them carefully and circle key words—that is, those that appear

most frequently and that seem to be most important. Based on your survey, what are the skills that employers seem to value most? Which ones are in greatest demand in your field? Compare your findings with those of others in your class; what implications can you draw for preparing your own job application?

2. Write your own Job Package, including a complete application letter, a résumé of your experience and education, a completed job application form, and a letter of recommendation.

   Letter of Application: Use the job descriptions provided by your instructor or answer an ad from the newspaper, placement office, or an Internet site. Type the letter in one of the formats you've learned.

   Résumé: You may complete this as though you have finished your current year of study. Review the section on résumé writing in this chapter and remember that your résumé must be correct, legible, and clear.

   Application Form: An application form appears on pages 291–294. Photocopy it, use the copy that your instructor provides for you, or download a comprehensive application form from the CACEE site at http://www.cacee.com/english/students/cacee_application.html. Fill out the application form completely, either by hand (in ink) or on the computer. Be sure to sign it where indicated.

   Recommendation Letter: Pretending to be a former employer or instructor, write a letter either for yourself or for someone else in your class. In either case, you will be graded for the letter you have written, not for the one someone else has written for you.

3. Your friend Gerry hopes to apply for the job described below, and has asked your advice on how to improve the content, organization, consistency, visual impact, and appeal of his résumé (see Figure 8.16). Using what you know about purpose, structure, audience, and self-presentation, identify at least five weaknesses in this résumé and explain 1) why these elements are weak and 2) how they might be improved. Write the results of your analysis in paragraph form, as if they might form the discussion segment of a short report.

FIGURE 8.15 Here is the job for which Gerry is applying.

Science Teacher

Canada World Academy, a private international high school with the highest academic standards, seeks a creative, energetic teacher for its science program. The ideal applicant will have a Bachelor of Education degree from a recognized institution, along with an undergraduate degree in natural sciences or biological sciences. We are looking for someone who can bring a strong understanding of scientific applications, and who can inspire enthusiasm and interest among students in upper level science classes. Ability to teach mathematics or computer science courses, along with courses in a natural science, would be a decided asset. A current teaching certificate is mandatory.

The individual we seek:

- is a committed, creative teacher;
- projects and inspires enthusiasm for science and mathematics;
- communicates positively and effectively with students, parents, colleagues, and administration;
- is willing to supervise extra-curricular activities in science and mathematics;
- can undertake some administrative duties when required

Please send your résumé, a cover letter, university transcripts, and the names of three people who can provide professional references to:

Dr. Richard Burton, Principal, Canada World Academy, PO Box 1984, Victoria, BC V5T 6Y3 <burton@admin.canacademy.ca>

FIGURE 8.16 Here is the résumé of Gerry Gervais.

<u>Résumé</u>

Gerry Francis Gervais   Height: 5'11"
12345–78 Street   Weight: 185 lbs.
Saskatoon, SK S5C 0C7   Date of Birth: May 9, 1977
Telephone: 306 467 3144
Marital Status: Single

EDUCATION:

1982–1995   M.E. LaZonte Composite High School
                 Matriculation degree

1995–97   Red Deer College
                 Red Deer, Alberta
                 First Two Years of General Science degree

1997–1998   University of Alberta Edmonton,
                 Alberta Bachelor of Engineering degree
                 Major: Bio-engineering
                 Minor: Biological sciences

1998–1999   University of Alberta
                 Faculty of Education
                 First year of two year after degree program
                 Major: Biological Sciences
                 Minor: General Sciences

2000   Bachelor of Education
                 Degree Professional Teaching Certificate

FIGURE 8.16 (Continued)

EMPLOYMENT HISTORY:

DATE	EMPLOYER	POSITION HELD
Summer 1995	Weston Bakery	Service Driver
Summers, 1996-98	Weston Bakery	Shipper/receiver
Winters, 1995-98	Weston Bakery	Order Desk (Saturdays)
Winter 1994	Shell Canada	Service Attendant
Summer 1984	Shell Canada	Mechanics Helper
Winter 1983	Shell Canada	Service Attendant
Summer 1983	Escher Construction	Labourer

PROFESSIONAL ORGANIZATIONS:

Member of The Baker, Confectionery and Tobacco Workers Union

Member of The Alberta Teachers Association

Member of the Canadian Wildlife Federation

RELATED WORK EXPERIENCE:

Summer 1999   College Preparatory Instructor
            Red Deer College

Duties:

Course and class preparation, instruction, examination preparation and laboratory preparation for Biology 120 (Biology 30 equivalent) for the months from May 7 to August 17, and Biology 75 (Biology 10/20 equivalent) for the months of July and August.

FIGURE 8.16 (Continued)

Summer 1999   Basic Computing Instructor
Grant MacEwan Community College
Millwoods Campus
Edmonton, Alberta

Duties:
Course and class preparation and instruction as well as program design. This course runs every Friday night for three hours for ten weeks. In addition to providing valuable experience in teaching in areas other than my major this course allows me to gain further proficiency in what might be considered my hobby.

I hope to one day be able to meld my professional occupation with my pastime hobby and produce software which instructs and quizzes students on areas related to the biological sciences area. It seems to me that part of the problem associated with the software available today is that it is written from a "programmers" point of view. While the program itself works properly, the content which it contains tends to stress irrelevant concepts, or quizzes the students on mere factual knowledge.

What is needed is a program that not only runs smoothly but is designed from the perspective of an educator.

To gain experience in the field of software teaching I have offered my services to a colleague/friend at Red Deer College who is attempting to transfer his Chemistry 250 course to web-based instruction.

His knowledge of computing is minimal, and so I will act as an advisor in this area while learning to sequence topics and questions to get the most knowledge onto the web and available to the students.

CHAPTER 9:

The Job Interview

Once you have submitted your application package, you will wait for anywhere from two weeks to two months for a response. In some cases, you will hear from the employer only if the company has decided to interview you. If they are interested in speaking with you regarding an interview, a representative of the company will likely telephone you.

Remember that the interview process begins when the employer calls you to set up a meeting. While you are conducting a job search, it's a good idea to think about the professional impression created by the message on your answering machine, or by those who may answer the phone at the contact number you have provided to employers. Caution everyone that you are seeking work, and that employers may telephone looking for you, and ensure that each person knows to take a message for you detailing the name of the caller and the company he or she represents, the telephone number where that person may be reached, and the time of the call. Ask them to record any other information that is provided, and to be sure to leave the message where you can readily find it so that you can respond promptly.

A Tip on Answering Machine Messages

An employer I recently interviewed had just telephoned a job applicant, only to find herself confronted with a lengthy and inane answering machine message. Like many busy employers, she resented the waste of her time. The effect on her opinion of the candidate could not have been worse, since it undermined the candidate's credibility by creating the impression that he lacked professional judgement. This employer did not leave a message; instead, she telephoned the second candidate on her list and offered that person an interview. I am sure that the first candidate still has no idea why he did not get the job.

A good rule to follow is this: a contact number left for work purposes should not waste the caller's time with "humorous" messages. At least while you are job hunting, keep the message on your answering machine simple and professional. Silly messages, especially if they are long, may make a negative impression on an employer who has not yet met you. Don't risk losing an interview by frustrating someone who calls for work reasons.

THE INTERVIEW

The interview is the employer's chance to get to know you in person, to determine whether you are right for the job. In the interview, you will want to maintain the positive impression you created with your résumé and application, since the employer's first impression of you strongly influences any decision to hire you. In many cases this decision is made in the first minute of the interview, during the employer's first reaction to you; if this impression is negative, the employer may spend the rest of the interview looking for faults to justify this dislike. It's obviously in your best interest to make the employer's first response to you a positive one. You should do all you can to prepare yourself, and thus give yourself an advantage.

A successful interview, like other effective professional communication, depends partly on your preparation. Before your interview, think carefully about the needs and interests of your audience—the person who may be paying your salary. What does that person want in an employee? What will the recruiter be looking for?

Although you cannot completely predict the interviewer's response to you, there are three aspects of that initial impression that you can control so as to make it a good one: your appearance, your attitude, and your background knowledge.

Appearance

An employer is seeking indications that you can demonstrate professional and appropriate judgement in a work situation. One of the easiest ways for the recruiter to assess your professionalism is to assess your appearance. Here are some tips for making a positive impression.

1. Wear appropriate clothes. For an interview, you should wear slightly more formal attire than you would wear on the job. For professional or office jobs, wear relatively conservative clothes: a suit is fine for both men and women, though for men a sports jacket and dress pants may serve the purpose, and for some jobs a woman may want to wear a dress and jacket. If you are applying for a labour or other blue-collar job, or for a technical position in the field, dress accordingly; in this situation, a suit might be considered overdressed. Your clothes should also be comfortable enough that you don't have to repeatedly adjust or fiddle with them. No matter what you're wearing, be sure you're neat and clean; avoid splashy colours or unusual hairdos, and leave the nose ring at home.

2. Be punctual. Arrive at the interview with a few minutes to spare, but not more than fifteen minutes early. Know how long it will take you to arrive by whatever means you're travelling and allow yourself enough time for delays. Be sure to take a watch.

   Occasionally, there may be a legitimate reason why you have to be late—car trouble, an accident, illness. If this happens to you, telephone the interviewer immediately to explain the situation and politely request a later interview. If you miss your interview, you should not expect automatically to be given a second chance; sometimes the employer will be unable or unwilling to reschedule, and you will just have to give up on that job. You should also be aware that even legitimate lateness may create a negative impression that damages your chances, so plan to be on time. Sleeping in or misjudging how long it takes to get to the interview are *not* acceptable reasons for being late.

3. Go alone to the interview. An interview is a formal meeting, not a social event, and bringing someone with you may cause the interviewer to question your maturity or your awareness of appropriate professional behaviour. A confident applicant is more likely to get the job, and you won't look confident if you bring someone else along.

4. Whether you are male or female, use a firm, confident grip when you shake hands. Don't let your hand hang limply, but be sure not to grip too tightly either. As silly as it may sound, it's a good idea to practise your handshake with a friend before going to your first interview.

5. Don't chew gum or smoke during the interview, though the interviewer may do either. If the interviewer smokes, don't do so unless you are invited to. Even then you may wish to refuse. And if you are a smoker, try not to smoke just before going in to the interview. Non-smoking interviewers can be put off by the reek of second-hand smoke on a candidate.

6. Make eye contact while you speak. Though some people avoid eye contact simply because they are nervous, this habit can make a very negative impression on an interviewer. It can suggest uncertainty or even dishonesty, neither of which will further your chances. Don't stare at the floor or ceiling or avoid meeting the interviewer's eyes; instead, make eye contact frequently and comfortably, but do look away occasionally to avoid appearing overly aggressive.

7. Speak clearly and use correct grammar. Employers do judge applicants' intelligence and education by the way they speak, and poor grammar is one of the indicators of weaknesses in either of these areas. Avoid such pitfalls as "I seen," "I done," "I did good in that course," or "between you and I" and keep away from slang expressions.

8. Watch your body language. Sit comfortably without slumping in your chair or hooking your feet around the chair legs. Don't fidget, tap your fingers, or fiddle with your clothes. You should appear self-controlled, and any of these is a clear message that you are unduly nervous or inexperienced. Excessive fidgeting can actually compromise your credibility because it signals that you are not in command of the situation or of yourself. Don't block your view of the employer, or the employer's of you, with a large briefcase or other unnecessary props. If you are carrying supporting materials in a briefcase, place the case on the floor beside you rather than in the way on the desk.

Attitude

Many interviewers agree that a good attitude is one of the most important things an applicant can bring to the interview. You should appear confident and positive. Although you will most likely feel, and the employer will expect, a little nervousness, you should try to be as relaxed and comfortable as you can. Be yourself—at your very best.

1. Avoid bragging or overstating your abilities. Employers, like everyone else, dislike arrogance. Be alert and attentive to questions, enthusiastic and

sincere in your answers. Show a willingness to learn and grow with the organization; no matter how much you feel you already know, there is always something else to learn.

2. Avoid one-word responses. At the same time, don't take over the interview with long, impossibly complicated replies. Watch the interviewer for cues that will signal when to stop speaking.

3. Show some interest in the profession or the firm and don't be vague about what you want from your career. Employers like someone who has thought about the future and can show some direction. Indicate a willingness to work hard and start at a reasonable level. You may display ambition, but don't give the impression that you expect to run the company.

4. Don't appear obsessed with money, benefits, or vacations. Remember that the goal of the interviewer is to find out what you can do for the company, not to hear what it can do for you. Don't stress how you can benefit from the position at the expense of showing how you can contribute to the company, and especially avoid talking about how much you need the job. Desperation in a candidate is a huge turn-off for the interviewer.

5. Be courteous at all times. Don't do or say anything that could be considered rude or discourteous. Remember this especially when you enter and are met by a receptionist or a secretary: rudeness to these people can cost you the job offer, since they often are part of the screening process. Among other things, the employer wants to know how well you will get along with other people in the organization, and one measure of this is how you treat people you meet on the way in. Always remember that first impressions count!

Knowledge

Employers are interested in discovering just how well you know the duties of the position you've applied for; they will want you to demonstrate the skills that you have claimed on your résumé. You should naturally be prepared to discuss your experience, always remembering to show how it is relevant to the position you are looking for.

But there's much more on the employer's mind. Although primarily interested in what you know about the duties of the job, your prospective employer will be impressed if you can demonstrate knowledge of the company's mission or mandate. Try to learn as much as you can about the firm before the interview, for example, how large it is and what products or services it offers. You can find out a little about the company by checking on the Internet or looking in your local library. Here are some questions you might consider answering for yourself before the interview.

1. What is the exact nature of the company—who are their clients, and what do they make or do?

2. Is it a local firm, a national company, or a multi-national?

3. How extensive is their client base?

4. How long has the company been in operation?

5. What is their organizational style?

An annual report will give you this information and more that might be useful. If you know someone who works for the organization, try to talk to that person before your interview. If the company has a Web site, visit it when you are preparing for the interview. It will provide you with important information about how the company sees itself. Find out as much as you can. Though you may not be asked such questions in the interview, the more you know when you go in, the more confident you will be and the better impression you will make.

Employers' Questions

In general, there are three things employers want to know: if you can do the job, if you are willing to do the job, and if you can fit in with the existing personnel and culture. Employers also want some indication that you are self-aware, that you know yourself and have thought about your goals, and that you have realistic expectations. They will also be interested to see if you can effectively solve the problems you are likely to face on the job. They will try to elicit this information through careful questioning.

Thus, though every interview is different, and interviewers have different focusses, there are some questions that occur regularly in one form or another. You will find that many interviews focus more on your personal qualities than on your particular specialized or technical qualifications, which are spelled out in the résumé and application materials you've submitted beforehand, and which are also guaranteed by your professional or technical training or education. Although it is possible that you may face an aptitude or skills test of some form in the interview process, employers are often more interested in using the opportunity of a face-to-face meeting to explore the kind of employee you will be than they are in assessing your technical knowledge. As a result, they will often employ questions designed to probe your attitudes and interest in the field.

As you read through the questions that follow, you should carefully consider the probable needs and interests of the employer, and try to tailor your response to fit the job you're applying for. Don't memorize a prepared answer and don't try to be something you're not; instead, think of strategies you would use for answering each question and consider specific examples from your experience that would serve as concrete illustrations of the qualities the employer is seeking. When you're at the interview, be sure to answer each question as fully and sincerely as possible, and be sure to provide an example from your background that supports your answer.

Finally, don't be fooled into thinking you won't face questions about philosophy, cultural interests, or pleasure reading if you're applying for a position in a technical field. Such questions offer a very effective way for an employer to determine how well-rounded you are and to gauge your ability to get along with others. In many cases, you will be interviewed by non-technical people whose job it is to determine such intangibles as attitude and ability to interact effectively with co-workers, management, and clients.

Here are some samples of favourite employer questions, with strategies for answering them. If you think about these, and about some potential answers, before you go to any interview, you will have a better chance of handling the questions effectively.

1. Tell me about yourself. Employers often like to begin the interview with this one, because it not only provides information about you, but it also shows something of your priorities. What you choose to discuss will tell them what you think is important. Be aware that the employer is interested primarily in information relevant to the position you've applied for. Avoid the temptation to deliver the epic of your life. A brief summary of educational and employment highlights, as they have prepared you for the job in question, will work best in answering this question.

2. What are your strengths? Don't be falsely modest. A mature person knows what his or her strong points are and can state them briefly without either bragging or understating them. Take some time to identify the things you do really well; go back over the skills section of your résumé and be ready with some examples.

3. What are your weaknesses? We all have weaknesses, and a mature adult is aware of his or her own. However, be careful when you're answering this question. First, don't identify weaknesses in such a way that they are likely to make an employer think twice about hiring you: such comments as "I have trouble finishing what I start" or "I can't get motivated in the morning" may put off a meticulous employer. If you have such weaknesses, of course you should be trying to correct them, but an interview is not the place to highlight them.

Second, avoid giving an overused, predictable response, such as "I'm a workaholic." Employers have heard this line from too many applicants to take it seriously any longer, even in cases where it may be accurate. Instead of relying on clichés, think about your own abilities and try to identify some concern from your own work habits, enough to show you are human, but not something that will make you sound unemployable. Use this strategy: identify a not-so-serious flaw and immediately balance it with a positive: "I sometimes find it hard to take criticism, but even though I find it difficult I usually benefit from constructive comments," or "I'm not really a morning person, but I find that using that time for routine tasks helps me to get focussed on the big jobs when I get to them later in the day." Avoid confessing really serious or negative traits, but don't say "I don't have any weaknesses" or "I don't know," and above all make sure you are sincere in your answers. An experienced interviewer can recognize phony or insincere answers and will reject an applicant she believes is misrepresenting himself.

4. Why do you want to work for this company? or Why do you want this job? Your research into the company will give you some information to use in answering this question; don't, however, identify pay or benefits as a primary reason for choosing this firm. Emphasize the position itself, and don't, as an acquaintance of mine did, answer by saying, "because I'm un-employed." Employers want some evidence that they are getting the best, not merely the most desperate, applicant.

This question might also appear as "In what ways do you think you can contribute to our mandate?" accompanied by "What can we do for you?" You can best prepare for these questions by thinking them through beforehand and planning how you will answer. Don't memorize an answer, however; doing so is bound to make you sound false or insincere.

5. What made you choose this field as a career? There are many acceptable answers to this question, and you will have to choose your own. However, "This is the field my father (or mother) works in," is not one of them. Neither is "I knew I'd earn lots of money." The employer wants to know that you've made a thoughtful choice, not just an expedient one.

6. What was your favourite subject in college, university, or school? You may wish to identify one or two subjects, but watch out for the other half of this question, which asks you to identify the subject(s) you disliked. You may indicate that there were some subjects you liked better than others, but you should avoid sounding like a complainer. Instead of identifying courses you couldn't stand, indicate that you learned something from all of them, even the ones you didn't particularly enjoy. A similar question is *What was the job you've liked least? the most?* It is an attitude question more than anything, and you should use a similar strategy for answering it.

7. How would your friends describe you? This is another way for an employer to find out a bit about your self-awareness. You should not be overly modest. Again, try to identify real strengths, but avoid sounding arrogant.

8. What would you like to be doing in five years? ten? Employers can find out how much thought you've given to your career with this question. They may also be looking for evidence of a commitment to the company. If you say that you're not planning to stay in this job, or in this field, an employer may not want to spend the time or money on training you for the position. Answering "I don't know" isn't a good idea either; an employer might consider it a sign of indecision or immaturity and it could cost you the job. Try to show that you have given some thought to the future, but that you are flexible and able to adjust your goals as well.

9. Do you have plans to further your education? Would you be interested in doing so? Your answer here could depend on whether the employer is wondering if you're open to more training or afraid you'll quit the job in six months to return to full-time study. You should never close any door on yourself. You could indicate that you are willing to take further training if necessary. Even if you don't think right now that you would ever go back to school, remember that with time you might change your mind. On the other hand, you might also want to reassure the employer that you're not going to quit this job as soon as you're trained in order to return to school.

10. The situation question. One type of question that is often favoured by employers is the situation question. The most common type is one in which the interviewer gives an example of an incident that might happen to you on the job and asks for your solution. This question is designed to test your awareness of the job's requirements, and the best preparation is to go over your experience in your memory. Think of difficult situations you have faced and how you handled them; analyze how you might have handled them better than you did. If you consider in advance how you might answer such a question, you will be better prepared for it. The question may also appear in another form, such as:

Give me an example of a situation in which you showed leadership.
Tell me about a situation in which you resolved a difficulty.
Tell me about a situation in which you initiated change.
Tell me about a situation in which you handled criticism.
Describe an achievement you are proud of.
Give me an example of a situation in which you faced a challenge.
Tell me about a situation in which you resolved an interpersonal conflict.
Describe a situation in which you faced and resolved an ethical dilemma.

11. How well do you...? These questions might also be phrased evaluatively. Instead of asking for an example, the interviewer might simply ask you to outline your capabilities, using such questions as:

How well do you handle pressure?
How do you take criticism?
Are you able to adapt to change?
What specific strategies do you use to organize your time?

You should always support your answers with brief, specific examples from your work or educational experience.

12. Attitude questions. The employer might also probe your attitude toward others with such questions as: *Are you usually right? Are other people's ideas as important as yours?* Of course, you will want to show some openness to others' views and not indicate that you believe only you are ever correct. Provide some balance in your answer: you're not always right, naturally, but you are not always wrong either. You must show that you are capable of making decisions and standing by them, but that you can also recognize good ideas that others put forward and can compromise. You might also be asked:

> Do you make mistakes?
> How do you handle those?
> Can you give me an example?

Naturally you will want to show that you are aware that you can make mistakes, but also that you can learn from them. If you say you do not make mistakes, the interviewer will reject you outright, because of course we all make errors from time to time, and the measure of us is in how we handle the errors we make.

13. What are the qualities of a good leader? *a good boss? a good employee? Are you a good employee? What qualities of yours make you the best candidate for the position?* An acquaintance of mine, unnerved by the interview situation, once answered the latter question with, "I don't know; I just do the job!" By now, of course, you know that's not an appropriate answer; a good answer should be thought out beforehand in a way that's relevant to your program or profession.

14. Probing questions. The employer may also solicit some specific details about your work or educational experience, and invite you to interpret them in the light of the job you are applying for. Some of these questions include:

> Describe the most challenging project you have ever worked on. What it a team project? What was your role?
> Describe your experience in making formal (or technical) presentations. Give an example.
> Why are your grades only average?
> What did you do to prepare for this interview?

As with previous example questions, you should think about possible answers to questions like these before you go to the interview, and always be ready with specific examples. While there are no "correct" answers to such questions, you should pay attention to the sincerity and credibility of your answers, and to the likely interests, needs, and expectations of the person who is asking the question.

15. Your outlook on life. Personal attitude can be probed using questions about your general approach to life. Some of these are:

What motivates you?

What is your personal philosophy?

What are you doing right now to improve yourself?

Describe the biggest risk you have ever taken. What did it teach you?

What was the last book you read purely for personal interest?

Tell me a joke.

What can you bring to this job that nobody else can?

If a situation arose in which you might have to violate company policy, would you do so?

Once again, these questions have no simple "right" answer, but the way in which you choose to respond will tell the employer much about the kind of person you are and the general attitudes of tolerance, cooperation, reliability, tenacity, and good will you bring to the job. Remember that the employer is looking for someone who is motivated, capable, and able to get along with other employees. You should try to respond in ways that will showcase your abilities to do those things, and always support your answers with specific examples from your experience.

In answering any of these questions, keep in mind the probable needs and interests of all employers: they will be looking for someone who is confident and capable, but not arrogant or self-absorbed. Admit to mistakes, but show that you have learned from them and can handle criticism effectively. Be honest and appropriately modest, but not overly so; be confident, but not arrogant. Show that you have both strengths and weaknesses, but that your weaknesses are not serious and can be overcome. Don't cite weaknesses or flaws in character that are likely to damage your employment chances.

What to Expect in the Interview

Interviews can vary not only in the type and number of questions that employers ask, but in other ways as well. There is no set pattern for interviews and no "right" way to conduct them with respect to length or number of screenings. Employers tend to decide for themselves what selection process best suits their needs, and the more interviews you go to, the greater variety you will see.

For example, the length of time you spend in an interview may be anywhere from twenty minutes to two or even three hours, depending on the type of position and the number of applicants. An acquaintance of mine recently attended an interview that lasted seven hours! The last one I went to stretched over two days, but I have been to interviews as short as twenty minutes. Often the person who telephones you to set up the interview will indicate how long it will take; if she or he doesn't volunteer the information, ask for an estimate of how long the interview will last. If you're not sure, allow yourself at least two hours, just to be safe.

You may be interviewed by one person, by two or three together, or even by a committee of five or more. Again, you may be told this ahead of time. But whether you are told or not, be prepared for the possibility that you may be interviewed by a committee; the more responsible the position, the more likely it is that there will be more than one interviewer. In some organizations or institutions, two or three people interview you separately, then compare impressions. These may take place on the same day (you may spend twenty to forty minutes with three different people successively) or on subsequent days. You may also be individually interviewed, or you may be part of a group of interviewees who are put through a formal process as a group. These differences are neither good nor bad; they merely show the employer's personal preference.

In any of these cases, don't be thrown off by interviewers taking notes while you speak. Remember that they have seen several different people in a short space of time and are merely interested in keeping track of what was said. The note-taking is really for your benefit—you wouldn't want an interviewer to forget you or confuse you with someone else.

Sometimes, depending on the employer and on the position, you may be asked to complete some form of testing. Occasionally these tests will be vocationally specific—you may be asked to explain some important concepts in the field, to give a presentation, to write a report or memo, or to solve a mathematical or technical problem. In these cases, the interviewer is interested in knowing that you really do have the level of skill needed for the job.

There are other kinds of tests that you may be asked to take: general aptitude or even psychological tests. These are tests you can't really prepare yourself for; they are thought to reveal your general intelligence, attitude, or aptitude for the position you are interested in. Like all trends, aptitude testing goes in and out of favour, and is used more in some fields than in others. Even when they are popular, such tests are not universally employed by all organizations. However, employers who choose to use aptitude testing generally believe it is useful.

Your best bet if you are asked to write one of these is to be as honest and forthright as you can. Most of the tests are designed to double-check your responses by asking several questions aimed at the same information, so keep in mind that it's difficult to try to second-guess the tests. Fudged answers can usually be identified by the cross-questions. Simply try to relax as much as possible and do your best. There is nothing to be frightened of, and you will do better if you can keep from being too upset.

There is much talk these days about poor writing skills among college and university graduates, and many employers have expressed concern over such weaknesses. As a result, occasionally an employer will ask applicants to write a piece of professional correspondence—a letter, memo, or short report—right on the spot, in response to a situation such as the ones given in earlier chapters of this book. You should be prepared to write if necessary;

to prepare, you might review the chapters on the letter and the memo before you go to the interview. In this case, the employer will be looking not only for proper letter or memo format, but also for correct grammar and sentence structure, and the other Seven Cs of professional writing.

The Unskilled Interviewer

In spite of all that has been said here, you should also know that not all interviewers will be equally skilled, and you should be prepared for the possibility that one of these people may interview you. While in many companies interviewing is typically conducted by skilled personnel specialists, this isn't always the case. Frequently, in fact, you may be interviewed by the immediate supervisor or even by the person who is vacating the position you are applying for. Since interviewing people is not their area of specialization, these people may be inexperienced at the task, sometimes enough so that you will be able to detect their uncertainty.

If you find yourself in such a situation, maintain your cheerful, positive demeanour and answer each question to the best of your ability. Don't let the interviewer's lack of experience make you overconfident or arrogant. Unskilled interviewers are likely to be more sensitive to such nuances than an experienced interviewer would be, and possibly more easily intimidated. They will not appreciate challenges to their authority, however subtle. Maintain your poise and do your best with the questions you are given.

Problem Questions

Employers are forbidden by law to ask an applicant questions about age, marital status, religious affiliation, ethnic background, sexual orientation, or family relationships. Nevertheless, occasionally you will encounter an interviewer who asks you such questions anyway, either because of inexperience or because of deliberate disregard for the law. You are obviously not obliged to answer such questions, but refusing to do so may be awkward and could cost you the position.

The decision to volunteer such information or not is a personal one, based on your own comfort level. For example, if you don't mind a question about marital status, you may wish to answer it even though, strictly speaking, it's not appropriate for the employer to ask. Sometimes such questions are unthinking expressions of other employer concerns; when this is the case, you will sometimes be able to determine the employer's train of thought from the context of the remark. An employer may really be thinking about overtime and may ask about your marital status because he or she feels overtime might be more difficult for a person with a family. You may choose to phrase your response to address the employer's concern directly. For example, a prospective employer may ask if you are married. You could answer: "If you're concerned about my willingness to work overtime, I am willing to put in all the time the job requires."

If, on the other hand, the interviewer's questions seem a bit too personal or make you very uncomfortable, you may wish to decline to answer. Doing so is tricky, though. If you simply refuse, saying that you don't see the relevance of the question to the job, you will probably turn the interviewer against you. You may try restating the question as in the example just given, but you should be prepared for the possibility that doing so might cost you the job. You need to be able to balance your need for the job against your willingness to field inappropriate questions. This is entirely a judgement call, and it's important to know for yourself how much is too much.

Most interviewers want to see you at your best and will try very hard to put you at ease. However, if you do run into a difficult situation, know how much of such behaviour is tolerable to you, and don't be afraid to leave if you have to. If the interview is that unpleasant, it's unlikely that you would want to work for this company anyway.

After the Interview

Following the formal interview, you may be taken to lunch or dinner. If this occurs, remember that it's still part of the interview process, even though the formal questions are over. Although you will by this point have established a good impression, you will want to maintain your professional demeanour. Be careful to order food that is easy to manage—stay away from messy sauces or sloppy finger food. Recognize that the conversation will continue during the meal; pause occasionally as you eat, so that you will be able to respond naturally.

While the occasion may appear to be purely social, remember that you are still subject to the judgement of those who may become your future employers. If you are at dinner, you may wish to order an alcoholic drink, but if you do so, be sure that you limit yourself to one, or at most two, drinks. Continue to behave courteously and professionally, and be careful not to say anything that may be judged inappropriate. You should be friendly in your manner, but not so much so that you appear to be presuming you have clinched a job offer.

The Letter of Thanks

After the interview process is complete, many people like to send the employer a polite letter of thanks. While this courtesy is not required, it provides a final gesture of professionalism and good will that contributes to your overall ethos. It need not be lengthy, but it should express your sincere appreciation for the effort that went into interviewing you. The letter of thanks should also express your continued interest in the position, but you should not appear to assume you've got a job you haven't yet been offered. You will need to strike a balance between friendly courtesy and unwarranted expectation; the best way to do this, as you have already learned, is to put yourself in the audience's place. Figure 9.1 shows a sample letter of thanks.

1234 5th St W
Fredericton, NB E8J 0S3

May 15, 2004

Sven Runkvist, Managing Director
East Coast Technical Services
PO Box 930
Halifax, NS B2G 0X0

Dear Mr. Runkvist:

Thank you very much for the opportunity to meet with you for an interview for the Quality Assurance Specialist position in your Halifax office. I enjoyed meeting all the staff, and appreciated their friendly courtesy during the whole process.

I particularly appreciated the chance to demonstrate my problem solving and design skills in developing plans for a pattern-cutting system for Ceejay–Maritime Toy Mfg. Whether or not I am successful in the competition for this position, I am grateful for the chance to speak with you and to learn more about the operation of a large full-service engineering firm. Thank you again for the very positive experience, and for the enjoyable staff luncheon.

Yours truly,

Shawn Horton

Why Didn't I Get the Job I Interviewed For?

Sixty to eighty percent or more of the people who are selected for an interview fail to secure a job offer. As you search for the ideal position, you will certainly find yourself in this position some of the time. Of course, there are many reasons why people who are interviewed don't get an offer, and many of these are beyond the candidate's control—internal competition, someone with slightly more experience, a transfer from another branch of the same company, the timing of the interview. However, some fail in the interview because of factors that need not have presented a problem; they fail because they have made one or more of the common mistakes that lead to rejection. These factors are ones you can bring under your control, and while managing these elements effectively won't guarantee an offer, doing so will help to ensure that you won't be dismissed before you have had a chance to demonstrate your suitability for the job.

In general, you should avoid any extreme behaviour, and in particular, be sure to respond openly and willingly to questions, since a pattern of unresponsiveness suggests an uncooperative, even hostile personality. Extremes of behaviour such as talking too much or not enough could cost you a job offer, because such extremes suggest you are not entirely in control of yourself or the situation; similarly, being either too animated or too stiff implies extreme nervousness out of proportion to the stress of an interview, and suggests an inexperienced or incompetent applicant.

An inability to stay focussed on one point or one question will also make a bad impression; do your best to concentrate on the questions you are being asked, and ask for clarification occasionally if you do not understand the direction the question is taking. Even something as simple as smiling too much or not at all can be a sign to an employer that you are ill at ease. Finally, any indication of disorganization beyond the norm, such as not being able to find essential papers or a pen in your briefcase, purse, or pocket, forgetting significant documents at home, or being unable to recall details of past employment, will make an employer hesitate.

The following are some of the specific reasons employers have given for rejecting people who, on paper, were appealing enough to be invited for an interview.

1. Arrived late to the interview
2. Was inappropriately dressed.
3. Was poorly groomed.
4. Was rude to the receptionist.
5. Refused a handshake when offered.
6. Seemed unduly nervous.
7. Fidgeted; did not appear relaxed or confident.

8. Lacked energy or enthusiasm.

9. Didn't answer questions fully, or rambled on too long or pointlessly.

10. Was unforthcoming in answering questions.

11. Could not provide examples to support claims in the résumé.

12. Was aggressive, arrogant, or self-important.

13. Attempted to dominate the interview.

14. Was dismissive of contributions of professionals in fields other than his or her own.

15. Appeared more interested in pay or benefits than in the job.

16. Criticized former employers or professors.

17. Chewed gum.

18. Was unable or unwilling to provide references.

19. Spoke poorly, with poor grammar or diction.

20. Had a limp handshake.

21. Knew nothing about the company or was uninformed about the profession or field.

22. Was unfamiliar with current trends in the profession.

23. Had exaggerated on the application or résumé.

24. Had ambitions far beyond abilities.

25. Had no clear goals or professional interests.

26. Appeared whiny or unmotivated.

27. Could not take responsibility for weaknesses or mistakes; blamed others.

28. Was defensive when answering questions.

29. Was unwilling to start at the bottom.

30. Had a poor school record.

31. Was insincere or glib.

32. Lied in the application or interview.

Interviewing, like everything else in this book, is a skill you can learn and polish. You can do this best by practising. Go to as many interviews as you can, even if you're not sure you would want the job. You can never get too much experience, and every interview you go to will make the next one easier. You might even find that a job that didn't appear very attractive on paper turns out to be just the position you were seeking.

SHARPENING YOUR SKILLS

1. Consider the job you applied for when you wrote your résumé package in Chapter Eight. From the list of common interview questions given in this chapter, and based on the posting or advertisement for that position, select five questions that you feel would be likely to come up in an interview for this job. Develop some effective strategies for answering these appropriately. In a memo to your instructor, or as part of a brief oral presentation, discuss the ways in which your strategies demonstrate the principles of effective communication.

2. Study the sample questions in this book and select the three that seem the most challenging to you. Exchange your list with a classmate, and then interview each other using these questions. Note carefully the strategies used by your partner, paying attention to strengths in the person's interview and to areas that might be improved. Be prepared to report your observations in written or oral form.

3. Your instructor may wish to organize mock interviews to take place in class; in this exercise, each person in the class is interviewed in turn, with the rest of the class looking on. Your instructor may choose questions at random, or allow you to identify five of your own choosing, selecting one or more of these for your mock interview. After your mock interview, give a brief assessment of your own performance, then listen to comments from your classmates and instructor. Be prepared to discuss the experience with the rest of the class.

4. Which of the interview questions seems to you most difficult to answer? Why? What do such questions teach us about the nature and importance of communication?

5. In what ways does the interview constitute a nearly perfect illustration of the nine axioms of communication that we learned in Chapter One? Prepare a short oral report that may be delivered to the class in which you identify these features and suggest some implications for our study of communication.

Communication Ethics and Etiquette

ETIQUETTE

Throughout this book we have emphasized the importance of recognizing and adapting to the audience's needs and concerns, of accommodating the constraints of each specific situation, and of establishing personal and professional credibility. We have learned that being an effective and skilled communicator means being able to assess the demands of each communicative situation and respond appropriately to those demands.

We have also learned that communication is inescapably social: it involves a dynamic relationship between people. Communication therefore always involves others' feelings, attitudes, experiences, and concerns, as well as our own expectations and the quality of what we say. For this reason, communicating effectively means paying careful attention not only to the content of our messages, but also to the relationships we are building with each interaction. As part of this attentiveness to relation, we have emphasized "courtesy" as one of the Seven Cs of effective communication, and we have made it an important part of our communication toolbox.

Because communicating effectively is a matter of making appropriate judgements and choices, it is closely linked to social responsibility and ethics, the principles or values that guide our actions and behaviour. Through the choices we invite our audiences to make, and through the strategies we use to influence them, our communication practices bring us face to face with ethics, mainly because our communication affects the interests and well-being of others. For this reason, we cannot fully consider our communicative choices without considering their ethical implications and effect.

WHAT IS ETHICS?

Ethical systems are systems of values or codes of behaviour that have evolved to help us make reasonable and effective choices as we live, work, and socialize with other people. Although there may be some variation in the specific precepts of various ethical systems, individuals do not invent their own ethical systems. Instead, our interactions with others are shaped by cultural norms and religious beliefs; they are determined by law, by religious doctrine, by social custom, by family tradition, by the code of a profession, even by standards of conduct within a given company. Whether based on religious, cultural, familial, or professional codes of behaviour, all ethical systems have a strong collective, or social, component; they are not simply a matter of individual whim. As well, despite variations in ethical systems, there are some principles that they all seem to share.

At the heart of all ethical codes is some form of the principle that we have been calling courtesy: respect and regard for others with whom we interact. This principle can be seen in the ancient precept known as the golden rule, which is most frequently stated as "do unto others as you would have them do unto you."[1] The best-known and longest-standing code of professional ethics, the Hippocratic oath of physicians, embodies this same principle in its admonition to "first, do no harm."

Codes of ethics are guidelines or principles that have to be applied using careful judgement and attention; they are not cookie-cutter rules. They are

---

[1] Jess Stein, Editor-in-Chief, *The Random House College Dictionary*, rev. ed. (New York: 1975) 566.

meant to guide, not replace, human assessment and choice. Because the situations in which we have to make judgements are not always clear-cut, because they are not entirely predictable, and because each is unique, we must draw on our understanding to make choices that will best fulfill our obligations to our listeners, our messages, and ourselves.

Communication, like other forms of human choice and action, can be ethical or unethical; it can be honest or dishonest; it can be helpful or harmful. But because it is essentially social, effective communication places regard for others at the centre of ethical practice. We communicate ethically when we choose a course of action that limits the unnecessary harm done to others, when we take responsibility for our choices and actions, and when we do not knowingly deceive anyone or misrepresent facts, beliefs, or actions. Like other human actions, communication is ethical insofar as we accord to our audience the same consideration that we hope to receive in return.

Ethical communication involves genuine esteem for the audience, a sincere commitment to what you are saying, and respect for the integrity of the speaker. Aristotle's treatment of ethos—the character of the speaker—can thus be understood as essentially an ethical system. The principles of good judgement, good will, and good character that make up a speaker or writer's ethos constitute an ethical stance because, like any code of ethics, they are both a manifestation of the social nature of human interaction and a result of personal choice. The ethical principles of communication—credibility, sincerity, and integrity—are anchored in the participatory process of building connections with an audience.

A code of ethics is a list of the rules of conduct that a society, religious group, family, or organization holds to be important. These rules are based on the values that the group or organization accepts as central to human interaction. As you have probably guessed, the Seven Cs of professional communication that we have applied throughout our discussion represent the beginning of an ethical code for communicators. Building on these principles, and drawing on the nine axioms of communication, we can develop a code of ethics that will help to guide our professional communication with clients, colleagues, and the public.

A Communication Code of Ethics

1. I will take responsibility for my words and my actions.

I will focus on my audience's need for understanding, information, and support before my own need for self-expression. I will hold myself responsible for how well my audience understands my message. I will keep my word. If I am mistaken, I will admit it. If I do not know the answer, I will acknowledge it. I will respect the contributions of others, recognizing that a communication challenge may have more than one solution.

A Communication Code of Ethics

1. I will take responsibility for my words and my actions.
2. I will take care not to misrepresent myself or my message.
3. I will avoid unnecessary hurt to others by my words or my tone.

2. *I will take care not to misrepresent myself or my message.*

I will use language and visuals with care and precision. I will not speak of what I do not know. I will strive to communicate clearly, simply, and directly. I will label inferences and judgements as such, and not present them as fact. I will not distort or misrepresent issues in order to win a point. I will present my thoughts in my own language, and will not present the ideas or thoughts of others as though they're my own. I will know the source of information that I use, and be able to explain how and why it's credible. I will not claim expertise I do not possess. I will recommend to others only actions that I have taken or genuinely would take. I will avoid making promises I cannot keep.

3. *I will avoid unnecessary hurt to others by my words or tone.*

I will pay attention to the relational element of all communication. I will be respectful and courteous to others. I will make reasonable requests. If I must criticize others, I will speak with tact and generosity of attitude. I will assume good will on the part of others, and act with good will in return. I will listen to others as carefully as I expect them to listen to me. Where an apology is due, I will apologize to others genuinely and sincerely. I will avoid repeating unfounded gossip.

ETHICS AND ETIQUETTE

Few people recognize the ethical dimension in daily communication interactions; however, since all messages involve an element of relation and of risk, questions of ethics are inescapable, even in the simplest of communication events. We have learned that ethical communication involves respect for others, particularly when we are asking for their cooperation in return. In fact, codes of behaviour known as "politeness" or "etiquette" have evolved to guide our communicative choices and reduce the element of face threat inherent in most communication exchanges. The rules of etiquette are far more highly codified than are ethical systems, and they offer specific rules, often in the form of "dos" and "don'ts," for behaviour in social situations.

In business and the professions, as in other situations, etiquette rules help ensure that we do not violate the principles of social interaction that are embedded in the nine axioms we have been studying in this book. Throughout our discussion, we have devoted a great deal of space to the principle of courtesy in written communication and in highly formal interactions such as public speaking and interviewing. However, there are some additional on-the-job communication situations in which etiquette is equally important: telephone interactions and face-to-face encounters with clients and the public. Below are some guidelines for these interactions based on the principles we have already discussed.

Telephone Etiquette

Making a Call

1. When the person answers the telephone, clearly identify yourself, by name and title, state the name of the company or firm you represent, and briefly explain the purpose of your call. Do this first, and in a pleasant tone, before asking "How are you?" or otherwise exchanging greetings. The person answering the telephone has a right to know right away to whom he is speaking. A delay in identifying yourself can cause the recipient of the call to respond defensively or even to hang up.

2. Never, ever begin by asking "Who is this?" Expecting the recipient of the call to identify herself by name to an unknown caller is the height of rudeness, and doing so invites rudeness in return. As the caller, you should know whom you called; identify yourself first, and ask by name for the party to whom you wish to speak.

3. Recognize that, unless you are calling a help line or customer service desk where the person's main job is to answer the telephone, an unexpected telephone call is always an interruption to the person who has answered; do not assume that the person has time to deal with your questions right then and there. Respect the fact that you are encroaching on that individual's work time, and be courteous and brief. Always ask if the person has time to deal with your requests; if he is busy, offer to call back at a more convenient time.

4. Remember that the people you are speaking to have other duties to take care of. Make your calls brief and avoid detaining people for too long on the phone.

5. If you are telephoning someone on a professional matter at home, you should consider that you are not only imposing on her personal time, but in a sense you are also invading her privacy. If you must telephone someone at home, acknowledge the intrusion and offer your apology. Even if the person welcomes the call, your polite recognition of the interruption will be appreciated.

7. Do not hang up without speaking if you realize that you have mistakenly dialled the wrong number. Instead, politely apologize and terminate the call.

8. If you must leave a message on an answering machine or voice mail, leave a short, business-like message, stating your name and contact number clearly and distinctly so that the person hearing the message can get it down in one pass. Repeat your name and number at the end of the message to assist the recipient in recording them correctly.

Receiving a Call

1. Although it is conventional to answer the telephone with a simple "hello," it is also correct in a business or professional setting to answer by identifying yourself or your department. You should state this information in a manner appropriate to the situation and your own style. A former boss of mine answers the phone by crisply stating his last name: "Tobias." Another supervisor sets an entirely different tone by answering "Hi, this is Ken." Still another business acquaintance says "Good morning. Coordinator of Finance, Roxanne speaking." In many workplaces where several people must share a phone, it is customary to answer by stating the name of your department or division: "Research and Development," or "Claims," or "Shipping." Any of these is appropriate, and should be used in accordance with the conventions of your workplace and with your own manner and style.

2. Answer the telephone politely. The tone of your voice can set the tone for the call. Try to cultivate a pleasant telephone manner, even if you aren't feeling particularly cheerful. Remember that the person on the other end will make inferences about your character and attitude based on how you sound.

3. If you are busy with a client, or otherwise occupied, offer to place the caller on hold or to return the call when you are less busy. Never place a caller on hold without asking whether the person can wait, and do not place someone on hold if the wait will be longer than three or four minutes. If the person is unwilling to be placed on hold, or if the wait will be several minutes or longer, take the caller's number and offer to return the call when you are free. Then do it.

4. Try to establish a policy of handling all your calls during the same period each day, if your schedule allows. Let your contacts know you'll be available to take calls during that time, and make it a practice to return all your calls in that time period. If you can establish this routine, your colleagues will get to know the best time to phone you, and you will be able to keep telephone interruptions to a minimum during the rest of the day.

5. As a general rule, do not give out information about yourself to callers who have not identified themselves to you. Unscrupulous people obtain information through a variety of ruses, one of which is to request information over the phone without identifying themselves first. If a caller who has not self-identified asks "Who is this?" or "Who's there with you?" do not respond; they may be fishing for a name or trying to find out whether you are alone at home or in the workplace. Instead, politely ask, "Who were you hoping to speak to?" or "Who is calling, please?" If you encounter a caller who is reluctant to identify himself, or

who uses cagey responses such as "Don't you know who this is?" or "It's a friend of yours," terminate the call immediately. You could have an obscene or harassing caller on the line.

6. While it is standard practice to provide a caller with a colleague's work number, or to transfer a call directly to the person, you should never give out someone else's home address or phone number to an unidentified caller. Doing so may violate company policy, or even provincial law, since it could subject your co-worker to harassment or worse, and you or your company could be liable if something were to happen as a result of your disclosure. If a caller requests personal contact information for a colleague, a better policy is to take a message and offer to pass it on to the person concerned. That way, if the request is legitimate, your colleague can contact the caller and good will will be retained.

Face-to-Face Etiquette

In some businesses or organizations, contact with clients rarely occurs on a drop-in basis; in others, counter service is a normal part of the working day. However, all organizations involve face-to-face contact at some level between clients and company representatives. The following tips are designed to ensure that the clients, colleagues, and other professional contacts are handled as effectively as possible.

1. If you are at the counter or reception desk, greet clients promptly in a pleasant and professional manner. Make eye contact as soon as the person approaches you, and smile. Ask "How can I help you?" or "May I help you?"

2. No matter how important your other tasks may be, never, ever leave an unacknowledged client standing at the counter or reception area while you take care of filing, telephone calls, or other administrative details. If you are genuinely busy with a task that must be completed immediately, acknowledge the client and say "I'll be right with you" or "Someone will be with you in a moment." If it is appropriate, ask the person to have a seat, and offer coffee if there is some available.

3. Never, ever ignore or interrupt a person who is with you in order to take a call; if answering the telephone is your job, do so, but put the caller on hold until you have completed your business with the human being in front of you. A person who has taken the trouble to meet with you personally should always take precedence over one who contacts you by telephone.

Handling Face-to-Face Complaints

If you are handling complaints, if you are dealing with long lines of customers who have had to wait for an extended period, or if you are on the reception desk, you may be faced with grumpy or frustrated people. People who have been waiting a long time, who are frazzled or annoyed, or who have a legitimate gripe about your products and service may take their feelings out on the nearest person: you. It is normal to respond to rudeness by being rude in return. However, as you know by now, responding in kind will only aggravate the situation and intensify the unpleasantness for both you and the client. Furthermore, the memory of the unpleasantness will remain even after the content of the exchange is forgotten, and may sour your relationship with the client or with the organization that he or she represents.

Keep your cool and try to be pleasant even when others are rude. If you can remain calm and respond in a genuinely polite way, you can usually keep a difficult situation from becoming impossible. Try to retain your positive attitude, and try not to take the client's anger personally. Remember that, though you may not be able to resolve the situation, you could well inflame it if you respond rudely to someone who has been rude to you. An upset client is unlikely to cooperate in finding a resolution to a difficulty; someone whose interests have been acknowledged is more likely to be interested in finding a reasonable compromise.

Most angry people simply want to feel that their concerns are being heard. Listen attentively; even if the client's position seems unreasonable, the feelings of frustration and anger are genuine and should be acknowledged. If you can keep your cool, you may be able to keep the situation from escalating; you may even be able to resolve the problems and preserve the client relationship.

POINTS TO REMEMBER

1. Being an effective and skilled communicator means being able to assess the demands of each communicative situation and respond appropriately to those demands.

2. Communication always involves the feelings, attitudes, experiences, and concerns of others, as well as our own expectations and the quality of what we say.

3. Because our communication affects the interests and well-being of others, it always involves questions of ethics.

4. Communication, like all our interactions with others, is shaped by ethical codes based on religious, cultural, familial, or professional values and customs.

5. All ethical systems have a strong collective, or social, component.

6. Ethical communication involves taking responsibility for our words and actions, avoiding misrepresentation of ourselves or our messages, and limiting unnecessary harm done to others.

7. People tend to respond in kind. Defensive or rude behaviour will invite similar behaviour from others; polite, considerate behaviour will invite polite behaviour in return. Keep your cool and try to be pleasant even when others are rude.

8. What people want more than anything is for their concerns to be heard. If you can learn to listen to relation as well as content, you can dramatically improve your communication with others.

SHARPENING YOUR SKILLS

1. In a recent interview, a senior student interviewing for his first professional position was asked the following question:

   If a situation arose in which you were asked to violate company policy, under what circumstances would you do so?

   Consider your response to the ethical issues this question raises, and your own potential answer to the question. Prepare to discuss with the class, or write a memo to your instructor, regarding the following issues:

   a) Would it matter what the policy was?
   b) What special circumstances would warrant breaking company rules? Do you think your boss would agree?
   c) How accurately does your hypothetical answer to the question reflect your past experience? Is a hypothetical answer ethical? Is the question itself ethical?

2. Surveys and anecdotal reports suggest that academic cheating among university students is widespread. However, there is some disagreement about what counts as cheating, or whether it should be considered a serious academic violation. Consider the following scenarios; which do you consider to be cheating? Which are clearly not cheating? Which are questionable, but permissible in some circumstances, or justified by the workload in your programme? Which are never acceptable in any situation? Rank these in order from acceptable to unacceptable, and be prepared to defend your answer in class discussion.

   ____ copying routine assignments from someone else

   ____ fudging lab results

   ____ changing the results of calculations to coincide with answers given in the book

   ____ buying assignments from a Web site

   ____ copying in an examination

   ____ studying from a pirated copy of last year's exam

   ____ secretly taking notes into an exam, though they have been specifically disallowed

   ____ handing in the same report in two different classes

   ____ having someone else edit your written work

   ____ hiring a tutor for help with mathematics or English

___ sharing the workload in a study group

___ writing on your own the project report that the entire group was supposed to write

___ handing in a report that a friend or relative wrote for the same class a couple of years before

___ downloading a report from the Web and submitting it for an assignment

___ storing information in your calculator to take into an exam

___ passing answers to a friend during an exam

___ lending your work to a friend who is doing the same assignment

___ copying from a friend's work without her knowledge

___ copying a report or assignment from a textbook or other published source

___ copying a technical design from another source

___ knowingly letting a friend copy your assignments

Compare your list with those of others in the class. How much consensus was there? What differences did you discover? Both individually and as a class, consider whether you would change your opinion if the course in question were: a required elective; a free elective; a required core course. How should your university or college deal with plagiarism and cheating? Should students be expelled for handing in the work of others as if it is their own? Would such measures curb cheating? Summarize the results of the class discussion into a code of ethics, and write them up in a short memo report for your instructor. Submit the report via e-mail if your instructor directs you to do so.

3. You've just graduated and are seeking a permanent job, but you have been offered a second summer internship with the company where you worked last summer. It's a great job, and you've been told that there's the serious possibility that this position will become full-time when the summer is over. This year, they will be hiring a second intern, and you have been invited to participate in screening the applications for the job. Your boss has let you know that your recommendation will strongly influence the decision about whom they will interview.

In the applicant pool, you discover a submission from someone else in your college programme. Judging from the applications, he is the best qualified candidate for the position. As well, you know from being in classes with him that he would do a very good job—as good as you can do yourself. Unfortunately, though you have nothing personal against the guy, you know that there will be only one permanent job when the summer ends, and you fear that, should they hire your acquaintance, he will be your chief rival for the position. Using the principles of ethical communication as a guideline, what should you tell your boss? In what sense is this scenario strictly a communication dilemma? In what sense is it an ethical dilemma? To what extent are the two mixed? Be prepared to discuss your reactions with the class or to submit your response to your instructor in a memo or e-mail.

4. In a recent high-profile case, a Canadian university professor was removed from her job for misrepresenting her academic qualifications and research. In interviews contained in the news coverage of the incident, the students in the department where she taught for two years expressed outrage at this deception, and said they feared that this act of misrepresentation by their instructor would compromise the value of their university education. Do you agree? How serious is this breach of communication ethics? Should this professor have lost her job for her action? Why or why not? Be prepared to defend your argument using what you have learned about communication in this course.

5. Go to the section of your college or university calendar or Web site that deals with student rights and responsibilities. Based on what we have discussed in this chapter, does the material contained there constitute a code of ethics? To what extent is it a guideline for ethical communication? Most codes of ethics balance a sense of duty or obligation with freedom of choice; which of the two (obligation and choice) does your college's behavioural code emphasize? How prescriptive is it? Write your analysis into a brief informal report and submit it to your instructor in hard copy or electronic form.

6. Turn again to your responses to problem #2 and problem #4. Did you apply the same ethical standard in both cases? If not, why not? What is the relationship between students who resort to plagiarism and faculty who do the same? Should one be punished more severely than the other? Why or why not?

CRITICAL READING

As you have learned, one tangible benefit of studying the principles of rhetorical information is that they can help improve the general quality of your written and spoken communication. However, the study of communication can expand your understanding in other ways as well.

In this section of the text, we have explored in greater detail some of the ethical and social dimensions of communication that grow out of the

Axioms with which we began our discussion in Chapter One. We know that communication always takes place in a social context, that it involves implicit, unspoken messages that are just as influential as the explicit content of what we say. We know also that communication creates and maintains relationships between individuals, and between individuals and their social or professional groups. We create credibility, in part, by the extent to which we can identify ourselves with elements in the culture that our audiences, and we ourselves, perceive as authoritative.

In the report that follows, Julian Demkiw explores the power of engineering culture by analysing one of the profession's central cultural expressions: the Ritual Calling of an Engineer. Demkiw, himself an engineer by training, hopes to uncover some the reasons for the strong sense of identity that engineers feel within their profession, and in so doing, to understand the strength of his own identification with engineering as a cultural framework. Demkiw's analysis also sheds light on the way communication contributes to our sense of personal and social identity through the symbols we embrace and the stories we tell. As you read, consider the ways in which Demkiw's analysis makes use of, and also expands upon, what you have learned throughout this book about the way communication functions.

AN ORPHANED "SON OF MARTHA": DEFINING THE ENGINEER

Julian Demkiw

In May of 1999, after six long years of circuit design and math equations, and many hours of studying, I graduated from the University of Saskatchewan with a degree in electrical engineering. Normally, the next logical step to take after these six arduous years would be to follow my classmates and find a nice engineering job, most likely somewhere in the oil fields of Alberta, make a healthy living and never look back. Instead, I took a different route. After being involved in student politics for four years, I had the opportunity to work in a public relations capacity for the university that had given me my degree. It was a far cry from my academic training, but well within my skill set and much more compatible with my future plans.

So I set off on a different journey from the one laid out for so many who graduate from the College of Engineering; however, although I am not working as a professional engineer, and although government-approved professional engineering guidelines state that I have no right to the title of "engineer," I find that I continue to identify with engineering culture and to define myself as an engineer. Although I am not eligible for membership in the Association of Professional Engineers and Geoscientists, somehow I continue to hold onto this identity. The question is, why?

Julian Demkiw is an Alumni and Development Officer for the University of Saskatchewan. A graduate of electrical engineering, he has worked in public relations and is now taking advanced study in the field of professional communication and rhetoric.

This report examines the process of professional enculturation in engineering as a communication process. In particular, I will examine some of the profession's cultural symbols that build such strong identification for engineering students, focussing on The Ritual of the Calling of an Engineer, a ceremony in which all Canadian engineering graduates take part, and which formally marks the entry of young engineers into the profession.

BACKGROUND

The Ritual of the Calling of an Engineer began in 1922 when a group of prominent engineers met in Montreal, Quebec to discuss a "concern for the general guidance and solidarity of the profession."[1] Their goal was to create an organization designed to bind more closely together all members of the engineering profession in Canada, and to imbue them with a strong sense of their responsibility toward society. To assist them in their goal, they enlisted the services of the poet Rudyard Kipling, whose military experience had given him a connection to engineers and who had caught the attention of the engineers through his writing about the profession.

Kipling designed an entire ceremony to achieve the goals of community-building and professional identification that the engineers indicated. In his own words,

> The Ritual of the Calling of an Engineer has been instituted with the simple end of directing the young engineer towards a consciousness of his profession and its significance, and indicating to the older engineer his responsibilities in receiving, welcoming and supporting the young engineers in their beginnings.[2]

The Ritual of the Calling of an Engineer, through the symbols connected with it, is the culmination of that sense of connection for graduating engineers. The Iron Ring, which each graduate receives at the ceremony, serves as a constant symbol of that connection to the profession and to all other obligated engineers across Canada.

The centrepiece of Kipling's ceremony is a poem he created specifically with engineers in mind, called "The Sons of Martha." It is my belief that an analysis of content and relation in this poem, of the ritual itself, and of the central symbol associated with the ritual, the iron ring, will help to demonstrate that the identity of an engineer is based not in law or regulation but in identification with culture and community. It is this powerful sense of identification that makes me continue to think of myself as an engineer, even though my career choices have taken me away from professional practice.

Kipling's poem[3] is central to the ceremony that marks the passage of a new graduate into the profession of engineering. Since this same poem is used in every ceremony in every school in Canada, it clearly functions as a key symbol of how the profession as a whole conceives of its role in and contribution to the social fabric. In the course of this analysis, I will show that Kipling's poem also functions as a symbolic identification of mythic, even Biblical, significance.

METHOD OF ANALYSIS

In conducting an analysis of Kipling's "The Sons of Martha," I will draw upon an understanding of communication as more than message-making. Instead, I will show how communication also functions to establish and maintain a sense of identity for a group. To do this, I will employ a method known as fantasy theme criticism,[4] a type of narrative analysis based on Ernest Bormann's theory of symbolic convergence. Fantasy theme criticism focusses on the dynamic process of group fantasizing. A group fantasy event involves the ritual repetition of a story that provides a common cultural focus, often through the invocation of mythical identification, for that specific group. Typically, the fantasy theme narrative features characters, sometimes real, sometimes fictitious, playing out a drama in a setting which, though removed from the present day transactions of the group, can be read as an allegory of the group's values, beliefs, and cultural mythology.

Symbolic convergence theory is based upon two assumptions; first, that symbols of rhetorical discourse do not merely represent reality as it is, but actually create the group's experience of the world through their power to establish a shared vision of meaning and value. Second, Bormann argues that, when individual conceptions of the fantasy theme, or vision, converge, group identification is assured. For Bormann, the message itself is not as significant as the act of sharing that message.

The fantasy theme is a shared story that expresses the value system and sense of identity of the group. It may depict either future or past, featuring either events that may be possible, or events that are said to have happened in the past. It is through such a ritualized, shared narrative that the group reaffirms its existence and interprets current experiences; it is through this shared narrative as well that the group welcomes new members into its community. By studying such a ritual fantasy theme, we can gain an understanding not only of how a specific group views its role in the world, but of how groups in general manifest their unique identity and their connection to society.

A key unit of analysis needed to understand the process that defines an engineer is the deliberative and descriptive language that sets the fantasy theme, both in the poem and in the terms used to describe the Ritual itself. The fantasy theme that is created interprets the world in a particular way for the group, and this analysis will also consider the setting of the fantasy, the characters placed in the setting, and the actions that are taken by the characters.

FEATURES OF THE RITUAL

The responsibility for the annual Ritual of the Calling of an Engineer is in the hands of seven Wardens, headquartered in Montreal, who represent the original seven prominent engineers who originally proposed the idea of the ceremony in 1922. Under their authority, Camps have been established throughout Canada to administer the Ritual. Although the Camps are often located near an accredited university, there is no necessary connection between the Camps and any institution or association of professional engineers.

A key feature of the Ritual is its privacy. There is no publicity associated with the ceremony, and admission is restricted—only graduating engineers who have reached eligibility by completing their programs and "obligated" engineers—those who have previously been through the Ritual—may attend. Each participant is required to "take an obligation," in return for which he or she is given a symbolic iron ring. According to the tradition of the Ritual, this obligation is not an oath as such, but a "solemn expression of the intention to adhere to the highest of ideals of the engineering profession."

The ring, another key feature of the Ritual, is a reminder to newly obligated engineers and a sign to others that they have taken the obligation. The ring is of unique importance to Canadian engineers, since Canada is the only nation which upholds the tradition of honouring and distinguishing our engineering graduates by awarding a ring in this fashion.[5] Although participation in the Ritual is voluntary, and choosing not to participate does not restrict the young engineer from any professional status, the Ritual represents the engineer's symbolic entry into the professional culture, an initiation rite that is further represented by the iron ring.

If, as Bormann suggests, rhetoric creates a shared cultural reality, particularly when it appears in ritualized narratives, we should expect to find in the ceremony, and consequently in Kipling's poem, a presentation of the shared mythic vision of the engineering profession—and indeed we do. As rhetorical theorist Sonja Foss explains, Bormann's theory assumes that such symbols "not only create reality for individuals but . . . individuals' meanings for symbols can converge to create a shared reality for the participants."[6] According to Foss, "the message itself is important, but the sharing of the message is seen as even more significant in the symbolic convergence theory."

As a result of symbolic convergence provided by the ritualized narrative, participants in the shared rhetorical vision develop a common consciousness that provides a basis for communicating with one another to discuss common experiences, achieve mutual understanding, and create community. The Ritual of the Calling of an Engineer is itself a culmination of such a convergence. It is a time and place where elder engineers come to educate the younger ones about what it means to be an engineer. The Ritual is designed to enact the community that defines the engineering culture. It has even been instituted so that it is shared and renewed at least once a year. As Foss says, the sharing of this Ritual is more significant than the actual message it relays.

THE SONS OF MARTHA

"The Sons of Martha" provides the vehicle for expressing the engineering fantasy theme known as the rhetorical vision. Its mythic power originates in part from its identification with Biblical narrative, which provides both precedent and interpretive frame. The principal characters of the narrative are the sons of Martha, the sons of Mary, and the Lord. The original narrative on which the poem is founded comes from the Gospel of Luke (10:38–42), where Jesus and his disciples visited the house of two sisters, Mary and Martha. While Martha worried about how to serve all the guests, Mary preferred to sit at Jesus' feet and listen to his teachings. When Martha implored Jesus to tell Mary to help out with the work, Jesus replied that Mary had chosen to do the better thing, listen to the word of the Lord. Because Mary had chosen the "better part," it was now Martha's burden to ensure that everything else was taken care of.

There can be no doubt that Kipling intends the reader to refer to this story, as he invokes it directly in his narrative. Given the centrality of Kipling's poem to the Ritual of the Calling of an Engineer, it becomes obvious that the "sons of Martha" are understood to be engineers, who take care of the tasks of living, while the sons of Mary are the non-engineers for whose benefit and comfort the engineers labour.

THE LANGUAGE OF THE FANTASY THEME

The very language used to describe the Ritual assists in creating the shared fantasy theme. The fact that engineering is referred to as a "calling" helps us gain a glimpse of how engineers see their profession. The word "calling" shows that, in the rhetorical vision, engineering is not chosen by the individual; instead, it is the profession itself that does the choosing. Further, because this terminology is normally used only for the "noble" professions such as the clergy or even medicine, its use here implies that engineering is similarly noble; rather than simply a profession, it is a "calling" to which the specially chosen few are beckoned by a higher entity.

The term "obligation" sets a similar tone because of its connotation of duty. Although an engineer does not take an oath to duty in quite the same fashion that police officers, doctors, lawyers and other professions do, they are nevertheless obligated to their duty by their "solemn expression of intention." In some sense, the engineer does not choose the burden of his obligation, though he accepts it with humility and gratitude; instead, it is laid upon him by powers beyond him—or such is the implication of Kipling's poem. Within the context of the poem, the "caller" of the engineer is therefore the Lord Himself, who obliges the engineer carry the burden of caring for Mary's sons, who are thus granted freedom from care. And while Mary's sons, having "cast their burden upon the Lord" (l. 32), can go about their business without worry, the sons of Martha in all humility and dedicated service assume the burdens that have in turn been laid upon them by the Lord.

READING THE FANTASY THEME

The poem can be read literally as well as symbolically. The stanzas describe, in turn, the various disciplines of engineering, at least the ones that existed in 1922. Each discipline is represented by the tasks and responsibilities most closely associated with it: references to "the gear" and "the switch" represent the mechanical engineer; the movement of mountains and damming of rivers, the civil engineer; "piecing the living wires," the electrical engineer; and the "secret fountains," the geological engineer.

Each characteristic displayed portrays a fantasy theme that contributes to the rhetorical vision. In each case, these sons of Martha are presented as careful and responsible for the mundane, but essential, tasks of day-to-day life, and in displaying these characteristics, they are shown to be true to the pattern set by their mother, who is distinguished by her "careful soul and troubled heart" (l. 2). Nevertheless, despite their care and the burdens they carry, the Sons of Martha are shown to be powerful and fearless, and are revealed to partake of some of the power of the Lord they serve. For example, they are able to achieve their deeds in much the way He did when creating the world He has now entrusted to the care of the Sons of Martha: "They say to mountains, 'Be ye removed.' They say to the lesser floods 'Be dry' (l. 9)." The *Sons of Martha* need not work to flatten the mountain or dam the water; instead, they merely say it, and it is so.

The engineer is ever watchful, diligently looking out for the sons of Mary. This vigilant watch is undertaken without complaint and is never questioned even when paid "with the blood some Son of Martha spilled" (l. 26). In addition to the power the Lord shares with them, the engineers are also equipped with the secret knowledge that they need in order to complete their service. This knowledge is enough to keep "Death at their gloves' end" (l. 13), to "hale him forth" and "goad him through the end of days." This is awesome power indeed, but it is all in the interest "simple service, simply given" by each Son of Martha "to his own kind in their common need" (l. 28).

Setting is another key component in developing a fantasy theme. While the settings in the poem are not geographically specific, they do take in the range and variety of the entire earth, and thereby show the extent of the engineer's watchfulness (l. 22). The Sons of Martha have command of mountains and floods, up high and down under the earth. Their influence can be found in "thronged and lighted ways," in the "dark and desert" (l. 23), in all the locations and situations of the world. Such descriptions help to establish the fantasy theme, that the influence of the Sons of Martha can be found "all around us, all the time,"[7]—leaving us to conclude that there is no one location where engineers are found, for they can be anywhere.

THE RHETORICAL VISION

The fantasy themes that permeate Kipling's ritual help to create the rhetorical vision. The rhetorical vision is the culmination of a group's separate fantasy themes, brought together to provide a shared experience and interpretation of reality. Collectively, through the Ritual of the Calling of an Engineer, engineers participate in a vision of their community that provides, for them, a mythical, even Biblical fantasy of what it means to be an engineer.

As the "Sons of Martha," engineers bear by birthright the burdens of care, which have been bestowed on them by the Lord himself (l. 32). Understanding the rhetorical vision means understanding that engineering is not simply a profession to engineers; instead, it is a calling to a higher purpose which, while it requires the chosen to bear burdens unknown to the rest of society, bestows upon them a special place at God's right hand. Born into this world with a purpose, engineers are called by their Maker to serve out their lives' obligation without complaint or special favour, "Not as a ladder from earth to Heaven, not as a witness to any creed,/ But simple service simply given to his own kind in their common need" (ll. 23–24).

In order to fulfill this obligation, the engineering culture creates fantasy themes surrounding the engineer's abilities. God-like powers, secret knowledge, and watchful ways are all parts of the fantasy that coalesce around the Biblical story. The universality of setting and multi-faceted nature of engineering discussed in the poem combine to illuminate the higher purpose of the engineering profession. These fantasy themes serve to support the Biblical rhetorical vision.

THE ROLE OF THE RHETORICAL VISION

Sonja Foss states that rhetorical visions "are always slanted and ordered in particular ways to provide compelling explanations for experiences."[8] The rhetorical vision of the Ritual of the Calling of an Engineer serves this purpose for those who are part of the profession. Despite strong ethical and ritualistic features in its internal culture, engineering enjoys comparatively less visibility and social prestige than the other "major" professions such as law and medicine, which have been glamourized in the public mind through prime time television, novels, movies, other popular culture venues. Although it could be argued that engineering has made contributions equal to those of the other professions, it has not, at least in recent history, achieved the same level of prestige in popular culture; indeed, the most visible manifestation of the engineer in current popular culture is Scott Adams's nerdy "Dilbert," who despite his command of technological advances is at the mercy of those who have surpassed him in the management hierarchy.[9] The engineer achieves transcendence through identification with a higher order, thus linking the profession to a higher purpose. The rhetorical vision created in and reinforced by the Ritual of the Calling of an Engineer enables the engineer to attribute greater significance to the profession while retaining as part of that vision an innate humility that explains its relatively low status in the cultural hierarchy.

The rhetorical vision represented by Kipling's poem does, as Bormann says, create a view of reality for its participants. Engineers have turned a potentially community-crushing reality into an attractive, even noble, experience. Accepting the fact that the profession itself is not considered glamorous, engineers have drawn upon the Biblical vision to transform the meaning of their social reality. No longer are engineers merely hard workers who get no recognition, as they are at the outset of Kipling's ritual; by the end, they have been transformed into "Angels" that no one can see. Where once they may have seemed "nerdy" and obscure, engineers now possess a "secret knowledge" that sets them apart, and ultimately, rather than being subservient to the "Sons of Mary" and doing their dirty work, the engineers are now assisting the Lord Himself by assuming responsibility for watching over His flock.

The rhetorical vision of Kipling's poem enables the engineer to transcend the ordinary to become part of an extraordinary community. Initially, the Sons of Martha are those who "wait upon" the sons of Mary and "deliver" them in their daily lives, almost in the manner of servants. But by the end of the poem, the Sons of Martha achieve symbolic transcendence through a transformation into "Angels," and the rhetorical vision is fulfilled. Despite their transcendent movement beyond the social hierarchy, the engineers of the poem do not act out of a desire to obtain "a ladder from earth to Heaven, [nor] as a witness to any creed" (l. 27). Instead, they are shown to accept willingly their burden of "simple service simply given" (l. 28), without rancour and without expectation of relief or assistance: "They do not preach that their God will rouse them a little before the nuts work loose./ They do not preach that His Pity allows them to drop their job when they damn-well choose." (ll. 21–22). Theirs is a charge that must be fulfilled, even when it is thankless or unnoticed by those who benefit from their labours.

Nevertheless, apparently contradictory themes appear; despite being bound to service, the engineers are endowed with great potency, akin to that of God himself. Indeed, they are revealed to enact their will through their words alone: "They say to mountains 'Be ye removed,' They say to the lesser floods 'Be dry'" (l.9). Such acts are by no means a mere "simple service." Thus, the engineer's humility is at the same time balanced by awesome power and responsibility; it may be a life devoted to service freely given, but it is also a life exalted by identification with the Creator of the universe.

Despite their apparent contradictions, these seemingly irreconcilable visions are united in the context of the rhetorical vision, as Foss predicts: "Actions that make little sense to someone outside of a rhetorical vision make perfect sense when viewed in the context of that vision, for the vision provides motive for action."[10] The idea that someone with God-like power would stoop so low as to be a servant, even spill blood as a servant, makes sense when viewed in the context of the rhetorical vision—the same framework within which Christ himself made the ultimate blood sacrifice. The parallels, which become evident on examination of the relational element of the message, contribute to the final transcendence of the engineers' rhetorical vision.

THE SYMBOLIC NATURE OF THE IRON RING

Although the fantasies themselves are important, true symbolic convergence happens only when those fantasies are shared in a ritual setting where the group creates its shared vision of reality. The Ritual of the Calling of an Engineer allows for that shared reality to be affirmed anew and passed on from generation to generation of engineers. Often in order to retain that shared memory, a group's fantasy themes are gathered into a phrase or slogan, or represented by a significant object, which may become what Barry Brummett describes as "a material manifestation of group identification."[11] The symbolic representation of the engineers' rhetorical vision, as enacted in the Ritual of the Calling of an Engineer, is of course the iron ring that each engineer receives upon taking up the "obligation" of the profession.

The iron ring is a symbol recognized nationally by engineers and non-engineers alike. Its symbolic meaning is literally laid out during the ceremony itself, and its symbolic power is represented in the ritual of the fantasy theme. The purpose of the ring is to remind the engineer of the ceremony just taken, and to link the individual to the shared vision of the community into which he or she has been introduced. The symbolic cue is necessary to the power of the fantasy theme as it consistently reinforces the emotions and meanings surrounding the theme.

CONCLUSIONS

Engineers may define themselves expressly through regulations and laws, but they also achieve an implicit identity through the unspoken elements of the rhetorical vision. The explicit values and expectations of the profession are thus deeply rooted in a symbolic, ritualized vision of the engineer's system of values and morals that is encoded in the Ritual of the Calling of an Engineer, and its centrepiece poem, "The Sons of Martha." Together they affirm that at its very core, being an engineer is not something that can be defined merely through the regulating bodies represented by professional associations or colleges of engineering. Instead, according to its own symbolic vision, engineering is more than simply a job or a profession; it is an obligated calling to a higher purpose.

The rhetorical vision enhances and reinforces the conceptual framework of social responsibility that is one of the fundamental principles in the Canadian Engineering Code of Ethics.[12] It also reminds engineers of their essential contributions to the infrastructures that support all aspects of today's civilization. A recent advertising campaign celebrating National Engineering Week featured the slogan "7-24-365: Engineers: all around us, all the time." Even as this slogan reminds the public of the ubiquity of engineering in society, it also takes for granted the naturalized conception of engineers as the watchful "Sons of Martha" who oversee our needs like guardian angels.

As this analysis has demonstrated, a shared rhetorical vision is as important to professional enculturation as laws and regulations, if not more so. The fantasy themes that combine in the Ritual of the Calling of an Engineer provide a framework of values for the naturalized assumptions of social obligation at the centre of an engineer's definition of professionalism. At the same time, they provide a link to the central mythical construct of the culture. The process of enculturation represented in the Ritual of the Calling of an Engineer of course takes much longer than the ceremony itself, but the Ritual crystallizes the vision and allows its participants to enact their identification symbolically. Such rites of passage inspire and reinforce powerful identifications, both for those entering the culture and for those who participate in the induction of new members.

Once "obligated," an engineer bears the "material manifestation" of the ritualized identification in the form of the iron ring, but while the ring is easy to remove, the process of enculturation that it represents remains. I am not currently a practising engineer and perhaps may never be one; however, I know now why I continue to identify myself with the professional culture that has shaped so much of my understanding. Although I am in some sense symbolically an orphan, like all orphans I retain the indelible marks of my parentage, which in the end I can neither repudiate nor deny.

NOTES   [1] J. Jeswie, "Information Relevant to the Iron Ring Ceremony."<http://conn.me.queensu.ca/dept/jj-ring.htm> (retrieved 13 March 2001)

[2] Jeswie, "Information Relevant to the Iron Ring Ceremony."

[3] The poem may be found on numerous engineering Web sites in Canada, including the Civil Engineering Department of the University of Alberta <http://web.cs.ualberta.ca/~hoover/cultinfo/sons_of_martha.htm> and on personal Web sites such as <http://www.conscoop.ottawa.on.ca/rgb/som.html>.

[4] Ernest G. Bormann, "Fantasy and Rhetorical Vision: The Rhetorical Criticism of Social Reality," *Quarterly Journal of Speech* 58 (1972): 396-407 and "Fantasy and Rhetorical Vision: Ten Years Later," *Quarterly Journal of Speech* 68 (1982): 288-305.

[5] For an indication of the importance of this symbol, and the legends that have grown up around it, see "The Iron Ring." Available at <http://engsoc.queensu.ca/tradition/ring/>

[6] Sonja Foss, "Fantasy Theme Criticism." *Rhetorical Criticism: Exploration and Practice*, 2e (Prospect Heights, IL: Waveland Press, 1996) 122.

[7] This expression is the slogan of National Engineering Week. See the National Engineering Week Web site <http://www.new-sng.com/maintemplate.htm>. Retrieved 10 March 2001.

[8] Foss, 122.

[9] See, for example, Scott Adams, *The Dilbert Principle : A Cubicle's-Eye View of Bosses, Meetings, Management Fads & Other Workplace Afflictions*, rpt. ed. (New York: HarperBusiness, 1997) and *Dogbert's Top Secret Management Handbook* (New York: HarperBusiness, 1996).

[10] Foss, 124.

[11] Barry Brummett, *Rhetoric in Popular Culture* (New York: St. Martin's Press, 1994) 17.

[12] The Code of Ethics for the Canadian Council of Professional Engineers can be accessed at <http://www.ccpe.ca/ccpe.cfm?page=ceqb_ethics>

Things to Consider

1. What is the purpose of Demkiw's report? What inferences can you make about its intended readers?

2. Like the article by Burton L. Urquhart on page 206, Demkiw's analysis studies the implicit, relational aspects of communication. However, despite this similarity, the two articles differ significantly. Identify three important differences in the way each writer presents his message, and use your understanding of rhetorical balance to suggest reasons for the difference.

3. How challenging was this critical reading, compared to others we have encountered in this book? Has your increased understanding of communication made these readings easier? How suitable is its chosen style—the language, the structure, the tone—to what its author hopes to communicate?

4. Re-read the article carefully, and write a summary of its findings that could be used if the article were presented in report form.

5. Is there any sense in which Demkiw's article can be understood as a study of ethical communication in the technical professions?

How Texts Communicate: Some Suggestions for Reading Critically

Reading critically means doing more than simply passing your eyes over words on a page. Instead, like other forms of communication, reading is an active process that requires careful attention to both verbal and nonverbal elements, to both content and relation. Becoming a better reader will make you a better communicator in general, will improve your critical skills, and will actually, in the long run, save you time and confusion.

The texts you read—from technical reports to your textbooks to newspapers and magazines—are as varied as the writers who produced them. They may be personal or impersonal, accessible or challenging, formal or informal. They may be informative, descriptive, narrative, or persuasive. But in spite of their differences, the words you read all share a common feature. No matter what their purpose, their intended audience, or their author, they are all the products of deliberate communicative choices made by a human being in order to bring about a desired result or achieve a desired goal. Reading carefully and attentively means that you pay attention not only to the content of the document but also to the relationship it is building with its audience, to its assumptions about the situation and the reader, and to the attitude it displays toward the audience, the situation, the subject, and writer herself.

Reading for pleasure and reading for work or study are really different activities. The stories we read for pleasure create their own narrative momentum, which doesn't require us to puzzle out the pattern of the writer's thinking or to remember details of an argument. They are typically organized chronologically, which is one of the easiest organizational patterns to process. The sorts of texts you read for school or work are rarely as accessible or pleasurable as novels or magazine articles, and are less likely to use the chronological pattern of organization. They are therefore typically more challenging to process and to remember.

In this segment we are going to focus on the reading you do for professional purposes, learning three levels of skills to help master these difficult kinds of reading. First, you will learn some "quick fixes" to help you read more critically, or analytically. These can be mastered immediately, and they will make your reading much more effective right away.

Second, we will consider a slightly more complex study-reading method that will help you manage large amounts of information efficiently. It is designed to help you with the kinds of texts you typically encounter on the job: reports, instruction manuals, policy and procedures documentation, and so on.

Finally, we will consider a set of more detailed questions designed to help you read more in-depth materials with greater care and attention. At times you may be asked not only to comment on the explicit content of a text, but also on its unexpressed agenda, its embedded purpose, and the ways in which it creates particular effects. As we know already, it is often the implicit, unspoken elements of a message that are the most influential and long-lasting. The in-depth questions are designed to guide you in uncovering what those meanings are and how they work.

> **Advantages of Reading Actively**
>
> 1 Allows you to process large amounts of reading material more efficiently and quickly.
> 2 Provides an overview of content and promote greater involvement in the interaction.
> 3 Increases understanding.

THREE QUICK FIXES FOR BETTER READING

How many times have you finished reading something for a class, only to find that you had no idea what you had just read? The time you put in reading the material was time wasted, since you understood very little and retained even less. If you wanted to retain the material you had to re-read the entire text, and even then you might have found that your eyes tended to wander across the page without taking in much of what you were reading. You might also find your mind straying from the task, and realize that you'd just read several pages without taking anything in: your eyes moved across the page, but you weren't really paying attention. The problem here is not necessarily with your ability to process and remember; the problem is with your reading strategy.

One of the mistakes made by most readers is to approach work-related texts as though they are novels: we start reading at the beginning and continue to the end. While this is a fine strategy for pleasure reading, it is ineffective for other kinds. Here are three "quick fixes" you can use immediately to improve both your understanding and your retention. They don't take much longer than a single reading, and they take far less time than reading a text twice.

> **Three Quick Fixes for Better Reading:**
>
> **Fix #1:** Preview the text first.
> **Fix #2:** Fill in the framework as you read.
> **Fix #3:** Ask the big question.

Fix #1: Preview the text first.

Before you ever read another document for work, preview the whole thing first. Read the summary or the introduction, and the conclusion. Make note of the headings and subheadings in the body of the text. If there are no headings or subheadings, scan the first and last sentence of major paragraphs or sections.

Taking a few moments to do this will actually save you time in the long run, because it will give you an overview of the purpose and shape of the text. It will be much easier to process and remember what the text says if you know what to expect before you start. Make it a habit *always* to take a few minutes to preview before you begin reading.

Fix #2: Fill in the framework as you read.

Once you have previewed the text, you will have a clearer idea of the framework around which the arguments are organized, and of the purpose of the chapter or report. As a result, you can read more knowledgeably and with an expectation about what is going to be presented. Because you know what to expect, you will be more engaged and alert as you read, and you can link the material in the reading to the framework you've already identified with your preview. You will even be able to take quicker and more efficient notes, because you've inferred an outline structure for the chapter.

Use a pencil rather than a highlighting pen to mark important ideas, since with a pencil you can jot down a brief outline as you read. Look for the main points and quickly list them in the margin. You should make your comments right in the book if you can, since your brief comments and notes will be right alongside what you've read and you'll easily be able to find them later. Writing your brief notes in the margins will also mean that you can work efficiently no matter where you are reading: you don't need to be sitting at a desk or in front of your computer.

Fix #3: Answer the big question.

Once you have previewed and then read the text, you are in a position to consider how it communicates on a relational, nonverbal level. One way to understand the relationship established or maintained in the text is to ask "Who is saying what to whom, and for what purpose?"

This question has three distinct parts. *who is speaking, to whom is she speaking,* and *what does she want?* There is no need, at the beginning, to answer these questions in detail, but you should think about what they can tell you about the dynamic being created or maintained. For example, consider the e-mail letter of congratulations shown in Figure 4.2, from Randy Alexander to Peter Holowaczok. *Who is saying what to whom, and for what purpose?* The explicit dynamic is of a more senior employee offering congratulations to a junior employee for his achievement in becoming employee of the month, and offering some helpful suggestions for his work.

However, it's clear that the explicit dynamic doesn't tell the whole story. *Who is saying what to whom?* Judging by the tone, emphasis, and language in the message, the dynamic that is being created is one of hostility and jealousy. The writer seems threatened by and jealous of the other man's success; he seems to feel that his own deserving work has not been recognized. However, instead of approaching those who could act on this problem, he chooses to vent his frustration and anger on a junior employee through condescending language, highhanded suggestions, and a sarcastic tone. Here, the nonverbal cues clearly contradict the verbal cues, making the analysis of the relational elements a little easier. What should Peter Holowaczok learn from this exchange? He should take care around Alexander, who appears to be a bitter, jealous person who is not above taking potshots at someone who is not in a position to defend himself.

Although it's unlikely that the relational elements of communication will always be as thinly veiled as they are in this example, there will always be some evidence in the text that communicates information about relation. Asking the big question of *Who is saying what to whom?* will help you to pay careful attention to this dynamic so that your own messages can effectively respond.

HOW THE PROS DO IT: STUDY-READING BY THE SQ3R METHOD

The "SQ3R" Study-Reading Method

S = Survey
Q = Question
R = Read
R = Recite
R = Review

During World War II, the American military developed an efficient reading method to assist recruits in quickly and efficiently reading the large amounts of training material facing them. Known as "SQ3R," it was made available to the public in Francis P. Robinson's 1946 book, *Effective Study*.[1]

The SQ3R method is similar to the quick fixes we have just discussed, but it is somewhat more elaborate. It was designed primarily for reading textbook materials which are organized for teaching, and which use headings, bullets, checklists, chapter summaries and other visual and organizational devices to aid the reader. In fact, a glance at several textbooks published in the first and second halves of the 20th century reveals that over the years since the study-reading method was popularized, textbook design has gradually evolved to accommodate some of its principles, making texts more reader-friendly.

The study-read method is a five-step process that enables you to read a text efficiently by making several quick passes through it rather than trying to read it all at once.

---

[1]Francis P. Robinson, *Effective Study* (New York: Harper and Bros, 1946) pp 29–35.

S=Survey

As in the preview step given above, the "survey" step in SQ3R involves glancing over the headings in the chapter to identify the few main ideas that the chapter discusses. In your quick survey, you should also read the final summary paragraph if the chapter has one. This survey should not take more than a minute and will reveal the three to six central ideas in the chapter. The survey will help you organize the ideas as you read them later.

Q=Question

The second step involves formulating a question out of each of the headings. If, for example, the heading is "Study-Reading by the SQ3R Method," your question might be "What does 'SQ3R' stand for?" Thinking about the main points in the text as questions will increase your active participation in what you are reading, and so increase your comprehension. Rephrasing the heading as a question will also make important points stand out, so they are easier to find. It doesn't take long to turn a heading into a question, but it does take conscious effort, and helps keep your attention focussed as you read to find the answer.

R=Read

Now read the text as you normally would, from beginning to end. However, as you read, look for the answers to the questions you posed. Reading to find answers to your own questions will keep you actively attentive and involved and prevent the kind of wandering of attention that sometimes occurs with unfocussed reading.

R=Recite

Stop when you have read the first headed section. Try to state the answer to your question out loud. Express your answer in your own words, and try to give an example. If you can do this, you will know you've actually learned what is in the section; if you can't, then take another look. You may want to jot down an outline of these main ideas as you go, but there is no need to rewrite the text. Brief notes should be enough. Continue this process until you have finished the entire reading.

R=Review

Once you have finished the reading, return to your brief notes to get an overview of the chapter's contents. You can check your memory by trying to recite the main ideas for each point once again.

The benefits of the SQ3R method are well known. It offers a way of processing large amounts of reading material fairly quickly and efficiently, and works well for reading technical materials and reports as well as textbooks. The first few times you try it, it will take a little longer than your normal approach to reading, but once you get the hang of it you will find that it takes no longer than a single read-through, but with the added benefit of increased understanding and better retention.

DETAILED QUESTIONS FOR CLOSE CRITICAL READING

In addition to reading a text or report to glean its content and general information, you will need to be able to read with some sensitivity to unspoken agendas, implications, and motives. The questions below are designed to help with the task of close critical reading in cases where what's going on in the communication interaction is not so immediately obvious, as has done in, for example, Julian Demkiw's essay "An Orphaned 'Son of Martha': Defining the Engineer" on page 338.

As you will recall from our Nine Axioms of communication, what is left unsaid can be as important, or even more important, than what is explicitly stated. It is in the nonverbal elements of the message that the relationship between speaker and audience is framed. However, these can also be much more difficult to analyse accurately and fully.

The questions below provide a kind of framework for thinking about those texts that you must understand more deeply. They will help you figure out the dynamics of the document you are reading; they will also provide you with a method for understanding how written communication in general influences its intended audiences. The questions have been divided into categories that reflect the major areas of consideration for written communication.

Subject, Purpose, and Meaning

The first step in understanding the meaning of any written material is to be sure you have grasped its topic and purpose. Think not only about what the message actually says, but about the way the tone and nonverbal qualities confirm or contradict the explicit content.

1. Define any special terms introduced or used in the reading.

2. What is the topic of this work?

3. Does this work have an explicit statement of purpose? If so, what is it? Where is it found?

4. Is there any distinction between its stated purpose and its actual purpose?

5. Is the audience being asked to undertake any specific action(s)? What are they?

6. What is the context or setting in which this message was created? How has the context shaped the message?

7. How is the selection organized? (That is, what pattern of organizations is employed?)

8. What evidence, or what kind of evidence, is provided to support the argument? Is anything significant left out?

9. Does the speaker misrepresent aspects of the problem?

10. Can you find examples of propaganda devices or fallacious reasoning?

The Character of the Speaker: Ethos

As you know, the speaker is an essential component of any communicative interaction; in fact, the speaker may well be the single most influential element of communication. The following questions are an aid to understanding what the speaker's character contributes to the meaning and purpose of the selection.

1. What kind of person is the speaker? How does she view herself?

2. Does the speaker display sincerity, authority, and credibility? How are these qualities made evident? Pick out some examples.

3. What is the speaker's stance, or attitude, toward the audience? Does the speaker address you as an equal? Does he seem to position himself as inferior or superior to you?

4. How personal or impersonal is this passage?

5. What common ground does the speaker create between herself and the audience? In other words, what values does the speaker seem to share with her intended reader?

6. Does the speaker appear sincere? How does he gain authority or credibility?

7. What is the tone of the selection?

Structure

The structure of a message refers to the number and arrangement of its significant parts. A typical report, for example, has five standard parts, arranged in the following order: summary, introduction, discussion, conclusion, recommendations, and appendices. However, what is most interesting from the point of view of message design is the internal structure of the discussion, which is where the choices of the writer become most evident and significant. Essentially, in analysing the structure of a message, you are looking for the types of sections or divisions contained in the discussion, the number of these sections, and the order into which they are arranged. The arrangement of these parts can make a difference in how the reader experiences the meaning of the message.

1. Into how many parts is the discussion divided? Is this division significant?

2. Where is the main purpose expressed? How many reasons are given to support the arguments in the discussion?

3. In what order are the reasons presented? What effect does this ordering of parts have on the passage?

4. What kind of evidence does the author use to support his position? Is it convincing?

5. What device does the author use to open the selection? To close it?

6. Does the author use any unusual sentence or paragraph structure (for instance, inversion)? Why might this be?

7. What contribution to the meaning or impact of the message, if any, is made by its physical layout (including paragraphing, numbering, headings, categories, illustrations, etc.)?

Style and Strategy

In a communication act, a strategy may be understood as any method or device that a communicator uses to influence an audience's attitudes, feelings, actions, or beliefs. Although they are usually linguistic or verbal devices, communicative strategies can also be visual or structural. Verbal strategies include such devices as specialized language or jargon, humour, irony, comparisons, name calling, and so on. Structural strategies or devices may include organization patterns, numbering of items, repetition, and parallel constructions. Visual elements may include the use of white space, pictures, borders, or special print features such as boldfacing or underlining.

In short, any method that can help a writer to sway an audience can be considered a communicative or rhetorical device.

1. What, if anything, is special about the language used in this message? What effect does this kind of language have on the audience who will be reading the message?

2. Explain the meaning of the title. Why do you think the writer chose this particular title? What effect does it have on the audience?

3. Does the message employ humour? If so, is it appropriate? How does humour help the writer to communicate her message?

4. What details does the author emphasize? What role do these play in making the message effective? What are they meant to communicate?

5. Is this message written in a formal, informal, or casual style? How can you tell? How suitable is its chosen style to the purpose and audience for which it was written?

6. Is the selection ironic? Is it satirical? Does it employ parody? How do you know? What is the effect of these devices?

7. Metaphors and similes are forms of figurative comparison that compare two things that are essentially unlike. Does the message include any such comparisons? How do they work? What is their effect on the intended audience?

8. A symbol is a thing that stands for some meaning, usually an abstraction, beyond itself. Does the author employ any symbols? What are they? What do they represent? Why do you think the author chose to use symbolism?

Speaker–Audience Relationship

In addition to providing some kind of content, all communication attempts to establish, or helps to maintain, a relationship between the speaker and the audience. Looking closely at the selection can help you to identify the kind of reader that the speaker or writer intends to address. These questions are meant to help you uncover this information.

1. Who are the intended readers of this passage? Are they general readers, or are they specialized? How can you tell?

2. Is any specific group excluded by this passage? Is this exclusion significant?

3. What is the speaker's attitude toward the audience? Are *you* part of the audience?

4. What assumptions does the writer make about the reader's needs, expectations, concerns, or prior experience?

5. What assumptions does the writer make about the audience's beliefs, fears, prejudices, or desires?

6. What attention is paid by the writer to the emotional impact of the discourse on the audience? In what direction does the writer attempt to move them emotionally?

7. How does the speaker appeal to the audience's self-interest?

8. What effect does the writer want the audience to experience?

These in-depth questions are intended to help you get the most out of your critical reading. They are a useful place to begin if you are asked to assess someone else's messages. If you can learn to think your way through any of your readings—whether for this class or another—using these terms, you will develop a critical habit of reading that will help you to understand much of the way everyday messages influence and engage you. Such critical skills are helpful in every class, and in every situation in which people use language to influence each other.

Reading actively using one or more of these methods will improve both your immediate comprehension and your overall understanding of the communication process. Like any skill, critical reading takes practice to master, but it also gets easier and more efficient with practice. As reading attentively becomes a habit, you will find that it also takes much less time and effort, and that your awareness of what is being communicated, and of the dynamic that is operating in any interaction, becomes much keener. As with other communication skills, improving your reading skill will pay handsome dividends in your professional and personal interactions.

Grammar Review

Writing effectively means choosing your words carefully and putting them into an understandable order so that your reader receives the message clearly and without ambiguity. Part of this process involves using correct grammar. Incorrect grammar can undermine the clarity of your writing, and may sometimes create unintended meanings. In a business or career situation this can mean lost revenue. It quite literally pays to give some attention to what you're really saying in your work.

As well, you will recall that the accuracy and command in your written work contribute to the way others judge your professionalism and competence. Poor grammar reflects badly on your professional abilities, making you appear negligent or unskilful, or both. Remember that even those who are weak in these areas themselves will notice someone else's mistakes, and they may be just as unforgiving—in some cases more so—than your professors.

As you know from Chapter One, one of the most common complaints that colleges hear from employers, no matter what the field, is that graduates can't communicate clearly and accurately. The most visible of writing problems are those arising from poor grammar, and most readers are able to spot these weaknesses in the writing of others. It pays to take care with your sentence structure and usage, and to proofread carefully.

Though this is not intended to be a grammar book, the subject is important enough that this appendix is included to provide you with a handy guide to avoiding the most common grammar errors. If you feel you need a more thorough review, there are plenty of excellent books available. Ask your communication instructor to recommend one.

COMMON SENTENCE ERRORS

Though there are many ways to mangle the language, the following six errors in grammatical structure seem to occur with frequency.

subject-verb agreement
sentence fragments
run-on sentences
pronoun, tense, and person agreement
modifier errors
faulty parallelism

In order to understand these faults and correct them in your writing, it is important first to have a clear understanding of how sentences are built.

WORD GROUPS

Language is constructed principally of words, which can be grouped according to some pretty basic rules. By following these rules we can make three types of word groups:

clauses
sentences
phrases

A **clause** is a group of words that contains a *subject*—a "do-er" or "be-er" of something (this word will usually be a noun or a noun substitute)—and a *verb*—what the subject does or is.

Joe teaches.

subject: Joe   verb: teaches

Birds fly.

subject: birds   verb: fly

Babies cry.

subject: babies   verb: cry

The class learns.

subject: class   verb: learns

If these are the only elements contained in the word group, it is considered to be an **independent clause**. This means that it is able to stand by itself, and its meaning is complete. Independent clauses are important, because they are the fundamental units from which sentences are made.

Clauses may also be made **dependent** by the addition of a joining word called a *subordinate conjunction*. This word reduces the clause to a lesser (or subordinate) role in a sentence; it is no longer the fundamental unit within the sentence. Some subordinate conjunctions are:

although	because	if
when	whenever	which
before	after	who/m
that	though	since

Let's see what these do to our independent clauses from above:

When Joe teaches . . .
If birds fly . . .
Because babies cry . . .
After the students learn . . .

Suddenly these simple clauses are no longer complete; they merely set the stage for the really important information that is to follow. In other words, they must now be joined to something else; in fact, another independent clause, to make a complete thought.

When Joe teaches, he uses many examples.
If birds fly, people should, too.
Because babies cry, they can get attention.
After the students learn, they may be tested.

When you begin to join clauses in this way, you are really building sentences. The rule for a **sentence** is simple: each sentence *must* contain at least one independent clause; though it may contain other things, the independent clause is an absolute necessity. A **simple sentence** contains only an independent clause. You can see that our first grouping of clauses, though made up of only two words each, is also a grouping of sentences. A **compound sentence** contains two or more independent clauses joined by *coordinate conjunctions* (and, but, or, nor, yet, and so). A **complex sentence** contains a combination of at least one independent clause with one or more dependent clauses. The examples above are complex sentences.

REMEMBER: if there is no independent clause, you do not have a sentence.

The third classification of word groups is easy to remember: anything that is not a clause or a sentence is a **phrase**. To put this another way, anything that does not contain a subject or a verb is a phrase. This is true no matter how long the group of words may be.

climbing up the hill
across the street
beside the store with the big sign out front
over between the drive and the garage
with my dad

Because these groups of words contain neither a subject nor a verb, they are phrases, even though some may be quite long.

JOINING CLAUSES

Independent clauses may be combined into longer sentences, which are usually built of related independent clauses, joined in any of several ways. The following sample illustrates the different methods that can be used to build sentences.

Two related independent clauses:

Dave frequently dribbles. He plays basketball.

These may be joined by:

- *using a coordinate conjunction* (*and, but, or, nor, yet, and so*):

 Dave frequently dribbles, *so* he plays basketball. (compound sentence)

- *using a subordinate conjunction* (such as those listed above):

 Because Dave frequently dribbles, he plays basketball. (complex sentence)

- *using a semicolon*:

 Dave frequently dribbles; he plays basketball.

The following are common *incorrect* ways of joining two independent clauses.

- no joining method, simply running two clauses together:

 ✗ Dave frequently dribbles he plays basketball.

- using a comma:

 ✗ Dave frequently dribbles, he plays basketball.

Be sure to use only one of the correct methods at a time. It is, for example, also incorrect to use the semicolon *and* a conjunction to join two clauses:

 ✗ Because Dave frequently dribbles; he plays basketball.

 ✗ Dave frequently dribbles; and he plays basketball.

SIX COMMON SENTENCE ERRORS

1. Subject-Verb Agreement

Subjects ("do-ers" of an action or "be-ers" of a state) agree with their verbs in person and number. This means that singular subjects always take singular verbs, and plural subjects take plural verbs. Singularity of the verb is determined by its subject and not by its "s" ending.

the _dog_ walks (but)

the _dogs_ walk

the _girl_ sings

the _girls_ sing

Aidan does (but)

Aidan and his dad do

_it_ is (but)

they are

Verbs must also agree with their subjects in "person" as well as in number, and this agreement applies only to pronouns. First person pronouns are those that refer to the speaker—that is, "I" or "we." Second person pronouns are used for direct address to another person—that is, "you." Third person pronouns are used to speak of others who are not directly addressed by the speaker—"he," "she," "it," and "they." To list the form of the verb that goes with each of these pronoun subjects is called "conjugating" the verb. Luckily, most verbs in English follow consistent rules and do not change in form to match their pronoun subjects. The exception is the verb "to be." It is conjugated as follows:

As you write, take care to match subjects with the appropriate verb forms, by person and number. This is easy enough to do when subject and verb occur together in the sentence, but when they are separated by phrases or other words it is more difficult. Fortunately, there are some easy-to-learn rules that can help:

a) Subjects compounded with "and" always take plural verbs.

Bob and Devon were with me when it happened.

b) Subjects compounded with "either . . . or," "neither . . . nor," and "or" take verbs that agree with the subject _closest_ to the verb.

Neither the Kennedys nor Sheila is happy with the result.

Neither Sheila nor the Kennedys are happy with the result.

c) Words ending in "-one," "-thing," and "-body" are always singular.

Everybody was present at last night's meeting.

Everything is all right.

d) Phrases such as "together with," "in addition to," "along with," "apart from," and "as well as" are not part of the subject and do not influence the choice of verb.

Joyce, along with her friends, is going to the movie.

e) Collective nouns may take singular or plural verbs, depending on their context. If the group, family, committee, or class acts in unison, it is singular.

> The <u>committee</u> <u>has made</u> a decision.

If they act individually, the verb is plural.

> The <u>committee</u> <u>have argued</u> about this issue for months.

f) The word "each" is always singular; "both" is always plural.

> <u>Each</u> of these <u>is</u> perfect for my sister.

> <u>Both</u> of them <u>have</u> advantages.

First, Second and Third Person Pronouns

	Singular forms	Plural forms
First	I am	we are
Second	you are	you are
Third	he, she, it, or one is	they are

All nouns are third person.

2. Sentence Fragments

A fragment is a part of a sentence that has been treated as a complete sentence. Remember that complete sentences always contain at least one independent clause. Do not punctuate phrases or dependent clauses as sentences.

> ✗ Running down the street and around the corner.

> ✗ The thing being that I don't like him.

> ✗ After I had finished the laundry and the cleaning.

> ✗ For example, scrubbing, polishing, and waxing.

To fix these, either:

• join them to an independent clause; or
• add whatever is missing.

In the case of the second example, the word "being" is not a complete verb; therefore, this word group cannot be a sentence. Change the "-ing" form (known as the *present participle*) to the simple present "is."

John was running down the street and around the corner.

The thing is that I don't like him.

After I had finished the laundry and the cleaning, I took a nap.

I hate household chores, for example, scrubbing, polishing, and waxing.

3. Run-on sentences

This mistake is created by trying to cram too much information into a single sentence without correctly joining the elements that make up the structure of the sentence; to correct it, break the elements up in some way. The most common run-on sentences are created by putting two independent clauses together with only a comma.

✗   I slept in, I missed the bus.

This is incorrect. As is explained above, the only permissible ways to join clauses are with a coordinate conjunction, as in the first example sentence, below; with a subordinate conjunction, as in the second example; and with a semicolon, as in the third example.

I slept in, so I missed the bus.

Because I slept in, I missed the bus.

I slept in; I missed the bus.

4. Pronoun, Tense, and Person Agreement

Always strive for consistency in pronouns, person, and tense. Jumping from one to another person or tense is confusing; using ambiguous or inaccurate pronouns is likewise.

✗   Ted asked the neighbour to move his car. (Whose car? Ted's or the neighbour's?)

✗   A person should mind their own business. ("A person" is only one; "their" is plural. Pronouns should always agree with their antecedents in number [singular–plural] and gender [he–she–it].)

✗   His piece of cake was bigger than hers, which made her angry. ("Which" must refer to a single noun antecedent; a pronoun should not be used to refer to a whole idea.)

Maintain consistent tense. Generally, there are three kinds of time you may refer to in writing: past, present, and future. We tend to write about events in the past or present, and the rule is the same for both tenses. If you're writing in the past tense, stay with the past tense unless the time references change. The same applies to writing in the present tense, and the future tense, too, if you happen to be using it.

✗ So he came up to me and says, "Who do you think you are?" (This is incorrect due to the switch from past "came" to present "says" when the time referred to has not changed.)

Keep person (I, you, he or she, we, you, they) consistent. The most common problem with person agreement is moving from the first person "I" to the second person "you," as shown below; if you stop to think about the meaning of this sentence, it really doesn't make sense.

✗ My apartment faces a busy highway, so when I'm trying to sleep in, the noise of the traffic keeps you awake. (Why would the traffic keep you awake if I'm the one sleeping?)

The second common error in person agreement is switching from the use of third person "one" or "a person" to second person "you."

✗ One should always keep your eyes open. (In a sentence such as this one, you may use "you" or "one," but don't mix them in the same sentence.)

5. Modifier Problems

Modifiers are words or groups of words that describe, explain, intensify, or negate other words or groups of words. The two kinds are *adjectives*, which modify nouns or noun substitutes, and *adverbs*, which modify verbs, adjectives, or other adverbs. Modifier errors occur when the modifier is either misplaced or "dangling." The best rule for correcting both of these problems is to place the modifier as close as possible to the word or phrase it modifies; if that item is not in the sentence, rewrite the sentence so the meaning is clear.

Misplaced Modifiers

In the case of this error, the modifier is in the wrong position in the sentence. Put the modifier as close as possible to the thing modified.

✗ I only ate half my dinner. (I only ate it; I didn't dance with it or take it to a movie! Probably I intend the "only" to modify the "half.")

✓ I ate only half my dinner.

✗ I almost earned fifty dollars this morning. (Unless I mean that I had a chance to earn this money, but instead earned nothing at all, I need to move the modifier "almost" so that it modifies the amount of money I earned.)

✓ I earned almost fifty dollars this morning.

Dangling Modifiers

The modified element, though implied, is not actually given in the sentence. Rewrite the sentence so that the modified element is clear.

✗ Running alongside the river, a treasure chest lay in the bushes. (Since the treasure chest can't run, this sentence doesn't make sense. Who saw the treasure chest? Who was running?)

✓ Running alongside the river, Ted spotted a treasure chest lying in the bushes.

✗ After drawing a picture, I understood what she meant. (This sentence makes it sound as though I drew the picture myself, whereas it is more likely that the other person explained her meaning by drawing a picture.)

✓ After Nancy drew a picture, I understood what she meant.

6. Faulty parallelism

This error occurs when you are using lists or series of items. Whenever you are speaking of more than one item, place them all in the same grammatical form. Use nouns with nouns, adjectives with adjectives, "-ing" words with "-ing" words, clauses with clauses.

✗ Professionals include doctors and people who practise law.

✗ I like running, jumping, and to sing.

✗ She's pretty, but has ambition too.

Replace such faulty parallelism with corrected forms:

✓ Professionals include doctors and lawyers.

OR

✓ Professionals include people who practise medicine and people who practise law.

✓ I like running, jumping, and singing.

OR

✓ I like to run, jump, and sing.

✓ She's pretty, but ambitious, too.

OR

✓ She has beauty and ambition, too.

This coverage of grammar is necessarily a brief overview. There are many more subtleties to good grammar than there is room to cover in this supplement. A good grammar handbook will give you more information, should you require it. Your communication teacher will be able to recommend one.

SHARPENING YOUR SKILLS

Section A: Sentence Errors with Answer Key

The following sentences contain errors of the types explained above. See if you can correct them. The answer key follows.

1. I only lived in Ottawa for four months.

2. Doing his homework, the TV was distracting.

3. The thing being that I really enjoy their company.

4. Take me with you, I'll miss you too much if I stay here alone. In this scary place with no phone.

5. Give me a bite of your sandwich, I haven't had lunch.

6. Darlene's answer almost was right.

7. Charlene, but not the others, are going camping.

8. I nearly told you a hundred times! Don't call me!

9. A person should mind their own business.

10. One should always keep your eyes open.

11. Doing a test is better than explanations, you can see the rules in action.

12. I didn't want his sympathy, I sent him away forever.

13. If a mosquito bites your face, it should be squashed.

14. On the table was his hat and gloves, so I knew he was home.

15. I only mailed half my cards at Christmas.

16. In college I took English, History, and a course in people and society.

17. If I want to do a good job, you should never overlook details.

18. There was three people on the bus: a student, a mail carrier, and a man who worked on cars.

19. I love lasagna. Even though it's fattening.

20. I don't want to go to the party with him. The reason being that I had a lousy time when I dated him before.

21. Jeff looks familiar to me; because I have a friend who looks just like him.

22. There are lots of things you can do in winter. Skiing, skating, and hikes are only three of them.

23. Running up the stairs, someone tripped Lucy and she fell on her arm, breaking it in two places.

24. Where is Donna's dictionary, she'll be lost without it.

25. Neither Bill nor his friends is willing to help.

26. If the dog sleeps in your bed, it should be disinfected.

27. Being too large a sandwich, she declined to eat it.

28. It really made me laugh. The day she told us about Ted.

29. He is a kind person who has generosity too.

30. If a person is nervous, they should try to relax more.

31. Lola likes to wear soft sweaters, eat exotic food, and taking bubble baths.

32. My sister's boyfriend is stingy, sloppy, and doesn't have much ambition.

33. I noticed a crack in the window walking into the house.

34. Eating a hot dog, mustard dropped onto my shirt.

35. What do you think of this, Red Deer is the fourth largest city in Alberta.

36. When you're stuck. You can use your dictionary for help.

37. Hallowe'en is my least favourite holiday, I'm afraid of ghosts.

38. Although I like Christmas, since I love all the sparkle and magic.

39. I nearly earned a hundred dollars last week.

40. Jerry invited only Millie and me, I guess you can't come.

Answer Key

1. I lived in Ottawa for <u>only</u> four months.

2. <u>As</u> <u>he</u> <u>did</u> his homework, the TV was distracting.

   OR Doing his homework, <u>Ted</u> <u>found</u> the TV distracting.

3. The thing <u>is</u> that I really enjoy their company.

   OR I really enjoy their company.

4. Take me with yo<u>u;</u> I'll miss you too much if I stay here alone in this scary place with no phone.

5. Give me a bite of your sandwich. I haven't had lunch.

6. Darlene's answer was <u>almost</u> right.

7. Charlene, but not the others, <u>is</u> going camping.

8. I told you <u>nearly</u> a hundred times! Don't call me!

9. A person should mind <u>her</u> <u>or</u> <u>his</u> own business.

    OR <u>People</u> should mind their own business.

10. One should always keep <u>one's</u> eyes open.

    OR <u>You</u> should always keep your eyes open.

11. Doing a test is better than explanations, <u>because</u> you can see the rules in action.

12. I didn't want his sympathy, <u>so</u> I sent him away forever.

13. <u>A</u> mosquito <u>that</u> bites your face should be squashed.

14. On the table <u>were</u> his hat and gloves, so I knew he was home.

15. I mailed <u>only</u> half my cards at Christmas.

16. In college I took English, History, and <u>Sociology</u>.

17. If I want to do a good job, <u>I</u> should never overlook details.

    OR If <u>you</u> want to do a good job, you should never overlook details.

18. There <u>were</u> three people on the bus: a student, a mail carrier, and a <u>mechanic.</u>

19. I love lasagna, <u>e</u>ven though it's fattening.

20. I don't want to go to the party with him. I had a lousy time when I dated him before.

    OR I don't want to go to the party with him, <u>because</u> I had a lousy time when I dated him before.

21. Jeff looks familiar to me because I have a friend who looks just like him.

22. There are lots of things you can do in winter. Skiing, skating, and <u>hiking</u> are only three of them.

23. Running up the stairs, <u>Lucy</u> <u>tripped</u> and fell on her arm, breaking it in two places.

24. Where is Donna's dictionary<u>?</u> <u>S</u>he'll be lost without it.

25. Neither Bill nor his friends <u>are</u> willing to help.

26. <u>Your</u> <u>bed</u> should be disinfected if the dog sleeps in <u>it</u>.

27. <u>Because</u> <u>the</u> <u>sandwich</u> <u>was</u> <u>too</u> <u>large,</u> she declined to eat it.

28. It really made me laugh <u>the</u> day she told us about Ted.

29. He is a kind <u>and</u> <u>generous</u> person.

30. If <u>people</u> <u>are</u> nervous, they should try to relax more.

    OR <u>People</u> <u>who</u> <u>are</u> <u>nervous</u> should try to relax more.

31. Lola likes to wear soft sweaters, eat exotic food, and <u>take</u> bubble baths.

32. My sister's boyfriend is stingy, sloppy, and <u>unambitious.</u>

33. <u>Walking</u> <u>into</u> <u>the</u> <u>house,</u> I noticed a crack in the window.

34. <u>As</u> <u>I</u> <u>was</u> eating a hot dog, mustard dropped onto my shirt.

35. What do you think of this<u>?</u> Red Deer is the fourth largest city in Alberta.

36. When you're stuck<u>,</u> you can use your dictionary for help.

37. Hallowe'en is my least favourite holiday, <u>because</u> I'm afraid of ghosts.

38. I like Christmas, since I love all the sparkle and magic.

39. I earned <u>nearly</u> a hundred dollars last week.

40. Jerry invited only Millie and me, <u>so</u> I guess you can't come.

Section B: More Sentence Errors

The following sentences, like those in Section A, contain errors of the type discussed above. Your instructor can provide you with an answer key.

1. He doesn't know what he's talking about. But that never stopped him before.

2. Can I borrow your calculator, I need to figure something out.

3. Looking over my shoulder, Tim was about a block behind me.

4. Allan is obsessive, and he also has compulsions.

5. I want to try that ride again. Even though I was really scared.

6. I forget, is it "where no one has gone before," or "where no man has gone before."

7. He only wanted a sandwich.

8. I almost bought a hundred copies.

9. If you want to go. I will go with you.

10. Dave, along with his brothers, were in the room.

11. Here comes the doctor, I wonder what he will say.

12. My friend is tall, slender, and looks beautiful.

13. Laura is doing her project, it's on *This Hour has 22 Minutes*.

14. There was a sandwich and a glass of beer on the counter.

15. Gwen went for an interview, she didn't get the job.

16. Can you believe it, Alex changed his name.

17. Standing on the corner, a car ran through a puddle and soaked me.

18. Qualified individuals need only apply.

19. An individual should protect their credit rating.

20. There was three big mistakes in his report, he didn't notice them.

21. I hate that commercial. The one where the dog talks.

22. My mom is a great baker, you should taste her pies.

23. Waiting in line, my briefcase got stolen.

24. Dave hates rap, he says it should be spelled with a "c."

25. Don't get upset, I was only joking.

26. I only finished half my dessert.

27. Devon dislikes him; because he is a jerk.

28. Cheryl is so rude, maybe it's because she is so unpopular.

29. There are lots of reasons for leaving that place: the people are surly, the weather is bad, and it's an ugly town.

30. If you knew her as well as I do, you would see what is wrong. That she is just running away from her problems.

31. I am going downtown tomorrow, I want to get a new cd.

32. In living common law, a child is considered illegitimate.

33. My first big crush occurred at the age of thirteen.

34. Upon entering the room, there is a large cupboard.

35. I couldn't believe the weather we had in October, it was just like July.

36. The lady on the corner by the bus stop, whose hat blew off in the wind yesterday.

37. Although I hate typing, if I wait for her to do it, and I'd never get it done.

38. Not all sentences contain mistakes, does this one?

39. When depressed or lonely, a dog is always beside you wagging his tail.

40. After studying all night, I only got 68 per cent on the exam.

42. Being a math major, the test was easy for Jane.

43. Bill swatted the wasp that stung him with a newspaper.

44. My friend said in May we will be taking a trip to Hawaii.

45. Clyde and Charlotte decided to have two kids on their wedding day.

46. We could see the football stadium driving across the bridge.

47. Garage doors can either be opened sideways or upwards.

48. Poker is a great game, it takes skill as well as luck.

49. How can you treat him that way? When he has always been fair to you.

50. When you get here on Friday night, you will get to see Barb. If she decides to show up.

51. When I went to the door, I saw that it was Norm, who else would it be at 2 am?

52. The students no longer like the math teacher who failed the test.

53. One of my friends used to joke about working overseas, and nobody believed him. Until he did it.

54. After Christmas, the number of gift exchanges are no surprise to us.

55. Two-thirds of his car were covered by a white powder.

56. I only miss my friends, nothing else about living there.

57. We saw the injury, it was far worse than we expected.

58. Harold passed high school English, he can't read or write.

59. As I was watching TV last night, I see that the economy is in trouble.

60. Neither of those answers are correct.

61. My friend Norm knows more about music than me.

62. After criticizing both my work and my attitude, I was fired.

63. A large group of students are going on the field trip.

64. As a scientist, the only way a consumer can be injured by a microwave is if they trip over it.

65. The obituary column lists the names of those who have died recently for a nominal fee.

66. You go see what he wants, I'll wait here for you.

67. I'll be glad to help when you need me I'll be at my phone.

68. The combination of alcohol and tranquillizers are dangerous.

69. Everybody was excited, we were allowed to use the pool.

70. His style is clear, engaging, and a delight to read.

71. When caught, the weapon was still in the suspect's hand.

72. I count the cash, it came to $300.

73. I removed the rugs in order to clean the floor and for coolness in the summer.

74. The guard wouldn't permit Alfred and I to enter the building.

75. She borrowed an egg from a neighbour that was rotten and smelled bad when cracked.

Punctuation

This review is designed to remind you of the basic uses of the most common punctuation marks. Students often find punctuation confusing, but it is important. It is a convention designed to help your reader understand the meaning and intention of your written work. Like road signs on a highway, punctuation marks help to direct the reader on the journey through your written work. Though style guides differ somewhat, the following general rules should see you through most basic punctuation needs. The important thing is that you remember to follow a consistent style throughout your document. For more detailed information, consult a style manual.

REVIEWING COMMON PUNCTUATION MARKS

There are three common forms of "full stop" punctuation: the period, the question mark, and the exclamation point.

The Period (.)

1. Use a period to mark the end of a sentence.

2. Use a period following an abbreviation: etc., Dr., Sask. Some common abbreviations (TV, VCR) don't require periods. Neither do the new two-letter provincial abbreviations used by the Post Office: NS, NB, PE, NF, QC, ON, MB, SK, AB, BC.

3. When an abbreviation falls at the end of the sentence, use only one period, omitting the abbreviation period.

   Since Jerry completed dentistry school, he loves to be called Dr.

The Question Mark (?)

1. Use a question mark only after a direct question.

    Where are you taking that box?

    Why did you bring him with you?

2. Never use a question mark for an indirect question.

    I wonder whether Shirley has a copy of this book.

    I asked him where he intended to take that box.

A question mark is end punctuation and should not be directly followed by any other form of punctuation—period, comma, or semicolon.

The Exclamation Point (!)

1. Use an exclamation point after an exclamatory word (interjection) or phrase:

    Wow!

    Hey!

    How about that!

2. Use an exclamation point for emphasis when a statement or question is meant to be read with force.

    What did you do that for!

    I just won the lottery!

In formal writing, and most professional writing, you should avoid the exclamation point. Occasionally it is useful in sales letters, for emphasis, but too many exclamation points will create an overly loud or hysterical impression.

    You can win!

    Act now and save!

    No Down Payment! No Interest!

The Semicolon (;)

1. Use a semicolon to separate two independent clauses that are closely related in meaning.

    I am tired of Shawn; I really wish he would go away.

    I built my first model two years ago; now I'm hooked.

    I have a sinus infection; I went swimming without nose plugs.

2. Use a semicolon before conjunctive adverbs such as "however," "therefore," "thus," and "consequently" when they are used to join two related clauses.

> I don't want to deal with those people ever again; however, they are my relatives.

> Ernesto didn't prepare well enough for his presentation; consequently, he felt like a fool in front of the class.

3. Use a semicolon to separate items in a series, but only if the individual items in the series already contain commas.

> I have invited Madhu, my cousin; Nancy, my best friend; and Ian, Nancy's brother.

> I have to replace my VCR, which is ten years old; my television, which is twenty years old; and my stereo, which is practically an antique.

The Colon (:)

1. Use a colon to introduce a list on two occasions: when the list is vertical, or when the introductory clause is independent.

> Please bring the following supplies with you:
>
>> camera;
>> film;
>> flash attachment;
>> batteries; and
>> lenses.

> I can't believe how many things I have to do this week: prepare my presentation for Professional Communication; organize my project for Technical Design; finish packing for my vacation, and get my hair cut.

2. Use a colon after an independent clause if what follows (a word, phrase, or another independent clause) explains or enlarges upon the first one.

> Wait until you hear what he gave me: a machine for making rubber stamps.

> Cameron got the highest grade in the class on that test: A+.

> Vernon is the perfect place to stay away from: my in-laws live there.

3. Use a colon to introduce a formal quotation.

> In *Second Words*, Margaret Atwood says: "A voice is a gift: it should be cherished and used, to utter fully human speech, if possible."

If the quotation is lengthy (over three lines), begin it on a new line following the colon and indent it.

4. Use a colon after the salutation of a business letter.

Dear Dr. Faried:

Dear Ms Bergen:

5. Use a colon between a title and a subtitle.

Inside Language: A Canadian Language Anthology

A Gift of Voice: The Role of Self-Projection in the Rhetorical Appeal of Margaret Atwood's Nonfiction

The Comma (,)

1. Use commas to separate items in a list.

Betty, Elaine, and Debbie are close friends.

I can bring my cat, my dog, or my fish.

The comma before the coordinate conjunction is optional; it may be used to avoid confusion.

✗ Be alert for inclement weather, falling rocks and wildlife. (Was the wildlife falling or only the rocks?)

✓ Be alert for inclement weather, falling rocks, and wildlife.

2. Use a comma following someone's name in a direct address to that person; do not use a comma when the name is the subject of the sentence.

✓ Bill, will you hand me that book on economics?

✗ Bill, handed me the book.

3. Use a comma after an introductory dependent clause.

✓ Because I was late, I missed the pizza.

✓ Whenever he participates, we have problems.

✓ If I want you for anything, I'll call.

4. Use a comma after a lengthy introductory phrase.

After the best day of my life, I was exhausted.

Before the graduation dance, Ely came down with the measles.

5. Use a comma before a coordinate conjunction (*and, or, nor, but, yet, so*) that joins two independent clauses.

I went downtown, and Roxanne joined me.

I went for the interview, but somebody else got the job.

You can omit the comma if the two clauses have the same subject.

I went for the interview and I got the job.

I went for the interview and got the job.

6. Use a comma after conjunctive adverbs such as "however," "therefore," "thus," "moreover," and "consequently."

I like him very much; moreover, I think he is an outstanding accountant.

7. Use commas to set apart phrases or clauses that interrupt a sentence between its subject and verb.

David, who has the office beside mine, has accepted a job with another firm.

Feral, the little rascal, didn't finish her homework.

8. Use a comma before a short, direct quotation.

Paul said, "I'd like to nail that guy."

Don't use a comma for indirect quotations.

Paul said that he'd like to nail that guy.

9. Use a comma to separate parts of an address, if they appear on the same line.

57 Campus Drive, Saskatoon, Saskatchewan

10. Use a comma to separate the day and month when identifying a date.

Saturday, November 13

The Apostrophe (')

1. Use an apostrophe to indicate possession in nouns, but not in pronouns.

Paul's book

David's minutes

Maureen's comments

whose book (**NEVER** *who's* book—that's "who is")

its colour (**NEVER** *it's* colour—that's "it is")

Singular nouns, as above, add an apostrophe and an "s"; plural nouns or others ending in "s" usually don't need an additional "s" added.

the Inghams' house

the students' grades

However, when the possessive word is pronounced as though it had an additional syllable, you may wish to add the second "s."

Chris's test results

Iris's new gloves

For plural nouns not ending in "s" add apostrophe and "s."

children's coats

criteria's validity

2. Use apostrophes in contractions.

can't	there's	won't	what's
they're	it's (it is)	wasn't	I've
he's	we're		

Quotation Marks (" " or ' ')

1. Use quotation marks for indicating direct speech, but not indirect speech.

   David asked, "Do you need a ride to the airport?"

   David asked if I needed a ride to the airport.

2. Quotation marks within quotation marks are usually indicated by single marks within double ones.

   She said, "You know what Joseph Campbell says: 'Follow your bliss.'"

3. Use quotation marks to indicate that a word is slang, an inappropriate usage, or someone else's wording, but be very careful not to overuse them in this way.

   I don't think "intimacy" is quite the right word for this concept.

   I should have realized the truth when he said he was "cool."

4. Use quotation marks for the titles of short works, such as poems, short stories, essays, articles, songs, or chapters in a book.

   "The Rime of the Ancient Mariner"

   "Home on the Range"

   "Chapter Eight: The Job Application"

Enclose commas and periods within the quotation marks; semicolons and colons are placed outside the quotation marks. Other end punctuation (question marks and exclamation points) falls outside unless the original phrase was a question or exclamation.

Title Treatments

1. Italicize (or <u>underline</u> if you can't *italicize*) the titles of book-length works wherever they appear in your writing.

   Hamlet

   Lovable Soft Toys

   Technical Communication: Engineering Messages

2. Place the titles of short works (essays, book chapters, poems, short stories, songs, recipes, or articles) in quotation marks.

   "Chapter Eight: The Job Application"

   "Introduction: Are You a Designer?"

   "Stopping by Woods on a Snowy Evening"

3. Do not underline the titles of your own reports or essays, or place them in quotation marks, on the title page of your paper.

   The Agony and the Exigence: A Rhetorical Analysis of Two Presentations

   Innovations in Training: A Proposal for Improving Our Report Writing Workshops

4. If your title contains the title of another work, you should treat the title of the other work appropriately, depending on whether it is a short or long work.

   Method or Madness? An Analysis of Motive in <u>Hamlet</u>

   OR

   Method or Madness? An Analysis of Motive in *Hamlet*

   Miles to Go Before I Sleep: Hypothermia in Frost's "Stopping by Woods on a Snowy Evening"

This brief refresher is not intended to be a comprehensive guide to punctuation usage, but should provide enough detail for everyday usage. For a more detailed guide to punctuation conventions, you may wish to consult a good English handbook or manual of style, such as *The Chicago Manual of Style* or Strunk and White's *Elements of Style*.

Index

INDEX

Credits

Photo

(For each chapter-opening photograph.)

Chapter 1 (page 1): PhotoDisc

Chapter 2 (page 24): PhotoDisc

Chapter 3 (page 54): Photo Edit/Michael Newman

Chapter 4 (page 81): Photo Edit/David Young-Wolff

Chapter 5 (page 112): PhotoDisc

Chapter 6 (page 158): Digitalvision

Chapter 7 (page 217): PhotoDisc

Chapter 8 (page 250): Photo Edit/David Young-Wolff

Chapter 9 (page 309): PhotoDisc

Epilogue (page 328): Digitalvision

Literary

(For each Critical Reading.)

Page 15: Wayne C. Booth, "The Rhetorical Stance." Reprinted by permission of the author.

Page 51: Allan T. Dolovich and Lisa Coley, "Assignment Submission Requirements for ME 316" reprinted by permission of Allan T. Dolovich and Lisa Coley.

Page 74: Lloyd Bitzer, "The Rhetorical Situation." Reprinted by permission of Penn State University Press.

Page 206: Burton L. Urquhart, "Rhetorical Strategies for the Technical Workplace: Footing and

the Notion of Rhetorical Balance." Copyright © Burton L. Urquhart. Reprinted by permission of the author.

Page 247: Curtis Olson, "Strategy Report for a Five-Minute Proposal Speech." Copyright © Curtis Olson. Reprinted by permission of the author.

Page 338: Julian Demkiw, "An Orphaned 'Son of Martha': Defining the Engineer." Copyright © Julian Demkiw. Reprinted by permission of the author.